# THE ACTIVE PRACTICE
# OF STATISTICS

# THE ACTIVE PRACTICE
# OF STATISTICS

## A text for multimedia learning

David S. Moore
*Purdue University*

W. H. FREEMAN AND COMPANY
*New York*

Acquisitions editor: Holly Hodder
Associate editor: Patrick Farace
Project editor: Diane Cimino Maass
Design coordinator: Blake Logan
Cover designer: Marjory Dressler
Illustration coordinator: Bill Page
Illustrations: Publication Services
Production coordinator: Paul Rohloff
Composition: Publication Services
Manufacturing: R.R. Donnelley & Sons Company

**Library of Congress Cataloging-in-Publication Data**

Moore, David S.
    The active practice of statistics : a textbook for multimedia
learning / David S. Moore.
        p.   cm.
    Includes index.
    ISBN 0-7167-3140-1 (alk. paper)
    1. Statistics—Computer-assisted instruction.   I.  Title.
QA276.18.M66   1997
519.5—dc21                                               97-21730
                                                         CIP

Printed in the United States of America

First printing, 1997

# CONTENTS

vi     Contents

# PREFACE

*The Active Practice of Statistics (APS)* is an introduction to statistics that emphasizes working with data and statistical ideas. It is intended to accompany the multimedia CD *ActivStats*, authored by Paul Velleman of Cornell University and published by Addison Wesley Interactive. This book is based primarily on my *Basic Practice of Statistics (BPS)* (W. H. Freeman, 1995), but with substantial changes that reflect its place as the print component of a multimedia system.

In planning a multimedia system for learning basic statistics, Paul Velleman and I quickly decided the following: (a) The modern "data and concepts" introductory course based on *BPS* or on my longer text with George McCabe, *Introduction to the Practice of Statistics* (3rd ed., W. H. Freeman, 1998) should be the intellectual basis for the multimedia system. *ActivStats* closely follows the style, content, and order of these texts. (b) A printed text is an important part of a multimedia system, both because large blocks of text belong in print rather than on a screen and because we found that students very much want a text. (c) The multimedia presentation via CD and computer should form the student's first exposure to each lesson, with the text serving as a more portable source for reinforcement, reference, and review.

*APS* is therefore an unusual text. It differs from *BPS* in these ways:

- The exposition here, while complete, is briefer. I assume that readers have *ActivStats* available for a first presentation and that they will take advantage of the interactive nature of *ActivStats* to go over rough spots repeatedly.

- It is shorter in ways that go beyond the terser exposition. I have omitted material not covered by *ActivStats*. Because students are assumed to make constant use of the software in *ActivStats*, *APS* does not dwell on computational details.

- The content is customized for students who have constant access to good software. Some methods not in *BPS*, such as normal probability plots, appear in *APS* because software makes them accessible. Exercises often require software.
- There are many new examples, exercises, and data sets. Many of the data sets from this book are available on the *ActivStats* CD. For the convenience of instructors, I have attached a list of these exercises below.

## Exercises contained on the *ActivStats* CD

| | | | |
|---|---|---|---|
| Lesson 2 | Exercises 1, 2, 4 | Lesson 14 | Exercises 1, 2, 3, 7, 8, 10, 12, 14 |
| Lesson 3 | Exercises 2, 4, 6, 7, 11, 13, 14, 15 | Lesson 15 | Exercises 1, 4, 12, 13 |
| Lesson 4 | Exercises 1, 5, 11, 12, 14, 17, 18, 19, 20 | Lesson 16 | Exercises 3, 4, 7, 8 |
| | | Lesson 17 | Exercises 4, 7, 9, 10, 13, 14 |
| Lesson 5 | Exercises 2, 3, 5, 7, 8, 9, 11, 12, 13, 17 | Lesson 18 | Exercises 5, 6, 7, 9, 12 |
| | | Lesson 19 | Exercises 3, 4, 5, 7, 10, 11, 13, 14, 16, 18 |
| Lesson 6 | Exercises 2, 3, 4, 7 | | |
| Lesson 7 | Exercises 1, 2, 3, 4, 5, 8 | Lesson 20 | Exercises 4, 6, 7, 10, 12, 13, 14, 15 |
| Lesson 8 | Exercises 2, 3, 4, 5, 6 | | |
| Lesson 9 | Exercises 1, 2, 9, 10, 12, 14, 15 | Lesson 21 | Exercises 2, 6, 8, 9, 10, 13, 16 |
| Lesson 10 | Exercises 3, 4, 9, 11, 15 | Lesson 22 | Exercises 1, 2, 5, 9, 10, 15 |
| Lesson 11 | Exercises 3, 5, 7, 10, 14 | Lesson 23 | Exercises 1, 2, 6, 8, 12, 13 |
| Lesson 12 | Exercises 1, 2, 7, 8, 9, 10 | Lesson 24 | Exercises 1, 2, 3, 4, 5, 10 |
| Lesson 13 | Exercises 1, 3, 8, 9 | | |

The lesson and part structure of this book mirrors that of *ActivStats*. For each of the six parts, I have added review lessons that offer students an outline-style review and a set of review exercises. Lesson 1 is a brief discussion of how to use *APS* along with *ActivStats*. You may wish to read that lesson for more comments on the interaction between CD and book, and to read the Introduction for a brief statement of the approach to basic statistics that both the CD and the book encourage.

## What about those other books?

This text shares the basic spirit of my longer texts, *The Basic Practice of Statistics* (*BPS*) and *Introduction to the Practice of Statistics* (*IPS*, with George McCabe). It offers a modern presentation of statistics that gives proper emphasis to analysis of data and design of data production in addition to formal probability-based inference, stresses the practical and conceptual links between these aspects of the subject, and concentrates on core concepts and strategies that carry over into more advanced statistical settings. Most of all, the exposition centers on understanding what data tell us about substantive problems. *APS* of course contains many new examples, exercises, and data sets. It also differs from the other texts in systematic ways.

The intent of *BPS* is to present the core concepts and methods of data-oriented statistics in a manner accessible to relatively non-quantitative students. *BPS* does not assume that students will use technology beyond a two-variable statistics calculator, and it deliberately reduces the usual presentation of probability in order to concentrate on core statistics. *IPS* is a longer book that presents some more advanced statistical material as well as a quite traditional treatment of elementary probability. Its discussions have greater depth and richness; this is a substantial bonus for some groups of students, but a barrier for others.

The efficiency of good software allows *APS* to present some material not present in *BPS* without raising barriers to students. Normal probability plots appear throughout. The presentation of probability is much more complete than in *BPS*; in particular, there is a careful introduction to conditional probability. On the other hand, the topics discussed in *APS* are strictly limited to those present in *ActivStats*. *BPS* adds an introduction to oneway analysis of variance. *IPS* goes on to present multiple comparisons, twoway ANOVA, and some basics of multiple regression.

It has been a pleasure to work with Paul Velleman in thinking about multimedia instruction, and I thank him for his careful reading of this manuscript. We have aimed at close coordination between CD and book, but readers should be aware that software can be revised much more quickly than a text. It has once again been a pleasure to work with W. H. Freeman's excellent staff, especially project editor Diane Cimino Maass; designer, Blake Logan; production coordinator, Paul Rohloff; and copy editor Pamela Bruton.

I very much hope that *The Active Practice of Statistics,* in company with *ActivStats,* will make a contribution to moving statistics instruction closer to statistical practice in an era when the automation of calculations makes data judgment and understanding of statistical principles more important than ever.

David S. Moore

# INTRODUCTION: WHAT IS STATISTICS?

## Why study statistics?

Statistics is the science of gaining information from numerical data. We study statistics because the use of data has become ever more common in a growing number of professions, in public policy, and in everyday life. Here are some examples of statistical questions raised by a viewer of the nightly news.

- The Bureau of Labor Statistics reports that the unemployment rate last month was 6.5%. The government didn't ask me if I was unemployed. How did they obtain this information? How accurate is that 6.5%?

- Another news item describes restrictions on smoking in public places. I hear that much of the evidence that links smoking to lung cancer and other health problems is "statistical." What kind of evidence is statistical evidence?

- A medical reporter cites a study claiming to show that regular physical exercise leads to longer life. But a doctor she interviews casts doubt on the study's usefulness. How can I tell whether the data from a study really support the conclusions that are announced?

- Here's a special report on the international competitiveness of American industry. The experts interviewed talk about improving quality and productivity through better management, new technology, and effective use of statistics. What can statistics do except keep score?

We can no more escape data than we can avoid the use of words. Like words, data do not interpret themselves but must be read with understanding. Just as a writer can arrange words into convincing arguments or incoherent nonsense, so data can be convincing, misleading, or just irrelevant. Numerical literacy, the ability to follow and understand arguments based on data, is important for everyone. The study of statistics is an essential part of a sound education.

## What is statistics?

For most users of statistics, and even for most professional statisticians, statistics provides tools and ideas for using data to gain understanding of some other subject. Statistics in practice is applied to study the effectiveness of medical treatments, the reaction of consumers to television advertising, the attitudes of young people toward sex and marriage, and much else. Although statistics has an impressive mathematical theory, we are concerned with the *practice* of statistics. We can divide statistics in practice into three parts:

1. **Data analysis** concerns methods and ideas for organizing and describing data using graphs, numerical summaries, and more elaborate mathematical descriptions. The computer revolution has brought analysis of data back to the center of statistical practice. Statisticians have responded with new tools and (more important) new organizing ideas for exploring data. Parts I and II of this book discuss data analysis.

2. **Data production** supplies methods for producing data that can give clear answers to specific questions. Basic concepts about how to select samples and design experiments are perhaps the most influential of all statistical ideas. These concepts are the subject of Part III.

3. **Statistical inference** moves beyond the data in hand to draw conclusions about some wider universe. Statistical inference not only draws conclusions but accompanies those conclusions with a statement about how trustworthy they are. Inference uses the language of probability, which we introduce in Part IV. Because we are concerned with practice rather than theory, we can function with a quite limited knowledge of probability. Part V discusses the reasoning of statistical inference and methods for practical inference in several simple settings. Part VI offers introductions to inference in some more complex settings.

## How will we approach the study of statistics?

The goal of statistics is to gain understanding from data. Data are numbers, but they are not "just numbers." *Data are numbers with a context.* The number 10.5, for example, carries no information by itself. But if we hear that a friend's new baby weighed 10.5 pounds at birth, we congratulate her on the healthy size of the child. The context engages our background knowledge and allows us to make judgments. We know that a baby weighing 10.5 pounds is quite large, and that it isn't possible for a human baby to weigh 10.5 ounces or 10.5 kilograms. The context makes the number informative.

Because data are numbers with a context, doing statistics means more than manipulating numbers. This book is full of data, and each set of data has some brief background to help you understand what the data say. Examples and exercises usually express briefly some understanding gained from the data. In practice, you would know much more about the background of the data you work with and about the questions you hope the data will answer. No textbook can be fully realistic. But it is very important to form the habit of asking "What do the data tell me?" rather than just concentrating on making graphs and doing calculations. This book tries to encourage good habits.

Nonetheless, statistics involves lots of calculating and graphing. You will use computer software to automate most calculations and graphs. Because graphing and calculating are automated in statistical practice, the most important assets you can gain from the study of statistics are an understanding of the big ideas and the beginnings of good judgment in working with data. Ideas and judgment can't (at least yet) be automated. They guide you in telling the computer what to do and in interpreting its output. This book tries to explain the most important ideas of statistics, not just teach methods. Some examples of big ideas that you will meet (one from each of the three areas of statistics) are "always plot your data," "randomized comparative experiments," and "statistical significance."

These, then, are the principles that should guide your learning of statistics:

- Try to understand what data say in each specific context. All the methods you will learn are just tools to help understand data.
- Let a computer do as much of the calculating and graphing as possible, so that you can concentrate on what to do and why.
- Focus on the big ideas of statistics, not just on rules and recipes.

But perhaps the basic principle of all learning is persistence. The main ideas of statistics, like the main ideas of any important subject, took a long time to discover and take some time to master. The gain will be worth the pain.

# HOW TO USE THIS BOOK

*The Active Practice of Statistics (APS)* is intended to be used in company with the multimedia *ActivStats* system. It is therefore somewhat different from other texts. *ActivStats* has important advantages as a tool for learning. In particular, you interact with the ideas and examples presented as you move through a lesson in *ActivStats*. The active learning encouraged by a good multimedia system is a better starting point than reading the same material in a book.

> We recommend that you begin your study of each lesson with *ActivStats*. Then turn to this book for reinforcement and a record of the lesson content.

The organization, notation, and basic content of *APS* exactly match *ActivStats*. Each of the 24 lessons in this book corresponds to an *ActivStats* lesson. Many of the exercises in this book (especially those accompanied by sets of data) are available within *ActivStats* through the homework icon  WORK in the tool bar of each lesson. In addition, *APS* offers a review outline of each part of *ActivStats,* with lists of the specific skills you should have acquired from each part and review exercises to test those skills.

The presentation of the content of each lesson in this book is quite brief. It is intended to give you a compact statement that contains all the essentials but assumes that you have made a first acquaintance with the material during your work on the *ActivStats* lesson. Seeing two different expositions (multimedia and print) will solidify your understanding. You will also find discussions of some ideas that do not lend themselves to multimedia presentation.

## SOFTWARE TOOLS

In practice, users of statistics always use software to automate their calculations and graphics. This book fits actual statistical practice in that it assumes that you will use software in almost all your work. *ActivStats* makes available to you software tools for doing the many calculations and graphs that modern statistics requires. These are of two kinds:

- The *Data Desk* statistics software, a modern menu-driven statistics software package. *Data Desk* is widely used for statistical analyses. Moreover, its organization is generally similar to that of other software for doing statistics, so that experience with *Data Desk* equips you to do statistical work in practice.

- Animated graphics tools that make it easy to do some things that do not commonly appear in statistics software. These tools are used in *ActivStats* to help you learn, and they are available in the lesson tool bars for you to use detached from the lessons.

> This book assumes that you will use the software tools made available in *ActivStats*. Exercises require use of these or similar tools. The text does not tell you how to do calculations and graphs by hand except where knowing how helps your understanding.

You will find that *ActivStats* itself will teach you how to use the *Data Desk* software and the animated tools. Because you should start with the *ActivStats* lessons before reading the corresponding lessons in this book, there is no instruction here on how to use the software. One benefit of doing the exercises is to ensure that you can use the software effectively outside the *ActivStats* lessons that teach you how to use it.

## WHY DO I NEED A TEXTBOOK?

Multimedia systems are in many respects more effective than print as aids to learning. But printed books retain some important advantages:

- Large blocks of text are much easier to read and work with in print than on the screen. *APS* complements *ActivStats* by offering detailed explanations, worked examples, and comments on statistical practice that are better presented in a book.
- A book is more portable than a CD because you don't need a computer on hand. You can study *APS* anywhere.
- Because of its portability and because it is easy to find your way around in a book, a permanent record of what you must know is more convenient in book form.

Think of the lessons in *APS* as detailed and permanent records of the content of the *ActivStats* lessons, lacking only software instruction.

# FLORENCE NIGHTINGALE

Florence Nightingale (1820-1910) won fame as a founder of the nursing profession and as a reformer of health care. As chief nurse for the British army during the Crimean War, from 1854 to 1856, she found that lack of sanitation and disease killed large numbers of soldiers hospitalized by wounds. Her reforms reduced the death rate at her military hospital from 42.7% to 2.2%, and she returned from the war famous. She at once began a fight to reform the entire military health care system, with considerable success.

One of the chief weapons Florence Nightingale used in her efforts was data. She had the facts, because she reformed record keeping as well as medical care. She was a pioneer in using graphs to present data in a vivid form that even generals and members of Parliament could understand. Her inventive graphs are a landmark in the growth of the new science of statistics. She considered statistics essential to understanding any social issue and tried to introduce the study of statistics into higher education.

In beginning our study of statistics, we will follow Florence Nightingale's lead. Part I stresses the analysis of data as a path to understanding. Like her, we will start by measuring outcomes and making graphs to see what data can teach us. Along with the graphs we will present numerical summaries, just as Florence Nightingale calculated detailed death rates and other summaries. Data for Florence Nightingale were not dry or abstract, because they showed her, and helped her show others, how to save lives. That remains true today.

PART **I**

# Understanding Data

# DATA AND MEASUREMENT

Statistics is the science of data. We therefore begin our study of statistics by mastering the art of examining data. Any set of data contains information about some group of *individuals*. The information is organized in *variables*.

---

INDIVIDUALS AND VARIABLES

**Individuals** are the objects described by a set of data. Individuals may be people, but they may also be animals or things.

A **variable** is any characteristic of an individual. A variable can take different values for different individuals.

---

Data for a study of a company's pay policies, for example, might include data about every employee. The employees are the individuals described by the data set. For each individual, the data contain the values of variables such as the age in years, gender (female or male), job category, and annual salary in dollars. In practice, any set of data is accompanied by background information that helps us understand the data. When you meet a new set of data, ask yourself the following questions:

1. **Who?** What **individuals** do the data describe? **How many** individuals appear in the data?

2. **What?** How many **variables** do the data contain? What are the **exact definitions** of these variables? In what **units of measurement** is each variable recorded? Weights, for example, might be recorded in pounds, in thousands of pounds, or in kilograms.

3. **Why? What purpose** do the data have? Do we hope to answer some specific questions? Do we want to draw conclusions about individuals other than the ones we actually have data for?

Some variables, like gender and job title, simply place individuals into categories. Others, like height and annual income, take numerical values for which we can do arithmetic. It makes sense to give an average income for a company's employees, but it does not make sense to give an "average" gender. We can, however, count the numbers of female and male employees and do arithmetic with these counts.

---

CATEGORICAL AND QUANTITATIVE VARIABLES

A **categorical variable** records which of several groups or categories an individual belongs to.

A **quantitative variable** takes numerical values for which it makes sense to do arithmetic operations like adding and averaging.

The **distribution** of a variable tells us what values it takes and how often it takes these values.

---

## EXERCISES

2.1    Here is a small part of the data set that a company keeps to record information about its employees:

| Name | Age | Gender | Race | Salary | Job type |
|------|-----|--------|------|--------|----------|
| ⋮ | | | | | |
| Fleetwood, Delores | 39 | Female | White | 52,100 | Management |
| Fleming, LaVerne | 27 | Male | Black | 37,500 | Technical |
| Foo, Ruoh-Lin | 22 | Female | Asian | 15,250 | Clerical |
| ⋮ | | | | | |

(a) What individuals does the complete data set describe?

(b) The data set records five variables in addition to the name. Which of these are categorical variables?

(c) Which of the variables are quantitative? Based on the data in the table, what do you think are the units of measurement for each of the quantitative variables?

(d) What might be some of the purposes that led the company to gather these data?

2.2    Here is part of a data set that describes 1997 model motor vehicles:

| Vehicle | Type | Where made | City MPG | Highway MPG |
|---------|------|-----------|----------|-------------|
| Acura 2.5TL | Compact | Foreign | 20 | 25 |
| Buick Skylark | Compact | Domestic | 22 | 32 |
| Audi A8 Quattro | Midsize | Foreign | 17 | 25 |
| Chrysler Concorde | Large | Domestic | 19 | 27 |

Identify the individuals. Then list the variables recorded for each individual and classify each variable as categorical or quantitative.

2.3    Data from a medical study contain values of many variables for each of the people who were the subjects of the study. Which of the following variables are categorical and which are quantitative?

(a) Gender (female or male)

(b) Age (years)

(c) Race (Asian, black, white, or other)

(d) Smoker (yes or no)

(e) Systolic blood pressure (millimeters of mercury)

(f) Level of calcium in the blood (micrograms per milliliter)

2.4 Here is a small part of a data set that describes public education in each of the 50 states:

| State | Region | Population | SAT verbal | SAT math | Percent taking | Dollars per pupil | Teachers' pay ($1000) |
|-------|--------|-----------|------------|----------|----------------|-------------------|------------------------|
| ⋮ | | | | | | | |
| CA | PAC | 29,760 | 419 | 484 | 45 | 4,826 | 39.6 |
| CO | MTN | 3,294 | 456 | 513 | 28 | 4,809 | 31.8 |
| CT | NE | 3,287 | 430 | 471 | 74 | 7,914 | 43.8 |
| ⋮ | | | | | | | |

What are the individuals in this study? What are the variables? Which of the variables are categorical and which are quantitative? In what unit of measurement is each quantitative variable measured?

# THE DISTRIBUTION OF ONE VARIABLE

Statistical tools and ideas can help you examine your data in order to describe their main features. This examination is called **exploratory data analysis.** Like an explorer crossing unknown lands, we want first to simply describe what we see. Here are two basic strategies that help us organize our exploration of a set of data:

- Begin by examining each variable by itself. Then move on to study the relationships among the variables.
- Begin with a graph or graphs. Then add numerical summaries of specific aspects of the data.

We will organize our learning the same way. This lesson concerns displays for a single variable, and Lesson 4 looks at numerical summaries for one variable. Later lessons present ideas and tools for studying relationships among variables.

## CATEGORICAL VARIABLES

The values of a categorical variable are just labels for the categories, such as "male" and "female." The distribution of a categorical variable lists the categories and gives either the **count** or the **percent** of individuals who fall in each category. For example, here is the distribution of marital status for all Americans age 18 and over:

| Marital status | Count (millions) | Percent |
|---|---|---|
| Single | 41.8 | 22.6 |
| Married | 113.3 | 61.1 |
| Widowed | 13.9 | 7.5 |
| Divorced | 16.3 | 8.8 |

The graphs in Figure 3.1 display these data. The **bar graph** in Figure 3.1(a) quickly compares the sizes of the four marital status groups. The heights of the four bars show the counts in the four categories. The **pie chart** in Figure 3.1(b) helps us see what part of the whole each group forms. For example, the "married" slice makes up 61% of the pie because 61% of adults are married. Bar graphs and pie charts help an audience grasp the distribution quickly. But neither kind of graph is essential to

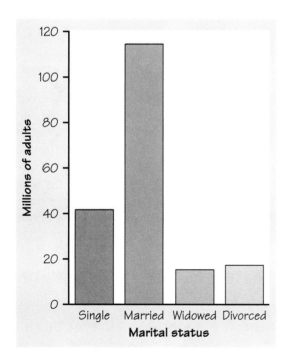

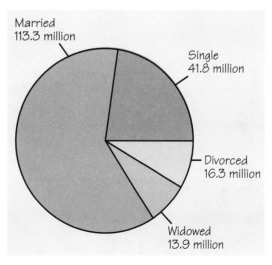

FIGURE 3.1(a)    Bar graph of the marital status of U.S. adults.

FIGURE 3.1(b)    Pie chart of the same data.

understanding the data. Categorical data on a single variable such as marital status are easy to describe without a graph. We can move on at once to graphs for quantitative variables.

## QUANTITATIVE VARIABLES: STEM-AND-LEAF PLOTS AND DOTPLOTS

For small data sets, a *stem-and-leaf plot* presents the distribution quickly and without losing information.

---

STEM-AND-LEAF PLOT

To make a stem-and-leaf plot:

1. Separate each observation into a **stem** consisting of all but the final (rightmost) digit and a **leaf**, the final digit.
2. Write the stems in a vertical column with the smallest at the top, and draw a vertical line at the right of this column.
3. Write each leaf in the row to the right of its stem, in increasing order out from the stem.

---

A **dotplot** is quite similar to a stem-and-leaf plot. Draw a number axis whose scale spans the range of your data. (You can make a dotplot either vertically like a stem-and-leaf plot or horizontally.) Mark each observation as a point on this number axis. *Data Desk* does exactly this, so that data sets containing many observations form dense clusters of points on the axis. Other software stacks the points for observations having the same value (after rounding off). The dotplot then looks very much like a stem-and-leaf plot with each leaf replaced by a dot.

---

**EXAMPLE 3.1**

Table 3.1 presents the percent of residents aged 65 years and over in each of the 50 states. The stem-and-leaf plot for these data appears in Figure 3.2. The stems are whole percents, and tenths of a percent form the leaves. Figure 3.3 is a horizontal dotplot of the same data.     ◄

**TABLE 3.1**  Percent of population 65 years old and over, by state (1991)

| State | Percent | State | Percent | State | Percent |
|---|---|---|---|---|---|
| Alabama | 12.9 | Louisiana | 11.2 | Ohio | 13.1 |
| Alaska | 4.2 | Maine | 13.4 | Oklahoma | 13.5 |
| Arizona | 13.2 | Maryland | 10.9 | Oregon | 13.7 |
| Arkansas | 14.9 | Massachusetts | 13.7 | Pennsylvania | 15.5 |
| California | 10.5 | Michigan | 12.1 | Rhode Island | 15.1 |
| Colorado | 10.1 | Minnesota | 12.5 | South Carolina | 11.4 |
| Connecticut | 13.7 | Mississippi | 12.4 | South Dakota | 14.7 |
| Delaware | 12.2 | Missouri | 14.1 | Tennessee | 12.7 |
| Florida | 18.3 | Montana | 13.4 | Texas | 10.1 |
| Georgia | 10.1 | Nebraska | 14.1 | Utah | 8.8 |
| Hawaii | 11.4 | Nevada | 10.8 | Vermont | 11.9 |
| Idaho | 12.0 | New Hampshire | 11.6 | Virginia | 10.9 |
| Illinois | 12.5 | New Jersey | 13.4 | Washington | 11.8 |
| Indiana | 12.6 | New Mexico | 10.9 | West Virginia | 15.1 |
| Iowa | 15.4 | New York | 13.1 | Wisconsin | 13.3 |
| Kansas | 13.9 | North Carolina | 12.3 | Wyoming | 10.6 |
| Kentucky | 12.7 | North Dakota | 14.5 | | |

SOURCE: *Statistical Abstract of the United States,* 1992.

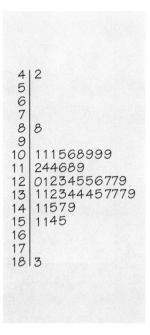

FIGURE 3.2   Stem-and-leaf plot of the percent of residents 65 years old and over in the 50 states, from Table 3.1.

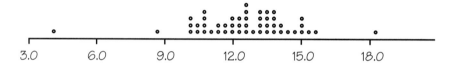

FIGURE 3.3    Dotplot of the age data from Table 3.1.

## QUANTITATIVE VARIABLES: HISTOGRAMS

Quantitative variables often take so many values that a graph of the distribution is clearer if nearby values are grouped together. The most common graph of the distribution of one quantitative variable is a **histogram**.

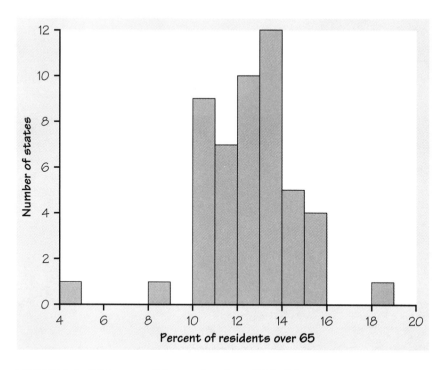

FIGURE 3.4    Histogram of the age data from Table 3.1.

EXAMPLE 3.2

Figure 3.4 is a histogram of the distribution of the percents aged 65 and over in the 50 states. Here are the steps that led to this histogram:

1. Divide the range of the data into classes of equal width. The classes in Figure 3.4 are

$$4.0 < \text{percent 65 and over} \leq 5.0$$
$$5.0 < \text{percent 65 and over} \leq 6.0$$
$$\vdots$$
$$18.0 < \text{percent 65 and over} \leq 19.0$$

Notice that each observation falls into exactly one class. A state with 5.0% of its residents aged 65 or older would fall into the first class, but 5.1% falls into the second.

2. Count the number of observations in each class. The counts are as follows:

| Class | Count | Class | Count | Class | Count |
|-------|-------|-------|-------|-------|-------|
| 4.1 to 5.0 | 1 | 9.1 to 10.0 | 0 | 14.1 to 15.0 | 5 |
| 5.1 to 6.0 | 0 | 10.1 to 11.0 | 9 | 15.1 to 16.0 | 4 |
| 6.1 to 7.0 | 0 | 11.1 to 12.0 | 7 | 16.1 to 17.0 | 0 |
| 7.1 to 8.0 | 0 | 12.1 to 13.0 | 10 | 17.1 to 18.0 | 0 |
| 8.1 to 9.0 | 1 | 13.1 to 14.0 | 12 | 18.1 to 19.0 | 1 |

3. Draw the histogram. The horizontal axis contains the scale for the variable whose distribution you are displaying. That's "percent of residents 65 and over" in this example. The vertical axis contains the scale of counts. Each bar represents a class. The base of the bar covers the class, and the bar height is the class count. There is no horizontal space between the bars unless a class is empty, so that its bar has height zero.    ◀

Our eyes respond to the *area* of the bars in a histogram. Because the classes are all the same width, area is determined by height and all classes are fairly represented. There is no one right choice of the classes in a histogram. Too few classes will give a "skyscraper" graph, with all values in a few classes with tall bars. Too many will produce a "pancake" graph, with most classes having one or no observations. Neither choice will give a good picture of the shape of the distribution. You must use your judgment in choosing classes to display the shape. Statistics software will choose the classes for you. The computer's choice is usually a good one, but you can change it if you want.

A stem-and-leaf plot looks like a histogram turned on end. The stem-and-leaf plot in Figure 3.2 is much like the histogram in Figure 3.4. The stem-and-leaf plot, unlike the histogram, preserves the actual value of each observation. Histograms are more flexible than stem-and-leaf plots because we can choose the classes. The classes (the stems) of a stem-and-leaf plot are given to us. Stem-and-leaf plots work poorly for large numbers of observations because each stem then carries too many leaves. Because of their flexibility, histograms are the most common display of the distribution of a quantitative variable.

## DESCRIBING DISTRIBUTION SHAPE

Making a statistical graph is not an end in itself. The purpose of the graph is to help us understand the data. After you make a graph, always ask, "What do I see?" Here is a general tactic for looking at graphs:

- Look for an **overall pattern** and also for striking **deviations** from that pattern.

In the case of a single quantitative variable, the overall pattern is the overall shape of the distribution. *Outliers* are an important kind of deviation from the overall pattern.

---

OUTLIERS

An **outlier** in any graph of data is an individual observation that falls outside the overall pattern of the graph.

Three states stand out in all of the graphs of the 65-and-over data. You can find them in Table 3.1 once a graph has called attention to them. Florida has 18.3% of its residents age 65 and over, and Alaska has only 4.2%. These states are clear outliers. You might also call Utah, with 8.8%, an outlier, though it is not as far from the overall pattern as Florida and Alaska. Whether an observation is an outlier is to some extent a matter of judgment. It is much easier to spot outliers in the graphs than in the data table.

Once you have spotted outliers, look for an explanation. Many outliers are due to mistakes, such as typing 4.0 as 40. Other outliers point to the special nature of some observations. Explaining outliers usually requires some background information. It is not surprising that Florida, with its many retired people, has many residents over 65 and that Alaska, the northern frontier, has few.

What about the *overall pattern* of the distribution in Figures 3.2 to 3.4? This distribution has an irregular shape that isn't easy to describe. It does have a single major peak, or **mode,** at about 13%. A distribution with one major peak is called **unimodal.** Some distributions have simpler shapes. Here are some shapes to look for.

---

SYMMETRIC AND SKEWED DISTRIBUTIONS

A distribution is **symmetric** if the right and left sides of the histogram are approximately mirror images of each other.

A distribution is **skewed to the right** if the right side of the histogram (containing the upper half of the observations) extends much farther out than the left side (containing the lower half of the observations). It is **skewed to the left** if the left side of the histogram extends much farther out than the right side.

---

EXAMPLE 3.3

Look at the histograms in Figures 3.5 and 3.6. Figure 3.5 comes from a study of lightning storms in Colorado. It shows the distribution of the hour of the day during which the first lightning flash for that day occurred. The distribution is *symmetric and unimodal*, with a single mode at noon.

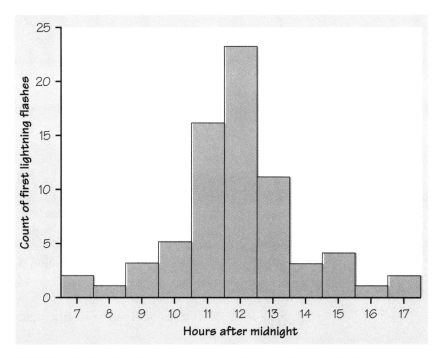

FIGURE 3.5    The distribution of the time of the first lightning flash each day at a site in Colorado.

Figure 3.6 shows the distribution of lengths of words used in Shakespeare's plays.[1] This distribution is also unimodal but is *skewed to the right*. That is, there are many short words (3 and 4 letters) and few very long words (10, 11, or 12 letters).    ◀

Notice that the vertical scale in Figure 3.6 is not the *count* of words but the *percent* of all of Shakespeare's words that have each length. A histogram of percents rather than counts is convenient when the counts are very large or when we want to compare several distributions.

The shape of a distribution is not changed by making a histogram of percents instead of counts. It is also not changed by a change in the unit of measurement. Measuring time in minutes instead of hours in Figure 3.5 would change the scale on the horizontal axis but would not change the shape of the histogram.

Some specific distribution shapes occur often enough to deserve special study. For example, the **normal distributions** are a family of symmetric, unimodal distributions that play an important role in statistics. We will meet the normal distributions in detail in Lesson 5.

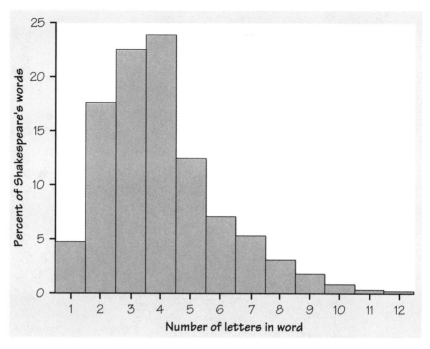

FIGURE 3.6   The distribution of lengths of words used in Shakespeare's plays.

## EXERCISES

3.1   Here are data on the percent of females among people earning doctorates in 1992 in several fields of study (from the 1995 *Statistical Abstract of the United States*):

| | |
|---|---|
| Computer science | 13.3% |
| Life sciences | 38.3% |
| Education | 59.5% |
| Engineering | 9.6% |
| Physical sciences | 21.9% |
| Psychology | 59.7% |

(a)  Present these data in a well-labeled bar graph.

(b)  Would it also be correct to use a pie chart to display these data? Explain your answer.

3.2   In 1992 there were 86,777 deaths from accidents in the United States. Among these were 40,982 deaths from motor vehicle accidents, 12,646 from falls, 3524 from drowning, 3958 from fires, and 7082 from poisoning. (Data from the 1995 *Statistical Abstract of the United States*.)

(a) Find the percent of accidental deaths from each of these causes, rounded to the nearest percent. What percent of accidental deaths were due to other causes?

(b) Make a well-labeled bar graph of the distribution of causes of accidental deaths. Be sure to include an "other causes" bar.

(c) Would it also be correct to use a pie chart to display these data? Explain your answer.

3.3    The Survey of Study Habits and Attitudes (SSHA) is a psychological test that evaluates college students' motivation, study habits, and attitudes toward school. A private college gives the SSHA to a sample of 18 of its incoming first-year women students. Their scores are

| 154 | 109 | 137 | 115 | 152 | 140 | 154 | 178 | 101 |
| 103 | 126 | 126 | 137 | 165 | 165 | 129 | 200 | 148 |

Make a stem-and-leaf plot or dotplot of these data. The overall shape of the distribution is irregular, as often happens when only a few observations are available. Is there a clear mode? Are there any outliers? About where is the center of the distribution (the score with half the scores above it and half below)? What is the spread of the scores (ignoring any outliers)?

3.4    Corn is an important animal food. Normal corn lacks certain amino acids, which are building blocks for protein. Plant scientists have developed new corn varieties that have more of these amino acids. To test a new corn as an animal food, a group of 20 one-day-old male chicks was fed a ration containing the new corn. A control group of another 20 chicks was fed a ration that was identical except that it contained normal corn. Here are the weight gains (in grams) after 21 days:[2]

| Normal corn | | | | New corn | | | |
|---|---|---|---|---|---|---|---|
| 380 | 321 | 366 | 356 | 361 | 447 | 401 | 375 |
| 283 | 349 | 402 | 462 | 434 | 403 | 393 | 426 |
| 356 | 410 | 329 | 399 | 406 | 318 | 467 | 407 |
| 350 | 384 | 316 | 272 | 427 | 420 | 477 | 392 |
| 345 | 455 | 360 | 431 | 430 | 339 | 410 | 326 |

To compare the two distributions of weight gains, make dotplots side by side. What do the plots show about the effect of the improved corn variety on weight gain?

3.5    Although the side-by-side dotplots of Exercise 3.4 are effective for comparing the two weight-gain distributions, they do not show clearly the shape of the distributions.

(a) Make separate histograms for the two groups. What do your histograms show about the effect of the improved corn on weight gain?

(b) Are the distributions of weight gains symmetric or skewed? Are they unimodal, bimodal (two distinct peaks), or multimodal (many peaks)?

3.6    Sketch a histogram for a distribution that is skewed to the left. Suppose that you and your friends emptied your pockets of coins and recorded the year marked on each coin. The distribution of dates would be skewed to the left. Explain why.

3.7    Environmental Protection Agency regulations require automakers to give the city and highway gas mileages for each model of car. Here are the highway mileages (miles per gallon) for 26 midsize 1997 car models:[3]

| Model | MPG | Model | MPG |
|---|---|---|---|
| Acura 3.5RL | 25 | Lexus GS300 | 24 |
| Audi A8 Quattro | 25 | Lexus LS400 | 25 |
| Buick Century | 29 | Lincoln Mark VIII | 26 |
| Cadillac Catera | 25 | Mazda 626 | 31 |
| Cadillac Eldorado | 26 | Mercedes-Benz E320 | 27 |
| Chevrolet Lumina | 29 | Mercedes-Benz E420 | 25 |
| Chrysler Cirrus | 30 | Mitsubishi Diamante | 26 |
| Dodge Stratus | 32 | Nissan Maxima | 28 |
| Ford Taurus | 28 | Oldsmobile Aurora | 26 |
| Ford Thunderbird | 26 | Rolls-Royce Silver Spur | 16 |
| Hyundai Sonata | 27 | Saab 900 | 26 |
| Infiniti I30 | 28 | Toyota Camry | 30 |
| Infiniti Q45 | 24 | Volvo 850 | 26 |

(a) Make a graph that displays the distribution. Describe its main features (overall pattern and any outliers) in words.

(b) The government imposes a "gas guzzler" tax on cars with low gas mileage. Which of these cars do you think are subject to the gas guzzler tax?

3.8    The total return on a stock is the change in its market price plus any dividend payments made. Total return is usually expressed as a percent of the beginning price. Figure 3.7 is a histogram of the distribution of total returns for all 1528 stocks listed on the New York Stock Exchange in one year.[4]

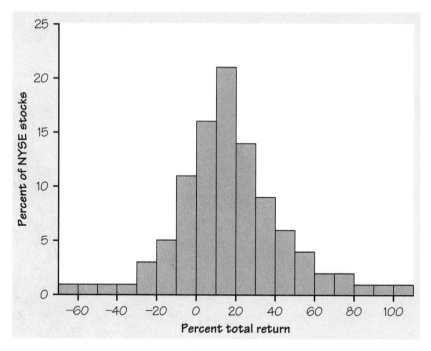

FIGURE 3.7    The distribution of percent total return for all New York Stock Exchange common stocks in one year, for Exercise 3.8.

(a) Describe the overall shape of the distribution of total returns.

(b) What is the approximate center of this distribution? (For now, take the center to be the value with roughly half the stocks having lower returns and half having higher returns.)

(c) Approximately what were the smallest and largest total returns? (This describes the spread of the distribution.)

(d) A return less than zero means that an owner of the stock lost money. About what percent of all stocks lost money?

3.9    Figure 3.8 is a histogram of the number of days in the month of April on which the temperature fell below freezing at Greenwich, England.[5] The data cover a period of 65 years.

(a) Describe the shape of this distribution. Are there any outliers?

(b) In what percent of these 65 years did the temperature never fall below freezing in April?

3.10    The distribution of the ages of a nation's population has a strong influence on economic and social conditions. The following table shows the age distribution of U.S. residents in 1950 and 2075, in millions of persons. The

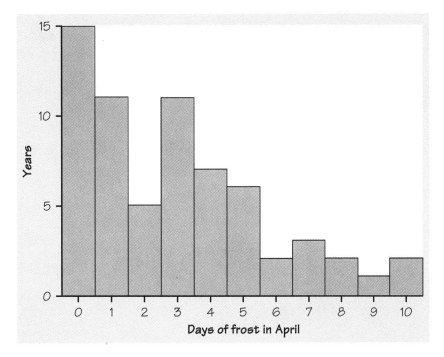

FIGURE 3.8   The distribution of the number of frost days during April at Greenwich, England, over a 65-year period, for Exercise 3.9.

1950 data come from that year's census. The 2075 data are projections made by the Census Bureau.

| Age group | 1950 | 2075 |
|---|---|---|
| Under 10 years | 29.3 | 34.9 |
| 10 to 19 years | 21.8 | 35.7 |
| 20 to 29 years | 24.0 | 36.8 |
| 30 to 39 years | 22.8 | 38.1 |
| 40 to 49 years | 19.3 | 37.8 |
| 50 to 59 years | 15.5 | 37.5 |
| 60 to 69 years | 11.0 | 34.5 |
| 70 to 79 years | 5.5 | 27.2 |
| 80 to 89 years | 1.6 | 18.8 |
| 90 to 99 years | 0.1 | 7.7 |
| 100 to 109 years | — | 1.7 |
| Total | 151.1 | 310.6 |

(a) Because the total population in 2075 is much larger than the 1950 population, comparing percents in each age group is clearer than comparing counts. Make a table of the percent of the total population in each age group for both 1950 and 2075.

(b) Make a histogram of the 1950 age distribution (in percents). Then describe the main features of the distribution. In particular, look at the percent of children relative to the rest of the population.

**TABLE 3.2**    Education and related data for the states

| State | Region* | Population (thousands) | SAT verbal | SAT math | Percent taking | Dollars per pupil | Teachers' pay ($1000) |
|-------|---------|------------------------|------------|----------|----------------|-------------------|------------------------|
| AL | ESC | 4,041 | 470 | 514 | 8 | 3,648 | 27.3 |
| AK | PAC | 550 | 438 | 476 | 42 | 7,887 | 43.4 |
| AZ | MTN | 3,665 | 445 | 497 | 25 | 4,231 | 30.8 |
| AR | WSC | 2,351 | 470 | 511 | 6 | 3,334 | 23.0 |
| CA | PAC | 29,760 | 419 | 484 | 45 | 4,826 | 39.6 |
| CO | MTN | 3,294 | 456 | 513 | 28 | 4,809 | 31.8 |
| CT | NE | 3,287 | 430 | 471 | 74 | 7,914 | 43.8 |
| DE | SA | 666 | 433 | 470 | 58 | 6,016 | 35.2 |
| DC | SA | 607 | 409 | 441 | 68 | 8,210 | 39.6 |
| FL | SA | 12,938 | 418 | 466 | 44 | 5,154 | 30.6 |
| GA | SA | 6,478 | 401 | 443 | 57 | 4,860 | 29.2 |
| HI | PAC | 1,108 | 404 | 481 | 52 | 5,008 | 32.5 |
| ID | MTN | 1,007 | 466 | 502 | 17 | 3,200 | 25.5 |
| IL | ENC | 11,431 | 466 | 528 | 16 | 5,062 | 34.6 |
| IN | ENC | 5,544 | 408 | 459 | 54 | 5,051 | 32.0 |
| IA | WNC | 2,777 | 511 | 577 | 5 | 4,839 | 28.0 |
| KS | WNC | 2,478 | 492 | 548 | 10 | 5,009 | 29.8 |
| KY | ESC | 3,685 | 473 | 521 | 10 | 4,390 | 29.1 |
| LA | WSC | 4,220 | 476 | 517 | 9 | 4,012 | 26.2 |
| ME | NE | 1,228 | 423 | 463 | 60 | 5,894 | 28.5 |
| MD | SA | 4,781 | 430 | 478 | 59 | 6,184 | 38.4 |
| MA | NE | 6,016 | 427 | 473 | 72 | 6,351 | 36.1 |
| MI | ENC | 9,295 | 454 | 514 | 12 | 5,257 | 38.3 |
| MN | WNC | 4,375 | 477 | 542 | 14 | 5,260 | 33.1 |
| MS | ESC | 2,573 | 477 | 519 | 4 | 3,322 | 24.4 |

(c) Make a histogram of the projected age distribution for the year 2075. Use the same scales as in (b) for easy comparison. What are the most important changes in the U.S. age distribution projected for the 125-year period between 1950 and 2075?

**TABLE 3.2** Education and related data for the states (*continued*)

| State | Region* | Population (thousands) | SAT verbal | SAT math | Percent taking | Dollars per pupil | Teachers' pay ($1000) |
|-------|---------|------------------------|------------|----------|----------------|-------------------|------------------------|
| MO | WNC | 5,117 | 473 | 522 | 12 | 4,415 | 28.5 |
| MT | MTN | 799 | 464 | 523 | 20 | 5,184 | 26.7 |
| NE | WNC | 1,578 | 484 | 546 | 10 | 4,381 | 26.6 |
| NV | MTN | 1,202 | 434 | 487 | 24 | 4,564 | 32.2 |
| NH | NE | 1,109 | 442 | 486 | 67 | 5,504 | 31.3 |
| NJ | MA | 7,730 | 418 | 473 | 69 | 9,159 | 38.4 |
| NM | MTN | 1,515 | 480 | 527 | 12 | 4,446 | 26.2 |
| NY | MA | 17,990 | 412 | 470 | 70 | 8,500 | 42.1 |
| NC | SA | 6,629 | 401 | 440 | 55 | 4,802 | 29.2 |
| ND | WNC | 639 | 505 | 564 | 6 | 3,685 | 23.6 |
| OH | ENC | 10,847 | 450 | 499 | 22 | 5,639 | 32.6 |
| OK | WSC | 3,146 | 478 | 523 | 9 | 3,742 | 24.3 |
| OR | PAC | 2,842 | 439 | 484 | 49 | 5,291 | 32.3 |
| PA | MA | 11,882 | 420 | 463 | 64 | 6,534 | 36.1 |
| RI | NE | 1,003 | 422 | 461 | 62 | 6,989 | 37.7 |
| SC | SA | 3,487 | 397 | 437 | 54 | 4,327 | 28.3 |
| SD | WNC | 696 | 506 | 555 | 5 | 3,730 | 22.4 |
| TN | ESC | 4,877 | 483 | 525 | 12 | 3,707 | 28.2 |
| TX | WSC | 16,987 | 413 | 461 | 42 | 4,238 | 28.3 |
| UT | MTN | 1,723 | 492 | 539 | 5 | 2,993 | 25.0 |
| VT | NE | 563 | 431 | 466 | 62 | 5,740 | 31.0 |
| VA | SA | 6,187 | 425 | 470 | 58 | 5,360 | 32.4 |
| WA | PAC | 4,867 | 437 | 486 | 44 | 5,045 | 33.1 |
| WV | SA | 1,793 | 443 | 490 | 15 | 5,046 | 26.0 |
| WI | ENC | 4,892 | 476 | 543 | 11 | 5,946 | 33.1 |
| WY | MTN | 454 | 458 | 519 | 13 | 5,255 | 29.0 |

SOURCE: *Statistical Abstract of the United States,* 1992.

*The census regions are: East North Central, East South Central, Middle Atlantic, Mountain, New England, Pacific, South Atlantic, West North Central, and West South Central.

3.11     The data below are the survival times in days of 72 guinea pigs after they were injected with tubercle bacilli in a medical experiment.[6] Survival times, whether of machines under stress or cancer patients after treatment, usually have distributions that are skewed to the right.

| | | | | | | | | |
|---|---|---|---|---|---|---|---|---|
| 43 | 45 | 53 | 56 | 56 | 57 | 58 | 66 | 67 |
| 73 | 74 | 79 | 80 | 80 | 81 | 81 | 81 | 82 |
| 83 | 83 | 84 | 88 | 89 | 91 | 91 | 92 | 92 |
| 97 | 99 | 99 | 100 | 100 | 101 | 102 | 102 | 102 |
| 103 | 104 | 107 | 108 | 109 | 113 | 114 | 118 | 121 |
| 123 | 126 | 128 | 137 | 138 | 139 | 144 | 145 | 147 |
| 156 | 162 | 174 | 178 | 179 | 184 | 191 | 198 | 211 |
| 214 | 243 | 249 | 329 | 380 | 403 | 511 | 522 | 598 |

Graph the distribution and describe its main features. Does it show the expected right skew?

3.12     Exercise 3.7 gives the fuel consumption of 26 midsize car models in miles per gallon. There are 3.785 liters in a gallon. Restate the data as miles per liter and make a histogram of the data in this new unit of measurement. What effect did the change of units have on the appearance of the histogram?

Table 3.2 presents data about the individual states that relate to education. The data for three states in Exercise 2.4 come from this table. Study of a data set with many variables begins by examining each variable by itself. Exercises 3.13 to 3.15 concern the data in Table 3.2.

3.13     Make a graph of the distribution of the percent of high school seniors who take the Scholastic Assessment Test (SAT) in the various states. Briefly describe the overall pattern of the distribution and any outliers.

3.14     Make a graph to display the distribution of average teachers' salaries for the states. Is there a clear overall pattern? Are there any outliers or other notable deviations from the pattern?

3.15     Make a graph of the distribution of dollars per pupil spent on education for the states. Is there a clear overall pattern? Are there any outliers or other notable deviations from the pattern?

# MEASURING CENTER AND SPREAD

A brief description of a distribution should include its shape, a number describing its center, and numbers describing its spread. **Shape, center, and spread provide a good description of the overall pattern of the distribution of a quantitative variable.** We describe the shape of a distribution based on inspection of a histogram, stem-and-leaf plot, or dotplot. Now we will learn specific ways to use numbers to measure the center and spread of a distribution. We can calculate these numerical measures for any quantitative variable. But to interpret measures of center and spread, and to choose among the several measures we will learn, you must think about the shape of the distribution and the meaning of the data. The numbers, like graphs, are aids to understanding, not "the answer" in themselves.

## MEASURING CENTER

A description of a distribution almost always includes a measure of its center or average. Three such measures are the *median,* the *mean,* and the *midrange.* The median is the "middle value," the mean is the "average value," and the midrange is "midway between the extremes." These are three different ideas for "center," and the three measures behave differently. Here are the precise recipes for median, mean, and midrange.

---

### THE MEDIAN

The **median** is the middle value of a distribution, the number such that half the observations are smaller and the other half are larger. To find the median of a distribution:

1. Arrange all observations in order of size, from smallest to largest.
2. If the number of observations $n$ is odd, the median $M$ is the center observation in the ordered list. Find the location of the median by counting $(n + 1)/2$ observations up from the bottom of the list.
3. If the number of observations $n$ is even, the median $M$ is the mean of the two center observations in the ordered list. The location of the median is again $(n + 1)/2$ from the bottom of the list.

---

### THE MEAN $\bar{y}$

To find the **mean** of a set of observations, add their values and divide by the number of observations. If the $n$ observations are $y_1, y_2, \ldots, y_n$, their mean is

$$\bar{y} = \frac{y_1 + y_2 + \cdots + y_n}{n}$$

or, in more compact notation,

$$\bar{y} = \frac{1}{n} \sum y_i$$

---

The $\sum$ (capital Greek sigma) in the formula for the mean is short for "add them all up." The subscripts on the observations $y_i$ are just a way of keeping the $n$ observations distinct. They do not necessarily indicate order or any other special facts about the data. The bar over the $y$ indicates the mean of all the $y$-values. Pronounce the mean $\bar{y}$ as "y-bar." This notation is very common. When writers who are discussing data use $\bar{x}$ or $\bar{y}$, they are talking about a mean.

---

**THE MIDRANGE**

The **midrange** of a set of observations is the average (mean) of the largest and smallest observations:

$$\text{midrange} = \frac{\text{max} + \text{min}}{2}$$

---

**EXAMPLE 4.1**

A study in Switzerland examined the number of hysterectomies (removal of the uterus) performed in a year by doctors. Here are the data for a sample of 15 male doctors:

27   50   33   25   86   25   85   31   37   44   20   36   59   34   28

The stem-and-leaf plot in Figure 4.1 shows that the distribution is skewed to the right and that there are two outliers on the high side.

The *mean* number of operations that these 15 doctors performed is

$$\bar{y} = \frac{27 + 50 + \cdots + 28}{15} = \frac{620}{15} = 41.3$$

◀

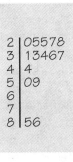

```
2 | 05578
3 | 13467
4 | 4
5 | 09
6 |
7 |
8 | 56
```

FIGURE 4.1   Stem-and-leaf plot of the surgery counts from Example 4.1.

To find the *median,* arrange the observations in increasing order:

20   25   25   27   28   31   33   **34**   36   37   44   50   59   85   86

The location of the median in the ordered list is

$$\frac{n+1}{2} = \frac{16}{2} = 8$$

That is, the median is the 8th observation in the list, $M = 34$. The Swiss study also looked at a sample of 10 female doctors. The numbers of hysterectomies performed by these doctors (arranged in order) were

5   7   10   14   **18**   **19**   25   29   31   33

Now $n$ is even, and the rule for locating the median in the list gives

$$\frac{n+1}{2} = \frac{11}{2} = 5.5$$

The location 5.5 means "halfway between the fifth and sixth observations in the ordered list." The median is $M = 18.5$.

The *midrange* for male doctors is the average of the largest and smallest observations:

$$\text{midrange} = \frac{86 + 20}{2} = 53$$

◀

One way in which these measures of center differ is in their **resistance** to the effect of a few extreme observations. The midrange uses only the two extremes, so it lacks resistance—only 3 male doctors performed more hysterectomies than the midrange value, 53. The mean uses all the data values, so it is also sensitive to outlying values, though less so than the midrange—5 of the 15 male doctors performed more than the mean number of hysterectomies. The median is very resistant, because it simply counts observations and ignores how far out they lie.

The median and mean are the most common measures of the center of a distribution. The mean and median of a symmetric distribution are close together. If the distribution is exactly symmetric, the mean and median are exactly the same. In a skewed distribution, the mean is farther out in the long tail than is the median. For example, the distribution of house prices is strongly skewed to the right. There are many moderately priced houses and a few very expensive mansions. The few expensive houses pull the mean up but do not affect the median. The mean price of new houses sold in mid-1994 was $152,000, but the median price for these same houses was only $127,000. Reports about house prices, incomes, and other strongly skewed distributions usually give the median ("middle value") rather than the mean ("arithmetic average value").

## MEASURING SPREAD: QUARTILES

Mean, median, and midrange provide different measures of the center of a distribution. A measure of center alone can be misleading. Two nations with the same median family income are very different if one has extremes of wealth and poverty and the other has little variation among families. A drug with the correct mean concentration of active ingredient is dangerous if some batches are much too high and others much too low. We are interested in the *spread* or *variability* of incomes and drug potencies as well as their centers. **The simplest useful numerical description of a distribution consists of both a measure of center and a measure of spread.**

Three numerical measures of spread are the *range,* the *interquartile range,* and the *standard deviation.* The range is the spread of all the data, and the interquartile range is the spread of the middle half of the data. The standard deviation is more complicated: it is based on the average of the squared distances of the observations from their mean. As in the case of measuring center, these measures of spread tell us different things and behave differently.

RANGE, QUARTILES, INTERQUARTILE RANGE

The **range** of a distribution is the distance between the largest and the smallest individual observations:

$$range = max - min$$

*(continued on next page)*

*(continued from previous page)*

The **quartiles** mark the middle half of the data. The **lower quartile** $Q_1$ is the 25% point, the value that is larger than one-quarter of the observations. The **upper quartile** $Q_3$ is the 75% point, the value that is larger than three-quarters of the observations. The second quartile is the median, which is larger than half of the observations.

The **interquartile range** is the distance between the upper and lower quartiles:

$$IQR = Q_3 - Q_1$$

This definition of the quartiles is not complete. We need a rule for exactly how to locate the value "larger than one-quarter of the observations." There are in fact several such rules, so different statistics software can give slightly different values for $Q_1$ and $Q_3$. The differences are always small, so just use the values that your software produces.

## EXAMPLE 4.2

Here again are the data on hysterectomies performed by 15 male Swiss doctors, arranged in increasing order:

20   25   25   **27**   28   31   33   **34**   36   37   44   **50**   59   85   86

The *range* is

$$range = 86 - 20 = 66$$

The quartiles appear in boldface in the list. The median is $M = 34$, and the lower and upper quartiles are $Q_1 = 27$ and $Q_3 = 50$. The *interquartile range* is therefore

$$IQR = 50 - 27 = 23$$

◀

There is a natural pairing of measures of center and spread: the range goes with the midrange, and the quartiles and interquartile range go with the median. The range, like its companion the midrange, is extremely sensitive to outlying observations. It says nothing about the spread of the body of the distribution. The quartiles, like the median, are *resistant.* They ignore the tails of the distribution. We can give information about both the body and the tails of a distribution by reporting the *five-number summary.*

---

THE FIVE-NUMBER SUMMARY

The **five-number summary** of a data set consists of the smallest observation, the lower quartile, the median, the upper quartile, and the largest observation, written in order from smallest to largest. In symbols, the five-number summary is

$$\text{Minimum} \quad Q_1 \quad M \quad Q_3 \quad \text{Maximum}$$

---

## MEASURING SPREAD: THE STANDARD DEVIATION

The five-number summary is not the most common numerical description of a distribution. That distinction belongs to the combination of the mean to measure center with the *standard deviation* as a measure of spread. The standard deviation measures spread by looking at how far the observations are from their mean.

---

THE STANDARD DEVIATION $s$

The **variance $s^2$** of a set of observations is the average of the squares of the deviations of the observations from their mean. In symbols, the variance of $n$ observations $y_1, y_2, \ldots, y_n$ is

$$s^2 = \frac{(y_1 - \bar{y})^2 + (y_2 - \bar{y})^2 + \cdots + (y_n - \bar{y})^2}{n-1}$$

or, more compactly,

$$s^2 = \frac{1}{n-1} \sum (y_i - \bar{y})^2$$

---

*(continued on next page)*

*(continued from previous page)*

The **standard deviation s** is the square root of the variance $s^2$:

$$s = \sqrt{\frac{1}{n-1}\sum(y_i - \bar{y})^2}$$

## EXAMPLE 4.3

A person's metabolic rate is the rate at which the body consumes energy. Metabolic rate is important in studies of weight gain, dieting, and exercise. Here are the metabolic rates of 7 men who took part in a study of dieting. (The units are calories per 24 hours. These are the same calories used to describe the energy content of foods.)

$$1792 \quad 1666 \quad 1362 \quad 1614 \quad 1460 \quad 1867 \quad 1439$$

The researchers reported $\bar{y}$ and $s$ for these men:

$$\bar{y} = 1600 \text{ calories} \quad s = 189.24 \text{ calories}$$

◀

Figure 4.2 is a dotplot of these data, with their mean marked by an asterisk (∗). The arrows mark two of the deviations from the mean. These deviations show how spread out the data are about their mean. Some of the deviations will be positive and some negative because some of the observations fall on each side of the mean. In fact, *the sum of the deviations of the observations from their mean will always be zero*. The sizes of the deviations, both positive and negative, show the spread of the data. Larger deviations mean greater spread. The variance is an average of the squares of the

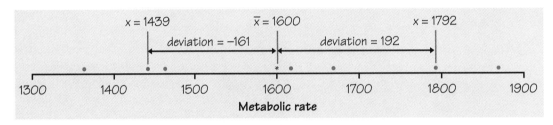

FIGURE 4.2   Metabolic rates for 7 men, with their mean (∗) and the deviations of two observations from the mean.

7 deviations from the mean, and the standard deviation is the square root of this average squared deviation.

Because the variance involves squaring the deviations, it does not have the same unit of measurement as the original observations. The variance of the metabolic rates, for example, is measured in squared calories. Taking the square root remedies this. The standard deviation $s$ measures spread about the mean in the original scale.

The idea of the variance is the average of the squares of the deviations of the observations from their mean. Why do we average by dividing by $n - 1$ rather than $n$? Because the sum of the deviations is always zero, the last deviation can be found once we know the other $n - 1$. So we are not averaging $n$ unrelated numbers. Only $n - 1$ of the squared deviations can vary freely, and we average by dividing the total by $n - 1$. The number $n - 1$ is called the **degrees of freedom** of the variance or standard deviation. Many calculators offer a choice between dividing by $n$ and dividing by $n - 1$, so be sure to use $n - 1$.

Here are the basic properties of the standard deviation $s$ as a measure of spread.

---

PROPERTIES OF THE STANDARD DEVIATION

- $s$ measures spread about the mean and should be used only when the mean is chosen as the measure of center.
- $s = 0$ only when there is *no spread*. This happens only when all observations have the same value. Otherwise $s > 0$. As the observations become more spread out about their mean, $s$ gets larger.
- $s$, like the mean $\bar{y}$, is not resistant. A few outliers can make $s$ very large.
- If a distribution is symmetric and unimodal, the middle 38% of the distribution is about one standard deviation wide.

---

The final property allows a rough estimate of $s$ from a histogram. It also allows us to compare the standard deviation $s$ with the interquartile range $IQR$, which is the spread of the middle 50% of the distribution. Do be aware that the 38% rule is very rough. It is not a substitute for calculating $s$. The real reason for the importance of the standard deviation is that it is the natural measure of spread for normal distributions. The 38% rule is exactly correct for normal distributions. We will meet the normal distributions in the next lesson.

## CHOOSING MEASURES OF CENTER AND SPREAD

How shall we choose between the five-number summary and $\bar{y}$ and $s$ to describe the center and spread of a distribution? A skewed distribution with a few observations in the single long tail will have a large standard deviation. The number $s$ does not give much helpful information in such a case. Because the two sides of a strongly skewed distribution have different spreads, no single number describes the spread well. The five-number summary, with its two quartiles and two extremes, does a better job.

---

CHOOSING SUMMARIES

The five-number summary is usually better than the mean and standard deviation for describing a skewed distribution or a distribution with strong outliers. Use $\bar{y}$ and $s$ only for reasonably symmetric distributions that are free of outliers.

---

Figure 4.3 is *Data Desk* output for the male Swiss doctors data, reporting both the five-number summary and $\bar{y}$ and $s$. Notice that the values given for the quartiles differ slightly from those in Example 4.2. This is an example of the effect of different detailed rules for calculating quartiles. The other summaries are as we have found them.

```
┌──────────────────────────────────────┐ ▢▣
│ Summary of     Hysterectomies        │
│ No Selector                          │
│                                      │
│ Percentile    25                     │
│                                      │
│             Count    15              │
│              Mean    41.3333         │
│            Median    34              │
│            StdDev    20.6074         │
│               Min    20              │
│               Max    86              │
│    Lower ith %tile   27.2500         │
│    Upper ith %tile   48.5000         │
│                                      │
└──────────────────────────────────────┘
```

FIGURE 4.3    *Data Desk* output for measures of center and spread.

Do remember that a graph gives the best overall picture of a distribution. Numerical measures of center and spread report specific facts about a distribution, but they do not describe its entire shape. Numerical summaries do not disclose the presence of multiple modes or gaps, for example. Exercise 4.16 gives an example of a distribution for which numerical summaries alone are misleading. **Always plot your data.**

## EXERCISES

4.1    The business magazine *Forbes* estimates (November 6, 1995) that the "average" household wealth of its readers is either about $800,000 or about $2.2 million, depending on which "average" it reports. Which of these numbers is the mean wealth and which is the median wealth? Explain your answer.

4.2    The NASDAQ Composite Index describes the average price of common stock traded over the counter, that is, not on one of the stock exchanges. In 1991, the mean capitalization of the companies in the NASDAQ index was $80 million and the median capitalization was $20 million. (A company's capitalization is the total market value of its stock.) Explain why the mean capitalization is much higher than the median.

4.3    Last year a small accounting firm paid each of its five clerks $25,000, two junior accountants $60,000 each, and the firm's owner $255,000. What is the mean salary paid at this firm? How many of the employees earn less than the mean? What is the median salary? What is the mode of the salaries? The midrange?

4.4    Which measure of center, the mean or the median, should you use in each of the following situations?

(a)    Middletown is considering imposing an income tax on citizens. The city government wants to know the average income of citizens so that it can estimate the total tax base.

(b)    In a study of the standard of living of typical families in Middletown, a sociologist estimates the average family income in that city.

4.5    This is a standard deviation contest. You must choose four numbers from the whole numbers 0 to 10, with repeats allowed.

(a)    Choose four numbers that have the smallest possible standard deviation.

(b)    Choose four numbers that have the largest possible standard deviation.

      (c) Is more than one choice possible in either (a) or (b)? Explain.

      (d) Would any of your answers to (a), (b), or (c) change if you used the range rather than the standard deviation to measure spread?

4.6    Here are the scores of 18 first-year college women on the Survey of Study Habits and Attitudes (SSHA):

$$154 \quad 109 \quad 137 \quad 115 \quad 152 \quad 140 \quad 154 \quad 178 \quad 101$$
$$103 \quad 126 \quad 126 \quad 137 \quad 165 \quad 165 \quad 129 \quad 200 \quad 148$$

      (a) Find the mean score from the formula for the mean. Then enter the data into your software and obtain the mean. Verify that you get the same result.

      (b) Find the midrange and the median of these data. A stem-and-leaf plot (Exercise 3.3) suggests that the score 200 is an outlier. How does this outlier explain the fact that for these data

$$\text{median} < \text{mean} < \text{midrange}$$

      (c) Find the mean, median, and midrange for the 17 observations that remain when you drop the outlier. Briefly describe how the outlier changes each measure of center.

4.7    Find the quartiles, the interquartile range, and the standard deviation for the SSHA scores in Exercise 4.6. Make a dotplot of the distribution, and mark the mean, median, and quartiles on your plot. Draw arrows from the median out to the quartiles (this spans $IQR$). Also draw arrows from the mean out $0.5s$ in either direction, spanning the standard deviation $s$. What feature of the data explains the fact that these two spans are about equal? The 38% rule for $s$ does not apply to these data.

4.8    Here are the number of home runs Babe Ruth hit in each of his 15 years with the New York Yankees:[7]

$$54 \quad 59 \quad 35 \quad 41 \quad 46 \quad 25 \quad 47 \quad 60 \quad 54 \quad 46 \quad 49 \quad 46 \quad 41 \quad 34 \quad 22$$

Roger Maris, who broke Ruth's single-year record, had these home run totals in his 10 years in the American League:

$$14 \quad 28 \quad 16 \quad 39 \quad 61 \quad 33 \quad 23 \quad 26 \quad 8 \quad 13$$

Make stem-and-leaf plots for both distributions, and find the five-number summaries. Use your results to compare Ruth and Maris as home run hitters.

4.9    The level of various substances in the blood influences our health. Here are measurements of the level of phosphate in the blood of a patient, in milligrams of phosphate per deciliter of blood, made on 6 consecutive visits to a clinic:

$$5.6 \quad 5.2 \quad 4.6 \quad 4.9 \quad 5.7 \quad 6.4$$

A graph of only 6 observations gives little information, so we proceed to compute the mean and standard deviation.

(a) Find the mean from its definition. That is, find the sum of the 6 observations and divide by 6.

(b) Find the standard deviation from its definition. That is, find the deviations of each observation from the mean, square the deviations, then obtain the variance and the standard deviation.

(c) Now enter the data into your software and obtain $\bar{y}$ and $s$. Do the results agree with your hand calculations?

4.10   Find the standard deviation for the data of Example 4.2. Then remove the two outliers and recalculate the midrange, interquartile range, and standard deviation for the remaining 13 doctors. Which measures change the most? Why?

4.11   Here are the percents of the popular vote won by the successful candidate in each of the presidential elections from 1948 to 1996:

| Year | 1948 | 1952 | 1956 | 1960 | 1964 | 1968 | |
|---|---|---|---|---|---|---|---|
| Percent | 49.6 | 55.1 | 57.4 | 49.7 | 61.1 | 43.4 | |

| Year | 1972 | 1976 | 1980 | 1984 | 1988 | 1992 | 1996 |
|---|---|---|---|---|---|---|---|
| Percent | 60.7 | 50.1 | 50.7 | 58.8 | 53.9 | 43.2 | 49.0 |

(a) Make a graph to display this distribution. What are its main features?

(b) What is the median percent of the vote won by the successful candidate in presidential elections?

(c) Call an election a landslide if the winner's percent falls at or above the upper quartile. Which elections were landslides?

4.12   In 1798 the English scientist Henry Cavendish measured the density of the earth with great care. It is common practice to repeat careful measurements several times and use the mean as the final result. Cavendish repeated his work 29 times. Here are his results (the data give the density of the earth as a multiple of the density of water):[8]

| | | | | | | | | | |
|---|---|---|---|---|---|---|---|---|---|
| 5.50 | 5.61 | 4.88 | 5.07 | 5.26 | 5.55 | 5.36 | 5.29 | 5.58 | 5.65 |
| 5.57 | 5.53 | 5.62 | 5.29 | 5.44 | 5.34 | 5.79 | 5.10 | 5.27 | 5.39 |
| 5.42 | 5.47 | 5.63 | 5.34 | 5.46 | 5.30 | 5.75 | 5.68 | 5.85 | |

Present these measurements with a graph of your choice. Does the shape of the distribution allow the use of $\bar{y}$ and $s$ to describe it? Find $\bar{y}$ and $s$. What is your estimate of the density of the earth based on these measurements?

4.13   A common criterion for detecting suspected outliers in a set of data is as follows:

1. Find the quartiles $Q_1$ and $Q_3$ and the interquartile range $IQR = Q_3 - Q_1$.
2. Call an observation an outlier if it falls more than $1.5 \times IQR$ above the upper quartile or below the lower quartile.

Find the interquartile range $IQR$ for the highway gas mileages in Exercise 3.7 (page 21). Are there any outliers by the $1.5 \times IQR$ criterion?

4.14   Exercise 3.11 (page 26) presents data on the survival times of guinea pigs in a medical experiment. Give a complete graphical and numerical description of the distribution of survival times, as well as a verbal summary of its most important features. Use the $1.5 \times IQR$ criterion (see the previous exercise) to check for outliers.

4.15   Table 3.1 (page 13) records the percent of people age 65 and older living in each of the states. Figure 3.4 (page 14) is a histogram of these data. Do you prefer the five-number summary or $\bar{x}$ and $s$ as a brief numerical description? Why? Calculate your preferred description. Does the $1.5 \times IQR$ criterion (Exercise 4.13) detect any outliers?

4.16   Table 3.2 (page 24) contains data on education in the states. We want to examine the distribution of median SAT math scores.

(a) Make a histogram of this distribution. Is it symmetric, skewed, or neither? Is it unimodal or bimodal? Are there any clear outliers?

(b) Give an appropriate numerical summary for this distribution.

(c) Your numerical summary does *not* show one of the most important features of the distribution. What feature is this? Remember to always start with a graph of your data—numerical summaries are not a complete description.

4.17   Table 4.1 shows the salaries paid to the members of the 1996 World Series champion New York Yankees baseball team (excluding bonuses).[9] Display this distribution with a graph and appropriate numerical measures, then give a description of its main features.

**TABLE 4.1**    1996 salaries for the New York Yankees baseball team

| Player | Salary | Player | Salary | Player | Salary |
|--------|--------|--------|--------|--------|--------|
| Cecil Fielder | 9,237,500 | Charlie Hayes | 1,750,000 | Darryl Strawberry | 560,000 |
| Paul O'Neill | 5,300,000 | Tony Fernandez | 1,500,000 | Mike Aldrete | 250,000 |
| Kenny Rogers | 5,000,000 | Tim Raines | 1,450,000 | Graeme Lloyd | 205,000 |
| David Cone | 4,666,666 | Ricky Bones | 1,425,000 | Andy Pettitte | 195,000 |
| Melido Perez | 4,650,000 | Jim Leyritz | 1,400,579 | Mariano Rivera | 131,125 |
| John Wetteland | 4,000,000 | Scott Kamieniecki | 1,100,000 | Derek Jeter | 130,000 |
| Bernie Williams | 3,000,000 | Dwight Gooden | 950,000 | Wally Whitehurst | 125,000 |
| Joe Girardi | 2,325,000 | Pat Kelly | 900,000 | Dave Pavlas | 120,000 |
| Tino Martinez | 2,300,000 | Jeff Nelson | 860,000 | Michael Figga | 109,000 |
| Pat Listach | 2,200,000 | Mariano Duncan | 845,000 | Dave Polley | 109,000 |
| Wade Boggs | 2,050,000 | Luis Sojo | 625,000 | Ruben Rivera | 109,000 |
| Jimmy Key | 1,750,000 | | | | |

4.18    At the time the salaries in Table 4.1 were announced, one dollar was worth 1.29 Swiss francs.

(a)  Convert the Yankees' salaries into Swiss francs. (What arithmetic operation did you do?)

(b)  Make a histogram of distribution in francs. How does this histogram compare with a histogram of the salaries in dollars?

(c)  Find the mean, median, and quartiles of the distribution in both dollars and francs. How are they related?

(d)  Find the standard deviation and interquartile range of the distribution in both dollars and francs. How are these measures of spread related?

(*Lesson*: Changing units by adding, subtracting, multiplying, or dividing has simple effects on our measures of center and spread.)

4.19    Exercise 3.7 (page 21) gives the highway gas mileages of 26 midsize car models.

(a)  Find the mean and standard deviation, and also the median and interquartile range, of these gas mileages.

(b)  There are 3.785 liters in a gallon. Use the results of the previous exercise to state the mean, median, standard deviation, and interquartile range if we change the unit of measurement to liters of gasoline per mile. (Make the change and do the calculations if you have any doubt about the result.)

4.20    Engineers prefer to record fuel consumption in gallons per mile rather than miles per gallon. Change the gas mileages in Exercise 3.7 into gallons per mile. (For example, 22 miles per gallon becomes $1/22 = 0.04545$ gallons per mile.)

(a) Make a histogram of the distribution in both miles per gallon and gallons per mile. Mark the position of the Rolls-Royce on both graphs. How did the shape of the distribution change?

(b) Calculate the mean and median of the distribution in both units. Each observation $y$ changed to $1/y$ when we changed units—do the mean and median change in this way?

(*Lesson*: Changes of units more complex than the basic arithmetic operations have complex effects.)

## LESSON 5

# NORMAL DISTRIBUTIONS

We now have a kit of graphical and numerical tools for describing distributions. What is more, we have a clear strategy for exploring data on a single quantitative variable:

- Start with a graph, usually a histogram.
- Look for the overall pattern and for striking deviations such as outliers.
- Choose a numerical summary to briefly describe center and spread.

Here is one more step to add to this strategy:

- Sometimes the overall pattern of a large number of observations is so regular that we can describe it by a smooth curve.

## DENSITY CURVES

Figure 5.1(a) is a histogram of the IQ test scores of 78 seventh-grade students in a rural school.[10] The distribution is a bit irregular but is unimodal

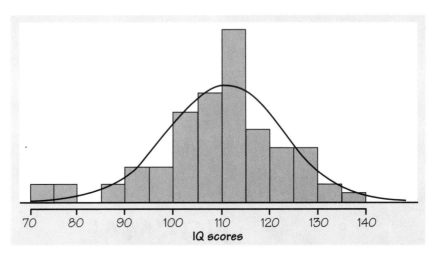

FIGURE 5.1(a)    Histogram of the IQ scores of 78 seventh-grade students, with a density curve that approximately describes the overall pattern of the distribution.

and roughly symmetric. We have drawn a smooth *density curve* through the histogram to give an idealized description of the distribution.

Areas in a histogram represent either counts or proportions of the observations. For example, the area of the shaded bars in Figure 5.1(b) represents the students whose IQ scores fall below 105. There are 24 such students, who make up the proportion $24/78 = 0.31$ of all the students. To make density curves easier to compare, we adjust the scale so that the

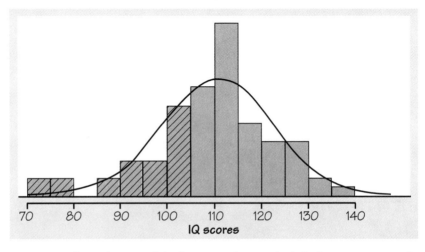

FIGURE 5.1(b)    The area of the histogram bars to the left of 105 represents the proportion of students with IQ scores below 105.

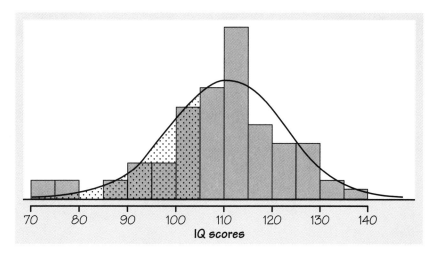

FIGURE 5.1(c)   The area under the density curve to the left of 105 represents the proportion of students with IQ scores below 105.

total area under any density curve is exactly 1. Areas under the curve then represent proportions of the observations. The shaded area under the density curve in Figure 5.1(c) is 0.38. You can see that areas under the curve are only a rough approximation to areas given by the histogram. On the other hand, the histogram depends on our choice of classes, and the density curve does not.

Density curves, like distributions, come in many shapes. Figure 5.2 shows two density curves: a symmetric density curve and a right-skewed curve.

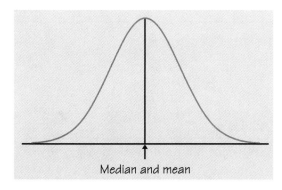

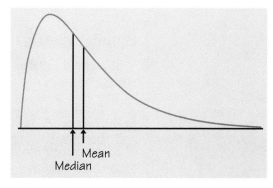

FIGURE 5.2(a)   A symmetric density curve, with its median and mean.

FIGURE 5.2(b)   A right-skewed density curve, with its median and mean.

DENSITY CURVE

A **density curve** is a curve that

- is always on or above the horizontal axis, and
- has area exactly 1 underneath it.

A density curve describes the overall pattern of a distribution. The area under the curve and above any range of values is the proportion of all observations that fall in that range.

Our measures of center and spread apply to density curves as well as to actual sets of observations, but only some of these measures are easily seen from the curve. A **mode** of a distribution described by a density curve is a peak point of the curve, the location where the curve is highest. Because areas under a density curve represent proportions of the observations, the median and quartiles divide the area under the curve into quarters.

What about the mean and standard deviation? The mean of a set of observations is their arithmetic average. If we think of the observations as weights strung out along a thin rod, the mean is the point at which the rod would balance. This fact is also true of density curves. The median and mean are marked on the density curves in Figure 5.2. It's hard to locate the balance point by eye on a skewed curve. There are mathematical ways of calculating the mean for any density curve, so we are able to mark the mean as well as the median in Figure 5.2(b). The standard deviation can also be calculated mathematically, but it can't be located by eye on most density curves.

MEDIAN AND MEAN OF A DENSITY CURVE

The **median** of a density curve is the equal-areas point, the point that divides the area under the curve in half.

The **mean** of a density curve is the balance point, at which the curve would balance if made of solid material.

The median and mean are the same for a symmetric density curve. They both lie at the center of the curve. The mean of a skewed curve is pulled away from the median in the direction of the long tail.

Because a density curve is an idealized description of the distribution of data, we must distinguish between the mean and standard deviation of the density curve and the mean $\bar{y}$ and standard deviation $s$ computed from the actual observations. The usual notation for the mean of an idealized distribution is $\mu$ (the Greek letter mu). We write the standard deviation of a density curve as $\sigma$ (the Greek letter sigma).

## NORMAL DISTRIBUTIONS

One particularly important class of density curves has already appeared in Figures 5.1 and 5.2(a). These density curves are symmetric, unimodal, and bell-shaped. They are called *normal curves*, and they describe **normal distributions**. All normal distributions have the same overall shape. The exact density curve for a particular normal distribution is described by giving its mean $\mu$ and its standard deviation $\sigma$. The mean is located at the center of the symmetric curve and is the same as the median. Changing $\mu$ without changing $\sigma$ moves the normal curve along the horizontal axis without changing its spread. The standard deviation $\sigma$ controls the spread of a normal curve. Figure 5.3 shows two normal curves with different values of $\sigma$. The curve with the larger standard deviation is more spread out.

The standard deviation $\sigma$ is the natural measure of spread for normal distributions. Not only do $\mu$ and $\sigma$ completely determine the shape of a normal curve, but we can locate $\sigma$ by eye on the curve. Here's how. As we move out in either direction from the center $\mu$, the curve changes from falling ever more steeply

to falling ever less steeply

The points at which this change of curvature takes place are located at distance $\sigma$ on either side of the mean $\mu$. You can feel the change as you run a pencil along a normal curve, and so find the standard deviation. Figure 5.3 shows $\sigma$ for two different normal curves.

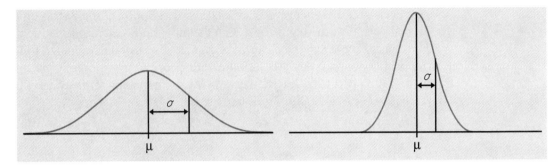

FIGURE 5.3    Two normal curves, showing the mean $\mu$ and standard deviation $\sigma$.

Although there are many normal curves, they all have common properties. Here are some of the most important.

---

### THE 68–95–99.7 RULE

In the normal distribution with mean $\mu$ and standard deviation $\sigma$:

- **68%** of the observations fall within $\sigma$ of the mean $\mu$.
- **95%** of the observations fall within $2\sigma$ of $\mu$.
- **99.7%** of the observations fall within $3\sigma$ of $\mu$.

### The 38% rule

- The middle 38% of any normal distribution is one standard deviation wide.

---

Figure 5.4 illustrates the 68–95–99.7 rule. By remembering these three numbers, you can think about normal distributions without constantly making detailed calculations.

## EXAMPLE 5.1

The distribution of heights of young women aged 18 to 24 is approximately normal with mean $\mu = 64.5$ inches and standard deviation $\sigma = 2.5$ inches.

Two standard deviations is 5 inches for this distribution. The 95 part of the 68–95–99.7 rule says that the middle 95% of young women are between $64.5 - 5$

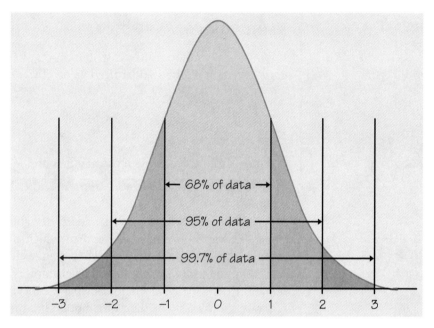

FIGURE 5.4    The 68–95–99.7 rule for normal distributions.

and 64.5 + 5 inches tall, that is, between 59.5 inches and 69.5 inches. This fact is exactly true for an exactly normal distribution. It is approximately true for the heights of young women because the distribution of heights is approximately normal.

The other 5% of young women have heights outside the range from 59.5 to 69.5 inches. Because the normal distributions are symmetric, half of these women are on the tall side. So the tallest 2.5% of young women are taller than 69.5 inches.    ◀

Because we will mention normal distributions often, a short notation is helpful. We abbreviate the normal distribution with mean $\mu$ and standard deviation $\sigma$ as $N(\mu, \sigma)$. For example, the distribution of young women's heights is $N(64.5, 2.5)$.

## STANDARDIZING OBSERVATIONS

As the 68–95–99.7 rule suggests, all normal distributions share many common properties. In fact, all normal distributions are the same if we measure in units of size $\sigma$ about the mean $\mu$ as center. Changing to these units is called *standardizing*. To standardize a value, subtract the mean of the distribution and then divide by the standard deviation.

STANDARDIZING AND $z$-SCORES

If $y$ is an observation from a distribution that has mean $\mu$ and standard deviation $\sigma$, the standardized value of $y$ is

$$z = \frac{y - \mu}{\sigma}$$

A standardized value is often called a **$z$-score**.

A $z$-score tells us how many standard deviations the original observation falls away from the mean, and in which direction. Observations larger than the mean are positive when standardized, and observations smaller than the mean are negative. For example, $z = -1.4$ indicates an observation 1.4 standard deviations below the mean. When we convert all observations into $z$-scores, their distribution has mean 0 and standard deviation 1.

## THE STANDARD NORMAL DISTRIBUTION

If the variable we standardize has a normal distribution, standardizing does more than give a common scale. It makes all normal distributions into a single distribution, and this distribution is still normal. Standardizing a variable that has any normal distribution produces a new variable that has the *standard normal distribution*.

STANDARD NORMAL DISTRIBUTION

The **standard normal distribution** is the normal distribution $N(0, 1)$ with mean 0 and standard deviation 1.

If a variable $y$ has any normal distribution $N(\mu, \sigma)$ with mean $\mu$ and standard deviation $\sigma$, then the standardized variable

$$z = \frac{y - \mu}{\sigma}$$

has the standard normal distribution.

Software and tables use $z$-scores and the standard normal distribution to find:

- The proportion of observations from a normal distribution that lie in a given interval of values.
- The value of a normal variable that marks off a given proportion of the observations either above it or below it.

Table A in the back of the book is a *standard normal table*. Table A also appears on the inside front cover. You can use Table A to do normal calculations, although software is usually more efficient.

---

THE STANDARD NORMAL TABLE

**Table A** is a table of areas under the standard normal curve. The table entry for each value $z$ is the area under the curve to the left of $z$.

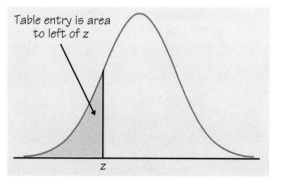

Table entry is area to left of z

$z$

---

EXAMPLE 5.2

*Problem:* What proportion of all young women are less than 68 inches tall? This proportion is the area under the $N(64.5, 2.5)$ curve to the left of the point 68. Because the standardized height corresponding to 68 inches is

$$z = \frac{y - \mu}{\sigma} = \frac{68 - 64.5}{2.5} = 1.4$$

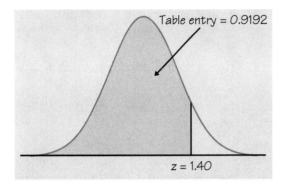

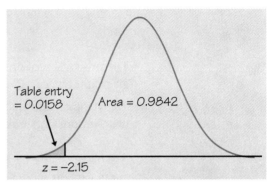

FIGURE 5.5(a)   The area under a standard normal curve to the left of the point $z = 1.4$ is 0.9192. Table A gives areas under the standard normal curve.

FIGURE 5.5(b)   Areas under the standard normal curve to the right and left of $z = -2.15$. Table A gives only areas to the left.

this area is the same as the area under the standard normal curve to the left of the point $z = 1.4$. Figure 5.5(a) shows this area.

*Solution:* To find the area to the left of 1.40, locate 1.4 in the left-hand column of Table A, then locate the remaining digit 0 as .00 in the top row. The entry opposite 1.4 and under .00 is 0.9192. This is the area we seek. Because $z = 1.40$ is the standardized value of height 68 inches, the proportion of young women who are less than 68 inches tall is 0.9192 (about 92%).

*Problem:* Find the proportion of observations from the standard normal distribution that are greater than $-2.15$.

*Solution:* Enter Table A under $z = -2.15$. That is, find $-2.1$ in the left-hand column and .05 in the top row. The table entry is 0.0158. This is the area to the *left* of $-2.15$. Because the total area under the curve is 1, the area lying to the *right* of $-2.15$ is $1 - 0.0158 = 0.9842$. Figure 5.5(b) illustrates these areas.

◀

The key to using either software or Table A to do a normal calculation is to sketch the area you want, then match that area with the areas that the table or software gives you.

## NORMAL PROBABILITY PLOTS

How can we assess whether a distribution is at least approximately normal? The histogram in Figure 5.1 only roughly resembles the normal density curve, yet these data are in fact approximately normal. The most useful tool for assessing normality is another graph, the **normal probability plot**.

A normal probability plot graphs the observed data values, arranged from smallest to largest, against the values that we would expect to find in a sample of the same size drawn from a standard normal distribution. The smallest observed value is graphed against what we would expect the smallest value to be in a sample of that size from a standard normal distribution, the next smallest against what we would expect the second smallest to be, and so on. If the data follow a normal distribution (with any mean and standard deviation), the resulting plot is a straight line.

**EXAMPLE 5.3**

Figure 5.6 is a normal probability plot of the 78 IQ scores whose histogram appears in Figure 5.1. The IQ scores are plotted vertically, and the corresponding values from a standard normal distribution (labeled "nscores") are plotted

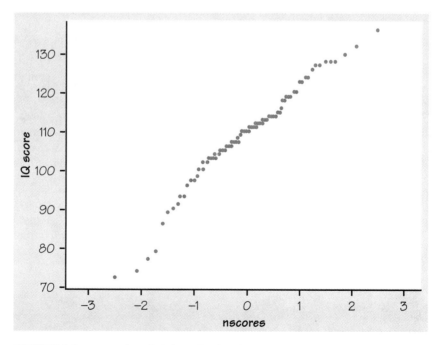

FIGURE 5.6    Normal probability plot for the IQ test scores shown in Figure 5.1. This distribution is approximately normal.

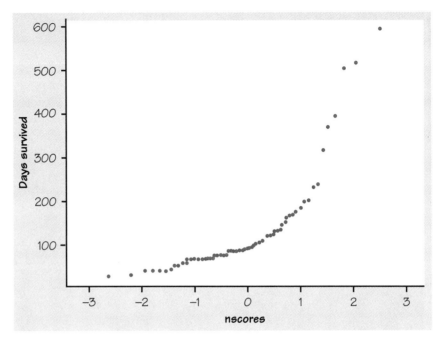

FIGURE 5.7   Normal probability plot for the guinea pig lifetimes from Exercise 3.11. This distribution is strongly skewed to the right.

horizontally. As the 99.7 part of the 68–95–99.7 rule suggests, the standard normal values lie between −3 and 3. The plot is roughly straight, so the distribution of IQ scores is roughly normal.

Figure 5.7 is a normal probability plot of the guinea pig survival times from Exercise 3.11 (page 26). The plot of the larger observations bends upward from a straight line. These data points are larger than we would expect in a normal distribution—that is, the right tail is long and the distribution is right skewed.

Returning to Figure 5.6, we can now see that the smallest IQ scores are a bit lower than they would be if they fell along the same straight line as the rest of the data. Use a straightedge to make this clear. So this distribution is slightly left skewed, as the histogram suggests. For practical purposes, we can still consider it roughly normal.   ◀

## EXERCISES

5.1   **(a)** Sketch a density curve that is symmetric but has a shape different from that of the curve in Figure 5.2(a).

**(b)** Sketch a density curve that is strongly skewed to the left.

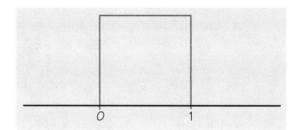

FIGURE 5.8    The density curve of a uniform distribution, for Exercise 5.2.

5.2    Figure 5.8 displays the density curve of a *uniform distribution*. The curve takes the constant value 1 over the interval from 0 to 1 and is zero outside that range of values. This means that data described by this distribution take values that are uniformly spread between 0 and 1. Use areas under this density curve to answer the following questions.

(a) What percent of the observations lie above 0.8?

(b) What percent of the observations lie below 0.6?

(c) What percent of the observations lie between 0.25 and 0.75?

5.3    What are the mean and the median of the distribution in Figure 5.8? What are the quartiles?

5.4    Figure 5.9 displays three density curves, each with three points marked on them. At which of these points on each curve do the mean, median, and mode fall?

5.5    The distribution of heights of adult men is approximately normal with mean 69 inches and standard deviation 2.5 inches. Use the 68–95–99.7 rule to answer the following questions.

(a) What percent of men are taller than 74 inches?

(b) Between what heights do the middle 95% of men fall?

(c) What percent of men are shorter than 66.5 inches?

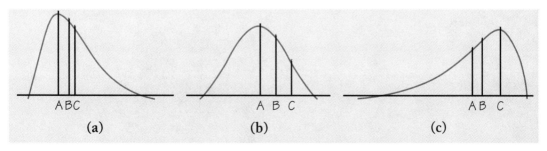

FIGURE 5.9    Three density curves, for Exercise 5.4.

5.6   Scores on the Wechsler Adult Intelligence Scale (a standard "IQ test") for the 20 to 34 age group are approximately normally distributed with $\mu = 110$ and $\sigma = 25$. Use the 68–95–99.7 rule to answer these questions.

(a) About what percent of people in this age group have scores above 110?

(b) About what percent have scores above 160?

(c) In what range do the middle 95% of all IQ scores lie?

5.7   The length of human pregnancies from conception to birth varies according to a distribution that is approximately normal with mean 266 days and standard deviation 16 days. Use the 68–95–99.7 rule to answer the following questions.

(a) Between what values do the lengths of the middle 95% of all pregnancies fall?

(b) How short are the shortest 2.5% of all pregnancies?

5.8   Eleanor scores 680 on the mathematics part of the Scholastic Assessment Test (SAT). The distribution of SAT scores in a reference population is normal with mean 500 and standard deviation 100. Gerald takes the American College Testing (ACT) mathematics test and scores 27. ACT scores are normally distributed with mean 18 and standard deviation 6. Find the $z$-scores for both students. Assuming that both tests measure the same kind of ability, who has the higher score?

5.9   Three landmarks of baseball achievement are Ty Cobb's batting average of .420 in 1911, Ted Williams's .406 in 1941, and George Brett's .390 in 1980. These batting averages cannot be compared directly because the distribution of major league batting averages has changed over the years. The distributions are quite symmetric and (except for outliers such as Cobb, Williams, and Brett) reasonably normal. While the mean batting average has been held roughly constant by rule changes and the balance between hitting and pitching, the standard deviation has dropped over time. Here are the facts:

| Decade | Mean | Std. dev. |
|--------|-------|-----------|
| 1910s | 0.266 | 0.0371 |
| 1940s | 0.267 | 0.0326 |
| 1970s | 0.261 | 0.0317 |

Compute the standardized batting averages for Cobb, Williams, and Brett to compare how far each stood above his peers.[11]

5.10    Scores on the Wechsler Adult Intelligence Scale (a standard "IQ test") for the 20 to 34 age group are approximately normally distributed with $\mu = 110$ and $\sigma = 25$.

(a) What percent of people age 20 to 34 have IQ scores above 100?

(b) What percent have scores above 150?

(c) How high an IQ score is needed to be in the highest 25%?

5.11    The rate of return on stock indexes (which combine many individual stocks) is approximately normal. Since 1945, the Standard & Poor's 500 index has had a mean yearly return of about 12%, with a standard deviation of about 16.5%. Take this normal distribution to be the distribution of yearly returns over a long period.

(a) In what range do the middle 95% of all yearly returns lie?

(b) The market is down for the year if the return on the index is less than zero. In what percent of years is the market down?

(c) In what percent of years does the index gain 25% or more?

5.12    The length of human pregnancies from conception to birth varies according to a distribution that is approximately normal with mean 266 days and standard deviation 16 days.

(a) What percent of pregnancies last less than 240 days (that's about 8 months)?

(b) What percent of pregnancies last between 240 and 270 days (roughly between 8 months and 9 months)?

(c) How long do the longest 20% of pregnancies last?

5.13    The quartiles of any density curve are the points with area 0.25 and 0.75 to their left under the curve.

(a) What are the quartiles of the standard normal distribution?

(b) How many standard deviations away from the mean do the quartiles lie in any normal distribution?

(c) What are the quartiles for the lengths of human pregnancies? (Use the distribution in the previous exercise.)

5.14    The *deciles* of any distribution are the points that mark off the lowest 10% and the highest 10%. The deciles of a density curve are therefore the points with area 0.1 and 0.9 to their left under the curve.

(a) What are the deciles of the standard normal distribution?

(b) The heights of young women are approximately normal with mean 64.5 inches and standard deviation 2.5 inches. What are the deciles of this distribution?

5.15    Use a normal probability plot to assess the normality of Cavendish's measurements of the density of the earth, from Exercise 4.12 (page 39). What do you conclude?

5.16    Make a normal probability plot of the percent of residents age 65 and over in the states, from Table 3.1. How do the outliers appear in your plot? Is the distribution roughly normal if we ignore the outliers? (It may help to compare your normal probability plot with the histogram in Figure 3.4 on page 14.)

5.17    Use software to generate 100 observations from the uniform distribution described in Exercise 5.2. Make a histogram of these observations. How does the histogram compare with the density curve in Figure 5.8? Make a normal probability plot of your data. According to this plot, how does the uniform distribution deviate from normality?

# PART I REVIEW

Data analysis is the art of describing data using graphs and numerical summaries. The purpose of data analysis is to describe the most important features of a set of data. Part I introduces data analysis by presenting statistical ideas and tools for describing the distribution of a single variable. To describe a distribution, we begin with a graph, add numerical descriptions of specific aspects of the distribution, and sometimes use a density curve to describe the overall pattern. Here is a review list of the most important skills you should have acquired from your study of Lessons 2 to 5.

## A. DATA

1. Identify the individuals and variables in a set of data.

2. Identify each variable as categorical or quantitative. Identify the units in which each quantitative variable is measured.

## B. DISPLAYING DISTRIBUTIONS

1. Make a bar graph or pie chart of the distribution of a categorical variable.

2. Make a histogram of the distribution of a quantitative variable.

3. Make a dotplot of the distribution of a quantitative variable.

4. Make a stem-and-leaf plot of the distribution of a quantitative variable.

## C. INSPECTING DISTRIBUTIONS (QUANTITATIVE VARIABLE)

1. Look for the overall pattern and for major deviations from the pattern.

2. Describe the overall pattern by giving numerical measures of center and spread and a verbal description of shape.

3. Assess from a stem-and-leaf plot or histogram whether the shape of a distribution is roughly symmetric, distinctly skewed, or neither. Assess whether the distribution is unimodal or has multiple modes.

4. Recognize outliers, using the $1.5 \times IQR$ criterion (Exercise 4.13) for guidance.

5. Decide which measures of center and spread are more appropriate: the mean and standard deviation (especially for symmetric distributions) or the five-number summary (especially for skewed distributions or when outliers are present).

## D. MEASURING CENTER

1. Find the mean $\bar{y}$ of a set of observations.

2. Find the median $M$ of a set of observations.

3. Understand that the median is more resistant (less affected by extreme observations) than the mean. Recognize that skewness in a distribution moves the mean away from the median toward the long tail.

## E. MEASURING SPREAD

1. Find the quartiles $Q_1$ and $Q_3$ for a set of observations.

2. Give the five-number summary of a distribution.

3. Find the standard deviation $s$ for a set of observations.

4. Know the basic properties of $s$: $s \geq 0$ always; $s = 0$ only when all observations are identical and increases as the spread increases; $s$ has the same units as the original measurements; $s$ is pulled strongly up by outliers or skewness.

## F. DENSITY CURVES

1. Know that areas under a density curve represent proportions of all observations and that the total area under a density curve is 1.

2. Approximately locate the median (equal-areas point) and the mean (balance point) on a density curve.

3. Know that the mean and median both lie at the center of a symmetric density curve and that the mean moves farther toward the long tail of a skewed curve.

## G. NORMAL DISTRIBUTIONS

1. Recognize the shape of normal curves and be able to estimate both the mean and the standard deviation from such a curve.

2. Use the 68–95–99.7 rule and symmetry to state what percent of the observations from a normal distribution fall between two points when both points lie at the mean or one, two, or three standard deviations on either side of the mean.

3. Find the standardized value ($z$-score) of an observation. Interpret $z$-scores and understand that any normal distribution becomes standard normal $N(0, 1)$ when standardized.

4. Given that a variable has the normal distribution with a stated mean $\mu$ and standard deviation $\sigma$, calculate the proportion of values above a stated number, below a stated number, or between two stated numbers.

5. Given that a variable has the normal distribution with a stated mean $\mu$ and standard deviation $\sigma$, calculate the point having a stated proportion of all values above it. Also calculate the point having a stated proportion of all values below it.

6. Make a normal probability plot and use it to assess approximate normality of a set of observations.

## REVIEW EXERCISES

**Review 1**   A professor who lives a few miles outside a college town records the time he takes to drive to the college each morning. Here are the times (in minutes) for 42 consecutive weekdays, with the dates in order along the rows:

| 8.25 | 7.83 | 8.30 | 8.42 | 8.50 | 8.67 | 8.17 | 9.00 | 9.00 | 8.17 | 7.92 |
|------|------|------|------|------|------|------|------|------|------|------|
| 9.00 | 8.50 | 9.00 | 7.75 | 7.92 | 8.00 | 8.08 | 8.42 | 8.75 | 8.08 | 9.75 |
| 8.33 | 7.83 | 7.92 | 8.58 | 7.83 | 8.42 | 7.75 | 7.42 | 6.75 | 7.42 | 8.50 |
| 8.67 | 10.17 | 8.75 | 8.58 | 8.67 | 9.17 | 9.08 | 8.83 | 8.67 | | |

(a) Make a graph of these drive times. Is the distribution roughly symmetric, clearly skewed, or neither? Are there any clear outliers?

(b) The data record shows three unusual situations: the day after Thanksgiving (no traffic on campus); a delay due to an accident; and a day with icy roads. Identify and remove these three observations, and give a numerical summary of the remaining 39 drive times.

(c) Are these data reasonably close to having a normal distribution when the unusual days are removed? Give numerical measures that summarize the professor's usual drive time (with outliers removed).

**Review 2**  The 1995 *Statistical Abstract of the United States* reports FBI data on murders for 1993. In that year, 56.9% of all murders were committed with handguns, 12.7% with other firearms, 12.7% with knives, 5.0% with a part of the body (usually the hands or feet), and 4.4% with blunt objects. Make a graph to display these data. Do you need an "other methods" category?

**Review 3**  *Consumer Reports* magazine (June 1986) presented the following data on the number of calories in a hot dog for each of 17 brands of meat hot dogs:

| 173 | 191 | 182 | 190 | 172 | 147 | 146 | 139 | 175 |
|-----|-----|-----|-----|-----|-----|-----|-----|-----|
| 136 | 179 | 153 | 107 | 195 | 135 | 140 | 138 | |

Make a graph of the distribution of calories in meat hot dogs and briefly describe the shape of the distribution. Most brands of meat hot dogs contain a mixture of beef and pork, with up to 15% poultry allowed by government regulations. The only brand with a different makeup was *Eat Slim Veal Hot Dogs*. Which point on your stemplot do you think represents this brand?

**Review 4**  A marketing consultant observes 50 consecutive shoppers at a grocery store, recording how much each shopper spends in the store. Here are the data (in dollars), arranged in increasing order for convenience:

| | | | | | | | |
|---|---|---|---|---|---|---|---|
| 3.11 | 8.88 | 9.26 | 10.81 | 12.69 | 13.78 | 15.23 | 15.62 |
| 17.00 | 17.39 | 18.36 | 18.43 | 19.27 | 19.50 | 19.54 | 20.16 |
| 20.59 | 22.22 | 23.04 | 24.47 | 24.58 | 25.13 | 26.24 | 26.26 |
| 27.65 | 28.06 | 28.08 | 28.38 | 32.03 | 34.98 | 36.37 | 38.64 |
| 39.16 | 41.02 | 42.97 | 44.08 | 44.67 | 45.40 | 46.69 | 48.65 |
| 50.39 | 52.75 | 54.80 | 59.07 | 61.22 | 70.32 | 82.70 | 85.76 |
| 86.37 | 93.34 | | | | | | |

Make a graph of these data. Then describe the shape, center, and spread of the distribution. Are there any clear outliers? Make a normal probability plot, and explain why it has the shape that it does.

**Review 5** Joe DiMaggio played center field for the Yankees for 13 years. He was succeeded by Mickey Mantle, who played for 18 years. Here is the number of home runs hit each year by DiMaggio:

29  46  32  30  31  30  21  25  20  39  14  32  12

Here are Mantle's home run counts:

13  23  21  27  37  52  34  42  31  40  54  30  15  35  19  23  22  18

Compute the five-number summary for each player, and make side-by-side dotplots of the home run distributions. What does your comparison show about DiMaggio and Mantle as home run hitters?

**Review 6** The rate of return on a stock is its change in price plus any dividends paid, usually measured in percent of the starting value. Table I.1 gives the monthly rates of return (in percents) for the stock of Wal-Mart stores for the years 1973 to 1991, the first 19 full years Wal-Mart was listed on the New York Stock Exchange. There are 228 observations.

(a) Give a complete graphical and numerical description of these data, and describe the main features in words. In what way (if any) do the data deviate from normality? Are there outliers by the $1.5 \times IQR$ criterion? What are the mean and median monthly returns? Why do they differ as they do?

(b) If you had $1000 worth of Wal-Mart stock at the beginning of the best month during these 19 years, how much would your stock be worth at the end of the month? If you had $1000 worth of stock at the beginning of the worst month, how much would your stock be worth at the end of the month?

| TABLE I.1 | Monthly percent return on Wal-Mart common stock from January 1973 to December 1991 (in order across rows) | | | | | | | | |
|---|---|---|---|---|---|---|---|---|---|
| −14.13 | 4.64 | −26.61 | −1.65 | −1.68 | −31.25 | 58.68 | −11.77 | 5.92 | 5.03 |
| −34.04 | −13.71 | 42.06 | −11.84 | −6.42 | 0.80 | 7.94 | 14.71 | 0.00 | −15.06 |
| −14.42 | −12.39 | −19.19 | −5.00 | 57.89 | −5.83 | 41.81 | −14.38 | 22.63 | 16.21 |
| 0.51 | −4.08 | 4.38 | 15.31 | 6.19 | −12.37 | 3.81 | 11.01 | 8.40 | −12.21 |
| −4.35 | 0.15 | −12.73 | 22.92 | 5.22 | −8.06 | 7.89 | 0.98 | −10.48 | −1.80 |
| 2.02 | −13.51 | 10.42 | 9.77 | 0.86 | 5.98 | 9.97 | −0.74 | 9.90 | 9.46 |
| −6.17 | 1.97 | −2.94 | 24.00 | 3.23 | −2.38 | 4.81 | 11.22 | 2.04 | −18.02 |
| 1.10 | −0.85 | −0.55 | 2.21 | 5.19 | 4.64 | 0.49 | −0.69 | 2.97 | 32.02 |
| −6.93 | −7.45 | 18.64 | −0.50 | 2.52 | −10.18 | −10.62 | 14.04 | 5.38 | 12.34 |
| 14.33 | 10.20 | 3.63 | 1.00 | 11.39 | 7.73 | −0.83 | 5.42 | 15.62 | 5.82 |
| 0.32 | −1.77 | −2.63 | −3.54 | 5.96 | 5.30 | 10.38 | −2.99 | −0.29 | −2.06 |
| 10.46 | 13.66 | −6.01 | 8.37 | 3.07 | 15.30 | 13.15 | 10.56 | 20.70 | 5.37 |
| −5.76 | 4.79 | 17.65 | 6.48 | 16.63 | 8.97 | 4.79 | −3.88 | 3.81 | −3.98 |
| 7.32 | −7.34 | −8.97 | −9.51 | 3.67 | 6.77 | 4.58 | 11.25 | −1.21 | 8.41 |
| −2.83 | −1.75 | −5.34 | −4.88 | 20.13 | 0.00 | −1.77 | 5.04 | 11.73 | 4.43 |
| −9.38 | 4.69 | −7.73 | 10.99 | 11.79 | 7.71 | 0.78 | 5.84 | 15.93 | 2.86 |
| 18.21 | 8.44 | −10.36 | 1.17 | −10.64 | 6.85 | 6.13 | −2.27 | 4.30 | 12.89 |
| 5.13 | −1.52 | 3.31 | 15.70 | 10.17 | 8.81 | −6.48 | −27.06 | −13.12 | 8.46 |
| 5.29 | 7.31 | −4.12 | −1.33 | 1.80 | 8.55 | 7.35 | −8.62 | 7.50 | −2.71 |
| −3.59 | 3.85 | 7.57 | −5.56 | 2.13 | 10.77 | 5.21 | 0.81 | 13.44 | −2.18 |
| −2.96 | 1.22 | 5.12 | 2.99 | −5.01 | 3.23 | 7.55 | 5.03 | 13.60 | 10.77 |
| 0.60 | −9.05 | −4.39 | −1.83 | 12.62 | 0.53 | 9.09 | 7.20 | 9.66 | 4.52 |
| 5.86 | −0.19 | 11.40 | 6.39 | −5.68 | −3.14 | 5.68 | 20.55 | | |

**Review 7**    The Acculturation Rating Scale for Mexican Americans (ARSMA) is a psychological test that measures the degree to which Mexican Americans are adapted to Mexican/Spanish versus Anglo/English culture. The range of possible scores is 1.0 to 5.0, with higher scores showing more Anglo/English acculturation. The distribution of ARSMA scores in a population used to develop the test is approximately normal with mean 3.0 and standard deviation 0.8. A researcher believes that Mexicans will have an average score near 1.7 and that first-generation Mexican Americans will average about 2.1 on the ARSMA scale. What proportion of the population used to develop the test has scores below 1.7? Between 1.7 and 2.1?

**Review 8**    The army reports that the distribution of head circumference among soldiers is approximately normal with mean 22.8 inches and standard deviation 1.1 inches. Helmets are mass-produced for all except the smallest 5% and the largest 5% of head sizes. Soldiers in the smallest or largest 5% get custom-made helmets. What head sizes get custom-made helmets?

**Review 9**   The ARSMA test is described in Review Exercise 7. How high a score on this test must a Mexican American obtain to be among the 30% of the population used to develop the test who are most Anglo/English in cultural orientation? What scores make up the 30% who are most Mexican/Spanish in their acculturation?

**Review 10**   How long do presidents live? The table below gives the ages at death of deceased U.S. presidents, in years. Give a complete graphical and numerical summary of this distribution. Report your conclusions, including an assessment of normality.

| Washington | 67 | Fillmore | 74 | Roosevelt | 60 |
|---|---|---|---|---|---|
| Adams | 90 | Pierce | 64 | Taft | 72 |
| Jefferson | 83 | Buchanan | 77 | Wilson | 67 |
| Madison | 85 | Lincoln | 56 | Harding | 57 |
| Monroe | 73 | Johnson | 66 | Coolidge | 60 |
| Adams | 80 | Grant | 63 | Hoover | 90 |
| Jackson | 78 | Hayes | 70 | Roosevelt | 63 |
| Van Buren | 79 | Garfield | 49 | Truman | 88 |
| Harrison | 68 | Arthur | 56 | Eisenhower | 78 |
| Tyler | 71 | Cleveland | 71 | Kennedy | 46 |
| Polk | 53 | Harrison | 67 | Johnson | 64 |
| Taylor | 65 | McKinley | 58 | Nixon | 81 |

**Review 11**   You are planning a sample survey of households in Indiana. You decide to select households separately within counties. To aid in the planning, here are the populations of the state's 92 counties (in thousands of persons) according to the 1990 census:

| 31 | 301 | 64 | 9 | 14 | 38 | 14 | 19 | 38 | 88 |
|---|---|---|---|---|---|---|---|---|---|
| 24 | 31 | 10 | 28 | 39 | 24 | 35 | 120 | 37 | 156 |
| 26 | 64 | 18 | 20 | 19 | 32 | 74 | 30 | 109 | 46 |
| 30 | 76 | 48 | 81 | 35 | 38 | 25 | 22 | 30 | 24 |
| 88 | 40 | 65 | 29 | 476 | 107 | 43 | 131 | 797 | 42 |
| 10 | 37 | 109 | 34 | 56 | 14 | 38 | 5 | 18 | 17 |
| 15 | 19 | 13 | 129 | 26 | 13 | 30 | 27 | 25 | 18 |
| 247 | 21 | 40 | 19 | 23 | 27 | 19 | 8 | 131 | 16 |
| 7 | 165 | 17 | 106 | 35 | 8 | 45 | 24 | 72 | 26 |
| 24 | 28 | | | | | | | | |

Examine the distribution of county populations both graphically and numerically, using whatever tools are most suitable. Then write a brief description of the main features of this distribution.

**Review 12**   Scholastic Assessment Test (SAT) scores are approximately normal with mean 500 and standard deviation 100. Scores of 800 or higher are reported as 800, so a perfect paper is not required to score 800 on the SAT. What percent of students who take the SAT score 800?

**Review 13**   Scores on the Wechsler Adult Intelligence Scale for the 20 to 34 age group are approximately normally distributed with mean 110 and standard deviation 25. Scores for the 60 to 64 age group are approximately normally distributed with mean 90 and standard deviation 25.

Sarah, who is 30, scores 135 on this test. Sarah's mother, who is 60, also takes the test and scores 120. Who scored higher relative to her age group, Sarah or her mother? Who has the higher absolute level of the variable measured by the test? At what percentile of their age groups are Sarah and her mother? (That is, what percent of the age group has lower scores?)

**Review 14**   In the early 1990s, the median award to people who won medical malpractice lawsuits was $350,000. The mean award in these same lawsuits was about $1.7 million. Explain how this great difference between two measures of center can occur.

**Review 15**   If two distributions have exactly the same mean and standard deviation, must their histograms have the same shape? If they have the same five-number summary, must their histograms have the same shape? Explain.

**Review 16**   A school system employs teachers at salaries between $25,000 and $55,000. The teachers' union and the school board are negotiating the form of next year's increase in the salary schedule.
(a) If every teacher is given a flat $1000 raise, what will this do to the mean salary? To the median salary? To the extremes and quartiles of the salary distribution?
(b) What will a flat $1000 raise do to the standard deviation of teachers' salaries?

(c) If, instead, each teacher receives a 5% raise, the amount of the raise will vary from $1250 to $2750, depending on the present salary. What will this do to the mean salary? To the median salary?

(d) A flat raise would not increase the spread of the salary distribution. What about a 5% raise? Specifically, will a 5% raise increase the distance of the quartiles from the median? Will it increase the standard deviation?

## JOHN W. TUKEY

He started as a chemist, became a mathematician, and was converted to statistics by what he called "the real problems experience and the real data experience" of war work during the Second World War. John W. Tukey (1915–) came to Princeton University in 1937 to study chemistry but took a doctorate in mathematics in 1939. During the war, he worked on the accuracy of range finders and of gunfire from bombers, among other problems. After the war he divided his time between Princeton and nearby Bell Labs, perhaps the world's most eminent industrial research group.

Tukey devoted much of his attention to the statistical study of messy problems with lots of complex data: the safety of anesthetics used by many doctors in many hospitals on many patients, the Kinsey studies of human sexual behavior, monitoring compliance with a nuclear test ban, and air quality and environmental pollution.

From this "real problems experience and real data experience," John Tukey developed exploratory data analysis. He invented some of the simple tools we have met, such as stem-and-leaf plots. More important, he changed the way we approach data by emphasizing the need for a flexible, exploratory approach that does not just seek to answer specific questions but asks, "What do the data say?" Part II, like Part I, follows Tukey's path by presenting ideas and tools for examining data.

PART **II**

# Understanding Relationships

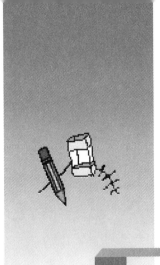

**LESSON 6**

# COMPARING
# GROUPS

Most statistical studies involve more than one variable. Sometimes we want
to compare the distributions of the same variable for several groups. For
example, we might compare the distributions of Scholastic Assessment Test
(SAT) scores among students at several colleges. "College" is a categorical
variable, and "SAT score" is a quantitative variable. We are examining the
relationship between a categorical and a quantitative variable. In other situations, we want to explore relationships among several quantitative variables measured on the same group of individuals. This lesson considers
comparing groups. Lessons 7 to 9 look at relationships between two quantitative variables.

A comparison should start with graphs. We can compare groups by
making separate histograms, stem-and-leaf plots, or dotplots for each
group. Use the same scales for all groups to make the comparison easier.
A *boxplot* is a type of graph designed for comparing the distributions of a
quantitative variable in several groups.

BOXPLOT

A **boxplot** is a graph of the five-number summary with suspected outliers plotted individually.

- A central box spans the quartiles.
- A line in the box marks the median.
- Observations more than $1.5 \times IQR$ outside the central box are plotted individually as possible outliers.
- Lines extend from the box out to the smallest and largest observations that are not suspected outliers.

EXAMPLE 6.1

Table 6.1 gives the city and highway gas mileages, as measured by the Environmental Protection Agency, for 1997 model midsize cars and four-wheel drive sports utility vehicles.[12] Figure 6.1 presents boxplots of the city mileage data.

**TABLE 6.1**    City and highway gas mileage for 1997 model vehicles

| Midsize cars | | | Sports utility vehicles | | |
|---|---|---|---|---|---|
| Model | City | Highway | Model | City | Highway |
| Acura 3.5RL | 19 | 25 | Chevrolet Blazer | 16 | 21 |
| Audi A8 quattro | 17 | 25 | Chevrolet Suburban | 12 | 16 |
| Buick Century | 20 | 29 | Chevrolet Tahoe | 13 | 17 |
| Cadillac Catera | 18 | 25 | Chrysler Town & Country | 15 | 22 |
| Cadillac Eldorado | 17 | 26 | Ford Expedition | 14 | 18 |
| Chevrolet Lumina | 20 | 29 | Ford Explorer | 14 | 19 |
| Chrysler Cirrus | 20 | 30 | Geo Tracker | 23 | 24 |
| Dodge Stratus | 22 | 32 | Isuzu Rodeo | 15 | 18 |
| Ford Taurus | 19 | 28 | Isuzu Trooper | 14 | 18 |
| Ford Thunderbird | 18 | 26 | Jeep Grand Cherokee | 15 | 20 |
| Hyundai Sonata | 20 | 27 | Kia Sportage | 19 | 22 |
| Infiniti I30 | 21 | 28 | Land Rover Defender | 14 | 15 |
| Infiniti Q45 | 18 | 24 | Land Rover Discovery | 14 | 17 |
| Lexus GS300 | 18 | 24 | Lexus LX450 | 13 | 15 |

*(continued)*

| Midsize cars | | | Sports utility vehicles | | |
| --- | --- | --- | --- | --- | --- |
| Model | City | Highway | Model | City | Highway |
| Lexus LS400 | 19 | 25 | Mazda MPV | 15 | 19 |
| Lincoln Mark VIII | 18 | 26 | Mitsubishi Montero | 16 | 19 |
| Mazda 626 | 23 | 31 | Nissan Pathfinder | 15 | 19 |
| Mercedes-Benz E320 | 20 | 27 | Range Rover | 13 | 17 |
| Mercedes-Benz E420 | 18 | 25 | Suzuki Sidekick | 21 | 24 |
| Mitsubishi Diamante | 18 | 26 | Toyota Landcruiser | 13 | 15 |
| Nissan Maxima | 21 | 28 | Toyota RAV4 | 22 | 26 |
| Oldsmobile Aurora | 17 | 26 | Toyota 4Runner | 18 | 22 |
| Rolls-Royce Silver Spur | 11 | 16 | | | |
| Saab 900 | 18 | 26 | | | |
| Toyota Camry | 23 | 30 | | | |
| Volvo 850 | 19 | 26 | | | |

**TABLE 6.1** City and highway gas mileage for 1997 model vehicles (*continued*)

SOURCE: Environmental Protection Agency, *Model Year 1997 Fuel Economy Guide.*

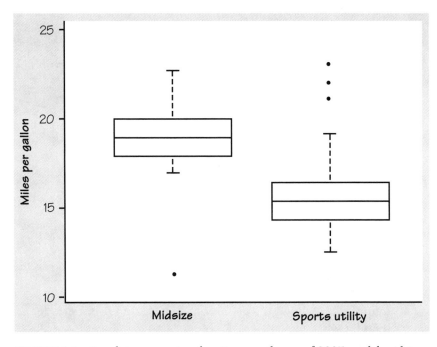

FIGURE 6.1    Boxplots comparing the city gas mileages of 1997 model midsize cars and sports utility vehicles.

We can see at once that sports utility vehicles as a group have markedly lower gas mileage than the midsize cars they often replace in our driveways. Several outliers appear as individual points in the graph. The Rolls-Royce is a low outlier in the midsize group, and three small four-wheel drive vehicles are high outliers in their group.

Boxplots allow rapid visual comparisons of more than two groups. Figure 6.2 shows the city and highway mileages for both groups of vehicles. We see that highway mileages are higher than city mileages, and that the gap between midsize cars and sports utility vehicles is somewhat greater on the highway than in the city. ◀

Boxplots are inferior to histograms and stem-and-leaf plots for displaying a single distribution because they hide details of its shape. They are primarily a tool for comparing distributions of the same variable or similar variables in two or more groups. For example, Figure 6.3 is a stem-and-leaf plot of the city gas mileages for sports utility vehicles. We now see a large cluster of vehicles that get 12 to 16 miles per gallon and five vehicles that get much better mileage. This detail is hidden in the boxplot.

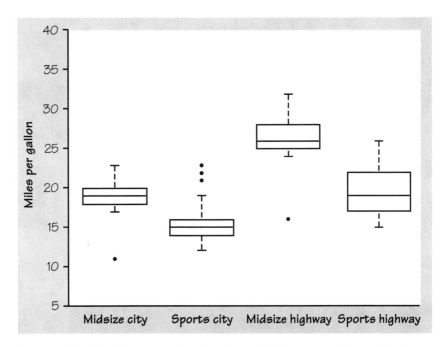

FIGURE 6.2   Boxplots comparing the city and highway gas mileages of 1997 model midsize cars and sports utility vehicles.

```
12 | 0
13 | 0000
14 | 00000
15 | 00000
16 | 00
17 |
18 | 0
19 | 0
20 |
21 | 0
22 | 0
23 | 0
```

FIGURE 6.3    Stem-and-leaf plot of the city gas mileages of sports utility vehicles.

Here is a general strategy for comparing distributions: Make histograms or stem-and-leaf plots of each individual distribution. Compare their shapes and identify outliers. On the basis of the shapes, choose numerical summaries of center and spread. Then make side-by-side boxplots for an overall visual comparison.

## EXERCISES

6.1    Exercise 4.8 (page 38) gives data on home runs hit by Babe Ruth and Roger Maris. Two more famous Yankees are Joe DiMaggio, who played center field for 13 years, and Mickey Mantle, who played for 18 years. Here is the number of home runs hit each year by DiMaggio:

29   46   32   30   31   30   21   25   20   39   14   32   12

Here are Mantle's home run counts:

13   23   21   27   37   52   34   42   31   40   54   30   15   35   19   23   22   18

Compare these four players as home run hitters, using boxplots and whatever other descriptions you think are helpful. What individual years are unusual performances by one of the players; that is, which years are outliers in a player's career?

6.2    We want to compare spending on public education in several regions of the country. Because states vary in population, we should compare dollars spent per pupil rather than total dollars spent. Table 3.2 (page 24) gives the dollars spent per pupil in each state. The Census Bureau groups the states into four regions:

Northeast    MA (Middle Atlantic) and NE (New England)
South    SA (South Atlantic), ESC (East South Central), and
WSC (West South Central)
Midwest    ENC (East North Central) and WNC (West North Central)
West    MTN (Mountain) and PAC (Pacific)

(a)    The District of Columbia is a city rather than a state. Is DC an outlier in the South region? We will omit DC because it is not truly a state.

(b)    Make numerical summaries and graphs to compare the four distributions. Write a brief statement of what you find.

6.3    How do the states in the four regions identified in the previous exercise compare in their median scores on the mathematics part of the SAT exam? Are there individual states that are outliers in their region? Make whatever comparisons you think are appropriate, and describe your findings.

6.4    Exercise 6.2 identifies the states in each Census Bureau region. Some people think that the West region has a younger population than the other regions. Compare the percent of residents age 65 and over in states in these regions (data in Table 3.1, page 13). Use boxplots and whatever other descriptions you think are helpful. What individual states are outliers in their regions?

6.5    Federal regulations allow three types of hot dogs: all beef, "meat" (mainly pork and beef, but regulations allow up to 15% poultry meat), and poultry. Because people concerned about health may prefer low-calorie, low-salt hot dogs, we ask: Are there any systematic differences among the three types in these two variables? Do poultry hot dogs as a group have an advantage?

Table 6.2 gives the results of laboratory tests done by *Consumer Reports* magazine on several brands of each type.[13] Do a careful comparison of the three distributions of calories per hot dog. Be sure to identify any outliers. For the 17 brands of meat hot dogs, comment on what you learn from a histogram or stem-and-leaf plot that a boxplot does not reveal.

6.6    Continue the previous exercise by comparing sodium content for the three varieties of hot dogs.

**TABLE 6.2**    Calories and sodium in hot dogs

| Beef hot dogs | | Meat hot dogs | | Poultry hot dogs | |
|---|---|---|---|---|---|
| Calories | Sodium | Calories | Sodium | Calories | Sodium |
| 186 | 495 | 173 | 458 | 129 | 430 |
| 181 | 477 | 191 | 506 | 132 | 375 |
| 176 | 425 | 182 | 473 | 102 | 396 |
| 149 | 322 | 190 | 545 | 106 | 383 |
| 184 | 482 | 172 | 496 | 94 | 387 |
| 190 | 587 | 147 | 360 | 102 | 542 |
| 158 | 370 | 146 | 387 | 87 | 359 |
| 139 | 322 | 139 | 386 | 99 | 357 |
| 175 | 479 | 175 | 507 | 170 | 528 |
| 148 | 375 | 136 | 393 | 113 | 513 |
| 152 | 330 | 179 | 405 | 135 | 426 |
| 111 | 300 | 153 | 372 | 142 | 513 |
| 141 | 386 | 107 | 144 | 86 | 358 |
| 153 | 401 | 195 | 511 | 143 | 581 |
| 190 | 645 | 135 | 405 | 152 | 588 |
| 157 | 440 | 140 | 428 | 146 | 522 |
| 131 | 317 | 138 | 339 | 144 | 545 |
| 149 | 319 | | | | |
| 135 | 298 | | | | |
| 132 | 253 | | | | |

SOURCE: *Consumer Reports,* June 1986, pp. 366–367.

6.7    Table 4.1 (page 41) gives the salaries for the 1996 World Series champion New York Yankees baseball team. The following members of this team are pitchers: Rogers, Cone, Perez, Wetteland, Key, Bones, Kamieniecki, Gooden, Nelson, Lloyd, Pettitte, Mariano Rivera, Whitehurst, Pavlas, and Polley. The others are "position players." Compare the distributions of salaries for pitchers and for position players and describe any important differences.

# SCATTERPLOTS

We have already examined the distribution of the percent of high school seniors in each state who take the SAT (Exercise 3.13) and the distribution of median SAT mathematics scores for the states (Exercise 4.16). Now we might ask how SAT math scores are related to the percent of a state's high school seniors who take the SAT. That is, we are interested in a relationship between two quantitative variables. The most effective way to display the relationship between two quantitative variables is a *scatterplot*.

---

SCATTERPLOT

A **scatterplot** shows the relationship between two quantitative variables measured on the same individuals. The values of one variable appear on the horizontal axis, and the values of the other variable appear on the vertical axis. Each individual in the data appears as the point in the plot fixed by the values of both variables for that individual.

If one of the variables influences or helps explain the other, always plot this variable on the horizontal axis (the *x* axis) of a scatterplot.

---

**EXAMPLE 7.1**

Joan is concerned about the amount of energy she uses to heat her home in the Midwest. She keeps a record of the natural gas she consumes each month over one year's heating season. Because the months are not all the same length, she divides each month's consumption by the number of days in the month to get the average number of cubic feet of gas used per day. Demand for heating is strongly influenced by the outside temperature. From local weather records, Joan obtains the average temperature for each month, in degrees Fahrenheit. Here are Joan's data:[14]

| Month | Oct. | Nov. | Dec. | Jan. | Feb. | Mar. | Apr. | May |
|-------|------|------|------|------|------|------|------|-----|
| Temperature $x$ | 49.4 | 38.2 | 27.2 | 28.6 | 29.5 | 46.4 | 49.7 | 57.1 |
| Gas consumed $y$ | 520 | 610 | 870 | 850 | 880 | 490 | 450 | 250 |

Figure 7.1 is a scatterplot of these data. Temperature helps explain gas consumption, so we plot temperature horizontally. Each month generates one point on the plot. October appears as the point (49.4, 520) above 49.4 on the $x$ axis and to the right of 520 on the $y$ axis.    ◀

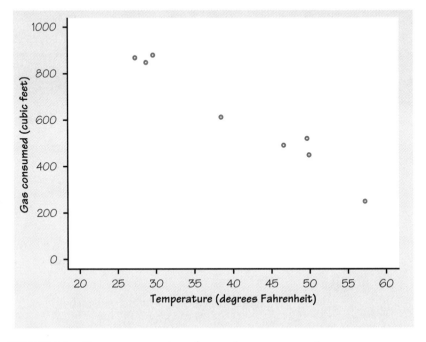

FIGURE 7.1    Home consumption of natural gas versus outdoor temperature.

# INTERPRETING SCATTERPLOTS

To interpret a scatterplot, look first for an overall pattern. **Form, direction, and strength provide a good description of the overall pattern of a relationship between two quantitative variables.**

---

**EXAMPLE 7.2**

Look again at Figure 7.1, the scatterplot of gas consumed versus average temperature. The *form* of the relationship is clear: the points lie close to a straight line. This is a *linear* form. The *direction* of the relationship is also clear: higher temperatures reduce gas consumption, so the plot shows a trend from upper left to lower right. This is a *negative association*. Because there is little scatter about the line, the relationship is a strong one.     ◄

---

POSITIVE ASSOCIATION, NEGATIVE ASSOCIATION

Two variables are **positively associated** when above-average values of one tend to accompany above-average values of the other and below-average values also tend to occur together. Two variables are **negatively associated** when above-average values of one accompany below-average values of the other, and vice versa.

---

Of course, not all relationships have a linear form. In fact, not all relationships have a clear direction that we can describe as positive association or negative association. Exercise 7.2 gives an example that is not linear and has no clear direction.

When your data contain several distinct groups of individuals (such as women and men), you can use a different plot symbol or color for each group. This allows a scatterplot to show the relationships among its two quantitative variables and a categorical variable that records which group an individual belongs to. Exercises 7.4 to 7.6 illustrate this idea.

After examining the overall pattern of a scatterplot, look for deviations from the pattern such as *outliers*.

OUTLIERS

An **outlier** in any graph of data is an individual observation that falls outside the overall pattern of the graph.

Figure 7.1 does not show any outliers—none of the points lies far from the overall linear pattern.

## EXERCISES

7.1    Table 6.1 gives the city and highway gas mileages for 1997 model midsize cars. We expect these mileages to be related to each other.
(a) Make a scatterplot to examine the relationship between city and highway fuel consumption for midsize cars. Plot city mileage on the $x$ axis.
(b) Describe the direction of the relationship. Are the variables positively or negatively associated?
(c) Describe the form of the relationship. Is it linear? Are there any outliers?
(d) Describe the strength of the relationship. Can highway mileage be predicted reasonably well from city mileage?

7.2    How does the fuel consumption of a car change as its speed increases? Here are data for a British Ford Escort. Speed is measured in kilometers per hour, and fuel consumption is measured in liters of gasoline used per 100 kilometers traveled.[15]

| Speed (km/h) | Fuel used (liters/100 km) | Speed (km/h) | Fuel used (liters/100 km) |
|---|---|---|---|
| 10 | 21.00 | 90 | 7.57 |
| 20 | 13.00 | 100 | 8.27 |
| 30 | 10.00 | 110 | 9.03 |
| 40 | 8.00 | 120 | 9.87 |
| 50 | 7.00 | 130 | 10.79 |
| 60 | 5.90 | 140 | 11.77 |
| 70 | 6.30 | 150 | 12.83 |
| 80 | 6.95 | | |

(a) Make a scatterplot. (Which variable should go on the $x$ axis?)

(b) Describe the form of the relationship. Why is it not linear? Explain why the form of the relationship makes sense.

(c) It does not make sense to describe the variables as either positively associated or negatively associated. Why?

(d) Is the relationship reasonably strong or quite weak? Explain your answer.

7.3    Some people use median SAT scores to rank state or local school systems. This is not proper, because the percent of high school seniors who take the SAT varies from place to place. Make a scatterplot to examine the relationship between the percent in a state who take the exam and the state median SAT mathematics score, using data from Table 3.2 (page 24). Which variable should we plot horizontally? Describe the direction, form, and strength of the relationship.

7.4    Metabolic rate, the rate at which the body consumes energy, is important in studies of weight gain, dieting, and exercise. The table below gives data on the lean body mass and resting metabolic rate for 12 women and 7 men who are subjects in a study of dieting. Lean body mass, given in kilograms, is a person's weight leaving out all fat. Metabolic rate is measured in calories burned per 24 hours, the same calories used to describe the energy content of foods. The researchers believe that lean body mass is an important influence on metabolic rate.

| Subject | Sex | Mass | Rate | Subject | Sex | Mass | Rate |
|---------|-----|------|------|---------|-----|------|------|
| 1  | M | 62.0 | 1792 | 11 | F | 40.3 | 1189 |
| 2  | M | 62.9 | 1666 | 12 | F | 33.1 | 913  |
| 3  | F | 36.1 | 995  | 13 | M | 51.9 | 1460 |
| 4  | F | 54.6 | 1425 | 14 | F | 42.4 | 1124 |
| 5  | F | 48.5 | 1396 | 15 | F | 34.5 | 1052 |
| 6  | F | 42.0 | 1418 | 16 | F | 51.1 | 1347 |
| 7  | M | 47.4 | 1362 | 17 | F | 41.2 | 1204 |
| 8  | F | 50.6 | 1502 | 18 | M | 51.9 | 1867 |
| 9  | F | 42.0 | 1256 | 19 | M | 46.9 | 1439 |
| 10 | M | 48.7 | 1614 |    |   |      |      |

(a) Make a scatterplot of the data, using different symbols or colors for men and women. It is often helpful to use different plot symbols or colors to distinguish different groups in a scatterplot.

**(b)** Is the association between these variables positive or negative? What is the form of the relationship? How strong is the relationship? Does the pattern of the relationship differ for women and men? How do the male subjects as a group differ from the female subjects as a group?

7.5   Return to the gas mileage data in Table 6.1. In Exercise 7.1, you made a scatterplot of highway versus city mileage for the midsize cars. Add the sports utility vehicles to your plot, using a different plot symbol or color to distinguish them. Do city and highway mileage have the same relationship for both vehicle types? For example, would you be willing to draw one line through the data to predict highway mileage from city mileage for both types? How do the differences between the two types appear on your graph?

7.6   Are hot dogs that are high in calories also high in salt? Make a scatterplot of the hot dog data from Table 6.2 (page 76), using a different symbol or color for each type of hot dog. Then describe the direction, form, and strength of the relationship between sodium and calories in hot dogs, as well as any differences among the groups in the nature of this relationship. Are there any outliers?

7.7   How much corn per acre should a farmer plant to obtain the highest yield? Too few plants will give a low yield. On the other hand, if there are too many plants, they will compete with each other for moisture and nutrients, and yields will fall. To find the best planting rate, plant at different rates on several plots of ground and measure the harvest. Here are data from such an experiment:[16]

| Plants per acre | Yield (bushels per acre) | | | |
|---|---|---|---|---|
| 12,000 | 150.1 | 113.0 | 118.4 | 142.6 |
| 16,000 | 166.9 | 120.7 | 135.2 | 149.8 |
| 20,000 | 165.3 | 130.1 | 139.6 | 149.9 |
| 24,000 | 134.7 | 138.4 | 156.1 | |
| 28,000 | 119.0 | 150.5 | | |

**(a)** Make a scatterplot of yield and planting rate. (Which variable should go on the $x$ axis?)

**(b)** Describe the overall pattern of the relationship. Is it linear? Is there a positive or negative association, or neither?

**(c)** Find the mean yield for each of the five planting rates. Plot each mean yield against its planting rate on your scatterplot and connect these five points with lines. This combination of numerical description and graphing makes the relationship clearer. What planting rate would you recommend to a farmer whose conditions were similar to those in the experiment?

7.8   Table 3.2 gives educational data for the states. We are interested in the relationship between how much states spend on education (dollars per pupil) and how much they pay their teachers (median teacher salaries, in thousands of dollars).

(a) Explain why you expect a positive association between these variables.

(b) We think that education spending helps explain teachers' pay. Make a scatterplot to display this relationship.

(c) Describe the relationship. Is there a positive association? Is the relationship approximately linear?

(d) On the plot, identify a state where teacher salaries are unusually high relative to the state's education spending. (This state is an outlier, though not an extreme outlier.) What state is this?

(e) How do the Mountain states compare with the rest of the country in education spending and teacher salaries? Mark the points for states in the MTN region with a different color on your scatterplot. Based on the plot, briefly answer the question.

7.9   (Optional) Data analysts often look for a simple *transformation* of data that simplifies the overall pattern. Here is an example of how transforming a variable can simplify the pattern of a scatterplot.

The population of Europe grew as follows between 1750 and 1950:

| Year | 1750 | 1800 | 1850 | 1900 | 1950 |
|------|------|------|------|------|------|
| Population (millions) | 125 | 187 | 274 | 423 | 594 |

(a) Make a scatterplot of population against year. Briefly describe the pattern of Europe's growth.

(b) Now take the logarithm of the population in each year. Plot the logarithms against year. What is the overall pattern of this plot?

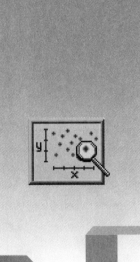

# 8

# CORRELATION

A scatterplot displays the form, direction, and strength of the relationship between two quantitative variables. Straight-line (linear) relations are particularly important because a straight line is a simple pattern that is quite common. We say a linear relationship is strong if the points lie close to a straight line, and weak if they are widely scattered about a line. Our eyes are not good judges of how strong a relationship is. The two scatterplots in Figure 8.1 depict exactly the same data, but the lower plot is drawn smaller in a large field. The lower plot seems to show a stronger relationship. Our eyes can be fooled by changing the plotting scales or the amount of white space around the cloud of points in a scatterplot.[17] We need to follow our strategy for data analysis by using a numerical measure to supplement the graph. *Correlation* is the measure we use.

---

CORRELATION

The **correlation** measures the strength and direction of the linear relationship between two quantitative variables. Correlation is usually written as $r$.

---

(continued on next page)

*(continued from previous page)*

Suppose that we have data on variables $x$ and $y$ for $n$ individuals. The values for the first individual are $x_1$ and $y_1$, the values for the second individual are $x_2$ and $y_2$, and so on. The means and standard deviations of the two variables are $\bar{x}$ and $s_x$ for the $x$-values, and $\bar{y}$ and $s_y$ for the $y$-values. The correlation $r$ between $x$ and $y$ is

$$r = \frac{1}{n-1}\sum\left(\frac{x_i - \bar{x}}{s_x}\right)\left(\frac{y_i - \bar{y}}{s_y}\right)$$

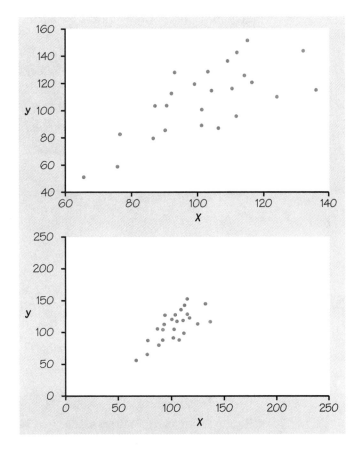

FIGURE 8.1    Two identical scatterplots; the straight-line pattern in the lower plot appears stronger because of the surrounding white space.

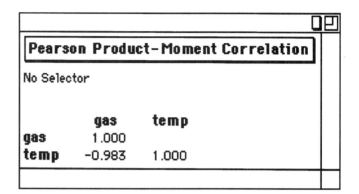

FIGURE 8.2   *Data Desk* correlation output.

As always, the summation sign $\sum$ means "add these terms for all the individuals." The formula for the correlation $r$ is a bit complex. It helps us see what correlation is, but in practice you should use software to obtain $r$. Figure 8.2 shows *Data Desk's* calculation of the correlation between natural gas consumption and outdoor temperature for Joan's house (Example 7.1).

## PROPERTIES OF CORRELATION

Here are the basic facts needed to interpret a correlation $r$:

- It makes no difference which variable you call $x$ and which you call $y$ in calculating $r$. Because $r$ uses the standardized values of the observations, $r$ does not change when we change the units of measurement of $x$, $y$, or both. Measuring height in centimeters rather than inches and weight in kilograms rather than pounds does not change the correlation between height and weight. The correlation $r$ itself has no unit of measurement; it is just a number.

- Correlation requires that both variables be quantitative, so that it makes sense to do the arithmetic required to find $r$. You can't calculate a correlation between the incomes of a group of people and what city they live in, because city is a categorical variable.

- Positive $r$ indicates positive association between the variables, and negative $r$ indicates negative association.

- The correlation $r$ is always a number between $-1$ and 1. Values of $r$ near 0 indicate a very weak linear relationship. The strength of the

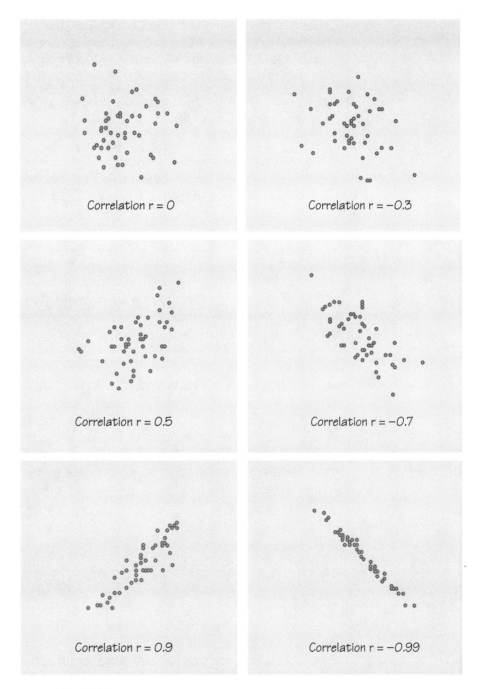

FIGURE 8.3 How the correlation coefficient measures the strength of linear association. Patterns closer to a straight line have correlations closer to 1 or −1.

relationship increases as $r$ moves away from 0 toward either $-1$ or 1. Values of $r$ close to $-1$ or 1 indicate that the points lie close to a straight line. The extreme values $r = -1$ and $r = 1$ occur only when the points in a scatterplot lie exactly along a straight line.

- Correlation measures the strength of only a straight-line relationship between two variables. Correlation does not describe curved relationships between variables, no matter how strong they are.

- Like the mean and standard deviation, the correlation is not resistant: $r$ is strongly affected by a few outlying observations. Use $r$ with caution when outliers appear in the scatterplot.

The correlation $r = -0.983$ in Figure 8.2 describes the strong negative linear relationship between outdoor temperature and gas consumption. The scatterplots in Figure 8.3 illustrate how values of $r$ closer to 1 or $-1$ correspond to stronger linear relationships. To make the essential meaning of $r$ clear, the standard deviations of both variables in these plots are equal and the horizontal and vertical scales are the same. In general, it is not so easy to guess the value of $r$ from the appearance of a scatterplot. Remember that changing the plotting scales in a scatterplot may mislead our eyes, but it does not change the correlation.

Remember also that *correlation is not a complete description of two-variable data*, even when the relationship between the variables is a straight line. You should give the means and standard deviations of both $x$ and $y$ along with the correlation. Because the formula for correlation uses the means and standard deviations, these measures are the proper choice to accompany a correlation.

## EXERCISES

8.1    *Archaeopteryx* is an extinct beast having feathers like a bird but teeth and a long bony tail like a reptile. Only six fossil specimens are known. Because these specimens differ greatly in size, some scientists think they are different species rather than individuals from the same species. If the specimens belong to the same species and differ in size because some are younger than others, there should be a positive linear relationship between the lengths of a pair of bones from all individuals. An outlier from this relationship would suggest a different species. Here are data on the lengths in centimeters of the femur (a leg bone) and the humerus (a bone in the upper arm) for the five specimens that preserve both bones:[18]

$$
\begin{array}{lccccc}
\text{Femur} & 38 & 56 & 59 & 64 & 74 \\
\text{Humerus} & 41 & 63 & 70 & 72 & 84
\end{array}
$$

(a) Make a scatterplot and describe the form, direction, and strength of the relationship.

(b) Find the correlation $r$ step-by-step. That is, find the mean and standard deviation of the femur lengths and of the humerus lengths. Then find the five standardized values for each variable and use the formula for $r$.

(c) Now enter these data into your software and use the software's correlation function to find $r$. Check that you get the same result as in (b).

8.2   Table 6.1 (page 71) gives the city and highway gas mileages for 26 mid-size car models. You made a scatterplot of these data in Exercise 7.1.

(a) Find the correlation between city and highway mileages. What does $r$ tell you about the relationship?

(b) The Rolls-Royce is an outlier—it has much lower gas mileage than any other car. Remove the Rolls-Royce from the data and again find the correlation. Explain why removing this one observation changes $r$ in the way that you observe.

8.3   Exercise 7.4 (page 81) gives data on the lean body mass and metabolic rate for 12 women and 7 men.

(a) Make a scatterplot. Use different symbols or colors for women and men. Do you think the correlation will be about the same for men and women or quite different for the two groups? Why?

(b) Find $r$ for women alone and also for men alone.

(c) Calculate the mean body mass for the women and for the men. Does the fact that the men are heavier than the women on the average influence the correlations? If so, in what way?

(d) Lean body mass was measured in kilograms. How would the correlations change if we measured body mass in pounds? (There are about 2.2 pounds in a kilogram.)

8.4   Exercise 7.2 (page 80) gives data on gas mileage versus speed for a small car. Make a scatterplot and find the correlation $r$. Explain why $r$ is close to zero despite a strong relationship between speed and gas used.

8.5    Changing the units of measurement can dramatically alter the appearance of a scatterplot. Consider the following data:

| $x$ | $-4$ | $-4$ | $-3$ | $3$ | $4$ | $4$ |
|---|---|---|---|---|---|---|
| $y$ | $0.5$ | $-0.6$ | $-0.5$ | $0.5$ | $0.5$ | $-0.6$ |

(a) Plot the data on $x$ and $y$ axes that extend from $-6$ to $6$.

(b) Form new variables $x^* = x/10$ and $y^* = 10y$, starting from the values of $x$ and $y$. Plot $y^*$ against $x^*$ on the same axes using a different plotting symbol. The two plots are very different in appearance.

(c) Find the correlation between $x$ and $y$. Then find the correlation between $x^*$ and $y^*$. How are the two correlations related? Explain why this is not surprising.

8.6    Consider again the correlation $r$ between the speed of a car and its gas consumption, from the data in Exercise 7.2 (page 80).

(a) Transform the data so that speed is measured in miles per hour and fuel consumption in gallons per mile. (There are 1.609 kilometers in a mile and 3.785 liters in a gallon.) Make a scatterplot and find the correlation for both the original and the transformed data. How did the change of units affect your results?

(b) Now express fuel consumption in miles per gallon. (So each value is $1/x$ if $x$ is gallons per mile.) Again make a scatterplot and find the correlation. How did this change of units affect your results?
    (*Lesson:* The effects of a linear transformation of the form $x_{new} = a + bx_{old}$ are simple. The effects of a nonlinear transformation are more complex.)

8.7    If women always married men who were 2 years older than themselves, what would be the correlation between the ages of husband and wife? (*Hint:* Draw a scatterplot for several ages.)

8.8    A college newspaper interviews a psychologist about student ratings of the teaching of faculty members. The psychologist says, "The evidence indicates that the correlation between the research productivity and teaching rating of faculty members is close to zero." The paper reports this as "Professor McDaniel said that good researchers tend to be poor teachers, and vice versa." Explain why the paper's report is wrong. Write a statement in plain language (don't use the word "correlation") to explain the psychologist's meaning.

8.9    Each of the following statements contains a blunder. Explain in each case what is wrong.

(a) "There is a high correlation between the sex of American workers and their income."

(b) "We found a high correlation ($r = 1.09$) between students' ratings of faculty teaching and ratings made by other faculty members."

(c) "The correlation between planting rate and yield of corn was found to be $r = 0.23$ bushel."

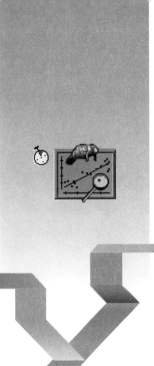

# 9

# LEAST-SQUARES REGRESSION

Correlation measures the direction and strength of the straight-line (linear) relationship between any two quantitative variables. If a scatterplot shows a linear relationship, we would like to summarize this overall pattern by drawing a line on the scatterplot. A regression line summarizes the relationship between two variables, but only in a specific setting: one of the variables helps explain or predict the other.

---

RESPONSE VARIABLE, EXPLANATORY VARIABLE

A **response variable** measures an outcome of a study. An **explanatory variable** attempts to explain the observed outcomes.

---

You will often see explanatory variables called **independent variables** and response variables called **dependent variables.** The idea is that the dependent variable depends on the independent variable. We will avoid these terms because "independence" has other meanings in statistics. When making a scatterplot of data on an explanatory variable and a response variable, always plot the explanatory variable on the horizontal axis. We therefore call the explanatory variable $x$ and the response variable $y$.

---

REGRESSION LINE

A **regression line** is a straight line that describes how a response variable $y$ changes as an explanatory variable $x$ changes. We often use a regression line to predict the value of $y$ for a given value of $x$. Regression, unlike correlation, requires that we have an explanatory variable and a response variable.

---

## EXAMPLE 9.1

Here again are Joan's data on average temperature (°F) and average natural gas consumed (cubic feet per day) for one heating season:

| Month | Oct. | Nov. | Dec. | Jan. | Feb. | Mar. | Apr. | May |
|---|---|---|---|---|---|---|---|---|
| Temperature $x$ | 49.4 | 38.2 | 27.2 | 28.6 | 29.5 | 46.4 | 49.7 | 57.1 |
| Gas consumed $y$ | 520 | 610 | 870 | 850 | 880 | 490 | 450 | 250 |

Temperature helps explain gas consumption, so temperature is plotted horizontally as the explanatory variable $x$. The scatterplot in Figure 9.1 shows a strong negative linear association: as temperature increases, gas consumption goes down because less gas is used for heating. The correlation is $r = -0.983$, so the points fall very close to a line. We have drawn a regression line for predicting gas consumption from temperature.

How much gas can Joan expect to use in a month that averages 30°F per day? Locate $x = 30$ on the horizontal axis and draw a vertical line up to the line and then a horizontal line over to the gas consumption scale. We predict that Joan will use slightly more than 800 cubic feet per day.  ◀

## THE LEAST-SQUARES REGRESSION LINE

Different people might draw different lines by eye on a scatterplot. This is especially true when the points are widely scattered. We need a way to draw a regression line that doesn't depend on our guess as to where the line should go. No line will pass exactly through all the points, but we want one

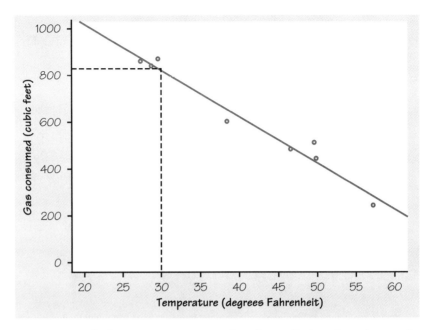

FIGURE 9.1    The least-squares regression line for predicting a home's natural gas consumption from outdoor temperature.

that is as close as possible. We will use the line to predict $y$ from $x$, so we want a line that is as close as possible to the points in the *vertical* direction. That's because the prediction errors we make are errors in $y$, which is the vertical direction in the scatterplot. If we predict 800 cubic feet per day for a month with average temperature 30° and the actual usage turns out to be 780 cubic feet, our error is

$$\text{error} = \text{observed} - \text{predicted}$$
$$= 780 - 800 = -20$$

A "good" regression line makes these prediction errors as small as possible. There are many ways to make "as small as possible" precise. The most common is the *least-squares* idea.

---

LEAST-SQUARES REGRESSION LINE

The **least-squares regression line** of $y$ on $x$ is the line that makes the sum of the squares of the vertical distances of the data points from the line as small as possible.

The equation of a line linking a response variable $y$ to an explanatory variable $x$ has the form

$$y = a + bx$$

The number $b$ is the **slope** of the line. The slope of a regression line is usually important for understanding the data. The slope is the rate of change—that is, the amount of change in the predicted $y$ when $x$ increases by 1. Think of slope as "$y$ units per $x$ unit" along the line. The number $a$ is the **intercept**, the value of $y$ when $x = 0$.

**EXAMPLE 9.2**

Figure 9.2 shows the *Data Desk* regression output for Joan's data. The equation of the least-squares line for predicting gas consumption $y$ from temperature $x$ is

$$\hat{y} = 1425.03 - 19.8719x$$

This line appears on the scatterplot in Figure 9.1. The slope of this line is $b = -19.8719$. This means that gas consumption goes down by about 19.9 cubic feet per day when the average outdoor temperature rises by 1 degree.

The intercept of the least-squares line is $a = 1425.03$. This is the value of the predicted $y$ when $x = 0$. Although we need the intercept to draw the line, it is

```
                                              ▢▣
Dependent variable is:    gas
No Selector
R squared = 96.6%    R squared (adjusted) = 96.0%
s = 46.38 with 8 - 2 = 6 degrees of freedom
```

| Source | Sum of Squares | df | Mean Square | F-ratio |
|--------|---------------|-----|-------------|---------|
| Regression | 362692 | 1 | 362692 | 169 |
| Residual | 12908.4 | 6 | 2151.40 | |

| Variable | Coefficient | s.e. of Coeff | t-ratio | prob |
|----------|-------------|---------------|---------|------|
| Constant | 1425.03 | 64.51 | 22.1 | ≤ 0.0001 |
| temp | -19.8719 | 1.530 | -13.0 | ≤ 0.0001 |

FIGURE 9.2  *Data Desk* regression output.

statistically meaningful only when $x$ can actually take values close to zero. An average temperature of $0°$ for a month never happens in Joan's location.

To use the equation for prediction, just substitute the value of $x$ and calculate $y$. For example, Joan's predicted gas consumption for a month with average temperature $30°$ is

$$\hat{y} = 1425.03 - (19.8719)(30)$$
$$= 829 \text{ cubic feet per day}$$

As is often the case, the output in Figure 9.2 contains other results that we ignore for now. Some of these will be important to us later.    ◀

We write $\hat{y}$ (read "y hat") in the equation of a regression line to emphasize that the line gives a *predicted* response $\hat{y}$ for any $x$. The predicted response will usually not be exactly the same as the actually *observed* response $y$.

# RESIDUALS

A regression line is a simple way to describe the overall pattern of a linear relationship between an explanatory variable and a response variable. Deviations from the overall pattern are also important. In the regression setting, we see deviations by looking at the scatter of the data points about the regression line. The vertical distances from the points to the least-squares regression line are as small as possible, in the sense that they have the smallest possible sum of squares. These distances are called *residuals*.

---

### RESIDUALS

A **residual** is the difference between an observed value of the response variable and the value predicted by the regression line. That is,

$$\text{residual} = \text{observed } y - \text{predicted } y$$
$$= y - \hat{y}$$

Data points with large residuals are *outliers*—they lie outside the overall pattern of the data. Try to learn if there is a reason for an outlier, such as an error entering data or an unusual occurrence.

## REGRESSION AND CORRELATION

Correlation and least-squares regression are closely related. We use software to obtain the regression line from data, so we don't need a recipe for the line. Nonetheless, an equation that relates regression and correlation helps us understand regression.

---

### EQUATION OF THE LEAST-SQUARES REGRESSION LINE

We have data on an explanatory variable $x$ and a response variable $y$ for $n$ individuals. The means and standard deviations of the sample data are $\bar{x}$ and $s_x$ for $x$ and $\bar{y}$ and $s_y$ for $y$, and the correlation between $x$ and $y$ is $r$. The equation of the least-squares regression line is

$$\hat{y} = a + bx$$

with **slope**

$$b = r\frac{s_y}{s_x}$$

and **intercept**

$$a = \bar{y} - b\bar{x}$$

---

Here are some basic facts about least-squares regression, most of which follow from study of this equation:

- Least-squares regression looks at the distances of the data points from the line only in the $y$ direction. So the two variables $x$ and $y$ play different roles in regression. The regression line for predicting gas consumption from temperature is not the same as the regression line for predicting temperature from gas consumption. There is, however, only one correlation $r$ between these variables.

- Along the regression line, a change of one standard deviation in $x$ corresponds to a change of $r$ standard deviations in the predicted response $\hat{y}$. As the correlation grows less strong, the prediction $\hat{y}$ moves less in response to changes in $x$.
- The least-squares regression line passes through the point $(\overline{x}, \overline{y})$ that marks the means of the two variables.
- The sum of the residuals from a least-squares regression line is always zero.
- Least-squares regression, like $\overline{y}$, $s$, and $r$, is not resistant. An extreme observation can move the line quite a bit. Points that are isolated in the $x$ direction—that is, points at the sides of the scatterplot—are particularly influential in fixing the position of the line.

Another connection between correlation and regression is also important. In fact, you can best interpret the numerical value of $r$ by thinking about regression. Here is the fact we need.

---

### $r^2$ IN REGRESSION

The **square of the correlation, $r^2$**, is the fraction of the variation in the values of $y$ that is explained by the least-squares regression of $y$ on $x$.

---

The idea is that when there is a linear relationship, some of the variation in $y$ is accounted for by the fact that as $x$ changes it pulls $y$ along with it. Look again at Figure 9.1. There is a lot of variation in the observed $y$'s, the gas consumption data. They range from a low of 250 to a high of 880. The scatterplot shows that most of this variation in $y$ is accounted for by the fact that outdoor temperature was changing and pulled gas consumption along with it. There is only a little remaining variation in $y$, which appears in the scatter of points about the line. The correlation for these data is $r = -0.983$, so the linear relationship explains $(-0.983)^2 = 0.966$, or about 97%, of the observed variation in gas consumption.

## EXAMPLE 9.3

Does the age at which a child begins to talk predict score on a test of mental ability taken several years later? A study of the development of young children recorded the age in months at which each of 21 children spoke their first word and their Gesell Adaptive Score, the result of an aptitude test taken much later. The data appear in Table 9.1.[19]

Figure 9.3 is a scatterplot, with age at first word as the explanatory variable $x$ and Gesell score as the response variable $y$. Children 3 and 13, and also Children 16 and 21, have identical values of both variables. We use a different plotting symbol to show that one point stands for two individuals. The plot shows a moderately linear negative relationship. That is, children who begin to speak later tend to have lower test scores than early talkers. The correlation describes the direction and strength of this linear relationship. It is $r = -0.640$.

The line on the plot is the least-squares regression line for predicting Gesell score from age at first word. Its equation is

$$\hat{y} = 109.8738 - 1.1270x$$

◄

The data in Example 9.3 illustrate the lack of resistance of correlation and regression. Children 18 and 19 are both unusual, but they are unusual in different ways. Child 19 lies far from the regression line. Child 18 is close to the line but far out in the $x$ direction.

Child 19 is an *outlier*, with a Gesell score so high that we should check for a mistake in recording it. In fact, the score is correct. The predicted

| TABLE 9.1 | Age at first word and Gesell score | | | | |
|---|---|---|---|---|---|
| Child | Age | Score | Child | Age | Score |
| 1 | 15 | 95 | 11 | 7 | 113 |
| 2 | 26 | 71 | 12 | 9 | 96 |
| 3 | 10 | 83 | 13 | 10 | 83 |
| 4 | 9 | 91 | 14 | 11 | 84 |
| 5 | 15 | 102 | 15 | 11 | 102 |
| 6 | 20 | 87 | 16 | 10 | 100 |
| 7 | 18 | 93 | 17 | 12 | 105 |
| 8 | 11 | 100 | 18 | 42 | 57 |
| 9 | 8 | 104 | 19 | 17 | 121 |
| 10 | 20 | 94 | 20 | 11 | 86 |
| | | | 21 | 10 | 100 |

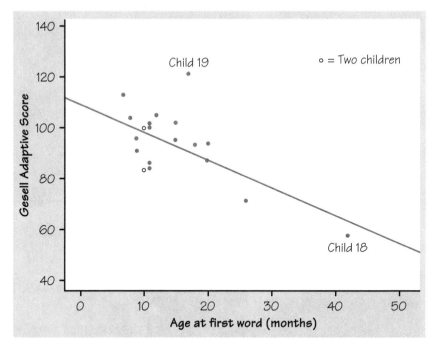

FIGURE 9.3   Scatterplot of Gesell Adaptive Score versus the age at first word for 21 children, from Table 9.1. The line is the least-squares regression line for predicting Gesell score from age at first word.

Gesell score for Child 19 is

$$\hat{y} = 109.8738 - (1.1270)(17) = 90.71$$

and the residual is

$$\text{residual} = \text{observed } y - \text{predicted } y$$
$$= 121 - 90.71 = 30.29$$

Despite this large residual, removing Child 19 would not move the least-squares line very much. This is because the several other children with similar $x$-values (points below Child 19 in Figure 9.3) anchor the line close to its present location.

Child 18 began to speak much later than any of the other children. *Because of its extreme position on the age scale, this point has a strong influence on the position of the regression line.* Figure 9.4 adds a second regression line, calculated after leaving out Child 18. You can see that this one point moves the line quite a bit.

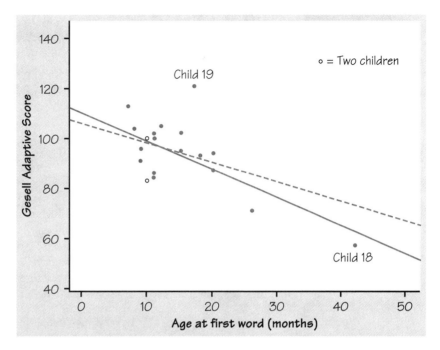

FIGURE 9.4    Two least-squares regression lines of Gesell score on age at first word. The solid line is calculated from all the data. The dashed line was calculated leaving out Child 18. Child 18 is an influential observation because leaving out this point moves the regression line quite a bit.

The original Gesell data have $r^2 = 0.41$. That is, the age at which a child begins to talk explains 41% of the variation on a later test of mental ability. This relationship is strong enough to be interesting to parents. But if we leave out Child 18, $r^2$ drops to only 11%. What should we do? We must decide whether Child 18 is so slow to speak that this individual should not be allowed to influence the analysis. If we exclude Child 18, much of the evidence for a connection between the age at which a child begins to talk and later ability score vanishes. If we keep Child 18, we need data on other children who were also slow to begin talking, so that the analysis no longer depends so heavily on just one child.

## ASSOCIATION IS NOT CAUSATION

When we study the relationship between two variables, we often want to show that changes in the explanatory variable *cause* changes in the response variable. A strong association between two variables is not enough to draw conclusions about cause and effect. Sometimes an observed association really does reflect cause and effect. Joan's house uses more natural gas in colder months because cold weather requires burning more gas to stay warm. In other cases, an association is explained by other variables that are present, and the conclusion that $x$ causes $y$ is either wrong or not proved.

## EXAMPLE 9.4

Measure the number of television sets per person $x$ and the average life expectancy $y$ for the world's nations. There is a high positive correlation—nations with many TV sets have higher life expectancies. Could we lengthen the lives of people in Rwanda by shipping them TV sets? No. Rich nations have more TV sets than poor nations. Rich nations also have longer life expectancies because they offer better nutrition, clean water, and better health care. There is no cause-and-effect tie between TV sets and length of life.                   ◄

### ASSOCIATION DOESN'T IMPLY CAUSATION

An association between an explanatory variable $x$ and a response variable $y$, even if it is very strong, is not by itself good evidence that changes in $x$ actually cause changes in $y$.

## EXERCISES

9.1  How well does a child's height at age 6 predict height at age 16? To find out, measure the heights of a large group of children at age 6, wait until they reach age 16, then measure their heights again. What are the explanatory and response variables here? Are these variables categorical or quantitative?

9.2  There appears to be a "gender gap" in political party preference in the United States, with women more likely than men to prefer Democratic candidates. A political scientist selects a large sample of registered voters, both women and men. She asks each voter whether they voted for the Democratic or for the Republican candidate in the last congressional election. What are the explanatory and response variables in this study? Are they categorical or quantitative variables?

9.3    (Review of straight lines) Fred keeps his savings in his mattress. He began with $500 from his mother and adds $100 each year. His total savings $y$ after $x$ years are given by the equation

$$y = 500 + 100x$$

(a) Draw a graph of this equation. (Choose two values of $x$, such as 0 and 10. Compute the corresponding values of $y$ from the equation. Plot these two points on graph paper and draw the straight line joining them.)

(b) After 20 years, how much will Fred have in his mattress?

(c) If Fred had added $200 instead of $100 each year to his initial $500, what is the equation that describes his savings after $x$ years?

9.4    (Review of straight lines) During the period after birth, a male white rat gains exactly 40 grams (g) per week. (This rat is unusually regular in his growth, but 40 g per week is a realistic rate.)

(a) If the rat weighed 100 g at birth, give an equation for his weight after $x$ weeks. What is the slope of this line?

(b) Draw a graph of this line between birth and 10 weeks of age.

(c) Would you be willing to use this line to predict the rat's weight at age 2 years? Do the prediction and think about the reasonableness of the result. (There are 454 grams in a pound. To help you assess the result, note that a large cat weighs about 10 pounds.)

9.5    Researchers studying acid rain measured the acidity of precipitation in a Colorado wilderness area for 150 consecutive weeks. Acidity is measured by pH. Lower pH values show higher acidity. The acid rain researchers observed a straight-line pattern over time. They reported that the least-squares regression line

$$pH = 5.43 - (0.0053 \times weeks)$$

fit the data well.[20]

(a) According to the regression line, what was the pH at the beginning of the study (weeks = 1)? At the end (weeks = 150)?

(b) What is the slope of the regression line? Explain clearly what this slope says about the change in the pH of the precipitation in this wilderness area. Is the association between pH and time positive or negative?

9.6    Concrete road pavement gains strength over time as it cures. Highway builders use regression lines to predict the strength after 28 days (when curing is complete) from measurements made after 7 days. Let $x$ be strength after 7 days (in pounds per square inch) and $y$ the strength after 28 days. One set of data gives this least-squares regression line:

$$\hat{y} = 1389 + 0.96x$$

(a) Draw a graph of this line, with $x$ running from 3000 to 4000 pounds per square inch.

(b) Explain what the slope $b = 0.96$ in this equation says about how concrete gains strength as it cures.

(c) A test of some new pavement after 7 days shows that its strength is 3300 pounds per square inch. Use the equation of the regression line to predict the strength of this pavement after 28 days. Also draw the "up and over" lines from $x = 3300$ on your graph (as in Figure 9.1).

9.7    Find the mean and standard deviation of the temperature and gas consumption data (Example 9.1). Use your results along with the correlation to find the equation of the regression line by hand. Verify that your work agrees (up to roundoff error) with the software result in Figure 9.2.

9.8    Return once more to Joan's data on outside temperature and natural gas consumption (Example 9.1). How much gas does the least-squares line predict Joan will use in a month with average temperature 90°? Why is this result impossible?

(Lesson: Using a regression line for prediction for values of $x$ outside the range of the data used to fit the line is called extrapolation. Extrapolation is risky, as this example illustrates.)

9.9    In Professor Smith's economics course the correlation between the students' total scores prior to the final examination and their final examination scores is $r = 0.6$. The pre-exam totals for all students in the course have mean 280 and standard deviation 30. The final exam scores have mean 75 and standard deviation 8. Professor Smith has lost Julie's final exam but knows that her total before the exam was 300. He decides to predict her final exam score from her pre-exam total.

(a) What is the slope of the least-squares regression line of final exam scores on pre-exam total scores in this course? What is the intercept?

(b) Use the regression line to predict Julie's final exam score.

(c) Julie doesn't think this method accurately predicts how well she did on the final exam. Calculate $r^2$ and use the value you get to argue that her actual score could have been much higher (or much lower) than the predicted value.

9.10   A study of class attendance and grades among first-year students at a state university showed that in general students who attended a higher percent of their classes earned higher grades. Class attendance explained 16% of the variation in grade index among the students. What is the numerical value of the correlation between percent of classes attended and grade index?

9.11 The discussion of Example 9.3 claims that Child 18 in the Gesell data of Table 9.1 has a strong influence on the least-squares line.

(a) Find the correlation and the equation of the least-squares line both with and without Child 18. Verify the correlation and equation given in Example 9.3 for the full data, and also verify the claim that $r^2$ drops to only 11% when we leave out Child 18.

(b) Now consider Child 19, who has a large residual. Find the least-squares regression line of Gesell score on age at first word, leaving out Child 19. Plot two regression lines—with and without Child 19—on a scatterplot of the data. Does Child 19 have more or less influence than Child 18 on the position of the regression line?

(c) How does removing Child 19 change the $r^2$ for this regression? Explain why $r^2$ changes in this direction when you drop Child 19.

9.12 Investors ask about the relationship between returns on investments in the United States and on investments overseas. Here are data on the total returns on U.S. and overseas common stocks over a 26-year period. (The total return is change in price plus any dividends paid, converted into U.S. dollars. Both returns are averages over many individual stocks.)[21]

| Year | Overseas % return | U.S. % return | Year | Overseas % return | U.S. % return |
|------|------|------|------|------|------|
| 1971 | 29.6 | 14.6 | 1984 | 7.4 | 6.1 |
| 1972 | 36.3 | 18.9 | 1985 | 56.2 | 31.6 |
| 1973 | −14.9 | −14.8 | 1986 | 69.4 | 18.6 |
| 1974 | −23.2 | −26.4 | 1987 | 24.6 | 5.1 |
| 1975 | 35.4 | 37.2 | 1988 | 28.5 | 16.8 |
| 1976 | 2.5 | 23.6 | 1989 | 10.6 | 31.5 |
| 1977 | 18.1 | −7.4 | 1990 | −23.0 | −3.1 |
| 1978 | 32.6 | 6.4 | 1991 | 12.8 | 30.4 |
| 1979 | 4.8 | 18.2 | 1992 | −12.1 | 7.6 |
| 1980 | 22.6 | 32.3 | 1993 | 32.9 | 10.1 |
| 1981 | −2.3 | −5.0 | 1994 | 6.2 | 1.3 |
| 1982 | −1.9 | 21.5 | 1995 | 11.2 | 37.6 |
| 1983 | 23.7 | 22.4 | 1996 | 6.4 | 23.0 |

(a) Make a scatterplot suitable for predicting overseas returns from U.S. returns.

(b) Find the correlation and $r^2$. Describe the relationship between U.S. and overseas returns in words, using $r$ and $r^2$ to make your description more precise.

(c) Find the least-squares regression line of overseas returns on U.S. returns. Add this line to your scatterplot.

(d) In 1993, the return on U.S. stocks was 10.1%. Use the regression line to predict the return on overseas stocks and find the residual. Are you confident that predictions using the regression line will be quite accurate? Why?

(e) Circle the point that has the largest residual (either positive or negative). What year is this? Are there any points that seem likely to be very influential, in the sense that the least-squares line would move if we omitted these points? If so, redo the regression without these points (one at a time) and see how much your prediction in part (d) changes.

9.13    Table 9.2 presents four sets of data prepared by the statistician Frank Anscombe to illustrate the dangers of calculating without first plotting the data.[22]

### TABLE 9.2    Four data sets for exploring correlation and regression

Data Set A

| x | 10 | 8 | 13 | 9 | 11 | 14 | 6 | 4 | 12 | 7 | 5 |
|---|----|----|----|----|----|----|----|----|----|----|----|
| y | 8.04 | 6.95 | 7.58 | 8.81 | 8.33 | 9.96 | 7.24 | 4.26 | 10.84 | 4.82 | 5.68 |

Data Set B

| x | 10 | 8 | 13 | 9 | 11 | 14 | 6 | 4 | 12 | 7 | 5 |
|---|----|----|----|----|----|----|----|----|----|----|----|
| y | 9.14 | 8.14 | 8.74 | 8.77 | 9.26 | 8.10 | 6.13 | 3.10 | 9.13 | 7.26 | 4.74 |

Data Set C

| x | 10 | 8 | 13 | 9 | 11 | 14 | 6 | 4 | 12 | 7 | 5 |
|---|----|----|----|----|----|----|----|----|----|----|----|
| y | 7.46 | 6.77 | 12.74 | 7.11 | 7.81 | 8.84 | 6.08 | 5.39 | 8.15 | 6.42 | 5.73 |

Data Set D

| x | 8 | 8 | 8 | 8 | 8 | 8 | 8 | 8 | 8 | 8 | 19 |
|---|----|----|----|----|----|----|----|----|----|----|----|
| y | 6.58 | 5.76 | 7.71 | 8.84 | 8.47 | 7.04 | 5.25 | 5.56 | 7.91 | 6.89 | 12.50 |

SOURCE: Frank J. Anscombe, "Graphs in statistical analysis," *The American Statistician,* 27 (1973), pp. 17–21.

(a) Without making scatterplots, find the correlation and the least-squares regression line for all four data sets. What do you notice? Use the regression line to predict $y$ for $x = 10$.

(b) Make a scatterplot for each of the data sets and add the regression line to each of the plots.

(c) In which of the four cases would you be willing to use the regression line to describe the dependence of $y$ on $x$? Explain your answer in each case.

(*Lesson*: ALWAYS PLOT YOUR DATA.)

9.14   Exercise 9.12 examined the relationship between returns on U.S. and overseas stocks. Investors also want to know what typical returns are and how much year-to-year variability (called *volatility* in finance) there is. Regression and correlation don't answer questions about center and spread.

(a) Find the five-number summaries for both U.S. and overseas returns, and make side-by-side boxplots to compare the two distributions.

(b) Were returns generally higher in the United States or overseas during this period? Explain your answer.

(c) Were returns more volatile (more variable) in the United States or overseas during this period? Explain your answer.

9.15   Someone says, "There is a strong positive correlation between the number of firefighters at a fire and the amount of damage the fire does. So sending lots of firefighters just causes more damage." Explain why this reasoning is wrong.

9.16   A study shows that there is a positive correlation between the size of a hospital (measured by its number of beds $x$) and the median number of days $y$ that patients remain in the hospital. Does this mean that you can shorten a hospital stay by choosing a small hospital?

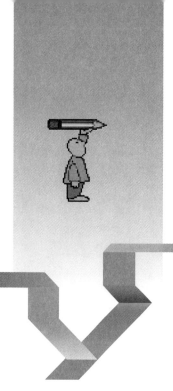

# PART II REVIEW

Part II continues our study of data analysis by presenting ideas and tools for examining relationships between two variables. Once again we start with a graph and then give selected numerical measures. Here is a review list of the most important skills you should have acquired from your study of Lessons 6 to 9.

## A. DATA

1. Recognize whether each variable is quantitative or categorical.
2. Identify the explanatory and response variables in situations where one variable explains or influences another.

## B. COMPARING GROUPS

1. Make side-by-side boxplots to compare the distributions of a quantitative response variable in several groups that are naturally different or have been treated differently.

## C. SCATTERPLOTS

1. Make a scatterplot to display the relationship between two quantitative variables. Place the explanatory variable (if any) on the horizontal scale of the plot.

2. Add a categorical variable to a scatterplot by using a different plotting symbol or color.

3. Describe the form, direction, and strength of the overall pattern of a scatterplot. In particular, recognize positive or negative association and linear (straight-line) patterns. Recognize outliers in a scatterplot.

## D. CORRELATION

1. Find the correlation coefficient $r$ between two quantitative variables.

2. Know the basic properties of correlation: $r$ measures the strength and direction of only linear relationships; $-1 \le r \le 1$ always; $r = \pm 1$ only for perfect straight-line relations; $r$ moves away from 0 toward $\pm 1$ as the linear relation gets stronger.

## E. REGRESSION

1. Find the least-squares regression line of a response variable $y$ on an explanatory variable $x$ from data.

2. Find the slope and intercept of the least-squares regression line from the means and standard deviations of $x$ and $y$ and their correlation.

3. Explain what the slope of a regression line says about the relationship between $x$ and $y$.

4. Use the regression line to predict $y$ for a given $x$. Recognize extrapolation and be aware of its dangers.

5. Use $r^2$ to describe how much of the variation in one variable can be accounted for by a straight-line relationship with another variable.

6. Recognize outliers and potentially influential observations from a scatterplot with the regression line drawn on it.

## F. LIMITATIONS OF CORRELATION AND REGRESSION

1. Understand that both $r$ and the least-squares regression line can be strongly influenced by a few extreme observations.

2. Understand that even a strong correlation does not mean that there is a cause-and-effect relationship between $x$ and $y$.

## REVIEW EXERCISES

**Review 1**  A food industry group asked 3368 people to guess the number of calories in each of several common foods. Here is a table of the average of their guesses and the correct number of calories:[23]

| Food | Guessed calories | Correct calories |
|---|---|---|
| 8 oz. whole milk | 196 | 159 |
| 5 oz. spaghetti with tomato sauce | 394 | 163 |
| 5 oz. macaroni with cheese | 350 | 269 |
| One slice wheat bread | 117 | 61 |
| One slice white bread | 136 | 76 |
| 2-oz. candy bar | 364 | 260 |
| Saltine cracker | 74 | 12 |
| Medium-size apple | 107 | 80 |
| Medium-size potato | 160 | 88 |
| Cream-filled snack cake | 419 | 160 |

(a) We think that how many calories a food actually has helps explain people's guesses of how many calories it has. With this in mind, make a scatterplot of these data.

(b) Describe the relationship. Is there a positive or negative association? Is the relationship approximately linear? Are there any outliers?

(c) Find the correlation $r$. Explain why your $r$ is reasonable based on the scatterplot. Find the least-squares regression line for predicting guessed calories from true calories and add it to your scatterplot.

(d) The guesses are all higher than the true calorie counts. How is this fact visible in either the correlation or the equation of the regression line? How would $r$ change if every guess were 100 calories higher? How would the slope of the regression line change if every guess were 100 calories higher?

(e) Circle the outliers on your scatterplot. Calculate $r$ for the other foods, leaving out the outliers. Explain why $r$ changed in the direction that it did. Then find the least-squares line leaving out the outliers. How much do these points influence the position of the line?

**Review 2**    Table II.1 gives the 1996–1997 salaries of the players on three National Basketball Association teams: the reigning champion Chicago Bulls, the runner-up Seattle Supersonics, and the New York Knickerbockers.[24] Compare the salary distributions for the three teams and report your conclusions.

(*Comment:* This example illustrates some limitations of our usual methods. You will have to remove some extreme outliers in order to

| TABLE II.1 | 1996 National Basketball Association player salaries | | | | |
|---|---|---|---|---|---|

| Chicago Bulls | | Seattle Supersonics | | New York Knicks | |
|---|---|---|---|---|---|
| Michael Jordan | $30,140,000 | Gary Payton | $10,212,000 | Larry Johnson | $5,004,000 |
| Dennis Rodman | 9,000,000 | Detlef Schrempf | 3,333,000 | Allan Houston | 5,000,000 |
| Toni Kukoc | 3,960,000 | Shawn Kemp | 3,300,000 | Patrick Ewing | 3,000,000 |
| Ron Harper | 3,840,000 | Jim McIllvaine | 3,000,000 | Chris Childs | 2,600,000 |
| Luc Longley | 2,790,000 | Sam Perkins | 2,730,000 | Charles Oakly | 2,600,000 |
| Scottie Pippen | 2,250,000 | Hersey Hawkins | 2,604,000 | John Starks | 1,913,000 |
| Randy Brown | 1,300,000 | Nate McMillian | 1,355,000 | Buck Williams | 1,193,000 |
| Dickey Simpkins | 1,040,000 | Greg Graham | 1,330,000 | Charlie Ward | 888,000 |
| Robert Parish | 1,000,000 | David Wingate | 560,000 | John Wallace | 785,000 |
| Bill Wennington | 1,000,000 | Craig Ehlo | 360,000 | Herb Williams | 757,000 |
| Steve Kerr | 750,000 | Steve Scheffler | 275,000 | Walter McCarty | 750,000 |
| Jason Caffey | 700,000 | Eric Snow | 247,500 | Dontae Jones | 691,000 |
| Jud Buechler | 500,000 | Larry Stewart | 247,500 | Scott Brooks | 247,500 |
| Thomas Hamilton | 247,500 | | | Chris Jent | 247,500 |
| | | | | Eric Leckner | 247,500 |

make useful graphs. Michael Jordan in particular is an "out-of-sight-lier." What is more, our usual descriptions don't include the measure that is most important to the team owner, who must pay the salaries. What is this measure, and how do the teams compare?)

**Review 3** Is wine good for your heart? Table II.2 gives data from a number of developed countries on wine consumption and deaths from

| TABLE II.2 | Wine consumption and heart attacks | | | | |
|---|---|---|---|---|---|

| Country | Alcohol from wine | Heart disease deaths | Country | Alcohol from wine | Heart disease deaths |
|---|---|---|---|---|---|
| Australia | 2.5 | 211 | Netherlands | 1.8 | 167 |
| Austria | 3.9 | 167 | New Zealand | 1.9 | 266 |
| Belgium | 2.9 | 131 | Norway | 0.8 | 227 |
| Canada | 2.4 | 191 | Spain | 6.5 | 86 |
| Denmark | 2.9 | 220 | Sweden | 1.6 | 207 |
| Finland | 0.8 | 297 | Switzerland | 5.8 | 115 |
| France | 9.1 | 71 | United Kingdom | 1.3 | 285 |
| Iceland | 0.8 | 211 | United States | 1.2 | 199 |
| Ireland | 0.7 | 300 | West Germany | 2.7 | 172 |
| Italy | 7.9 | 107 | | | |

heart disease.[25] Specifically, the explanatory variable $x$ is yearly liters of alcohol from drinking wine, per person, and the response variable $y$ is yearly deaths from heart disease per 100,000 people.

(a) Make a scatterplot and find the correlation and the equation of the least-squares regression line. Add the line to your plot. Describe the relationship between national wine consumption and heart disease deaths. Do any countries stand far outside the overall pattern?

(b) Explain clearly what the numerical value of the slope of the least-squares line says about the relationship between national wine consumption and heart disease deaths.

(c) Does the relationship we observe give good evidence that drinking wine *causes* a reduction in heart disease deaths? Why?

**Review 4**   Here are data on 18 people who fell ill from an incident of food poisoning.[26] The data give each person's age in years, the incubation period (the time in hours between eating the infected food and the first signs of illness), and whether the victim survived (S) or died (D).

|  | \multicolumn{9}{c}{Person} |
|---|---|---|---|---|---|---|---|---|---|
|  | 1 | 2 | 3 | 4 | 5 | 6 | 7 | 8 | 9 |
| Age | 29 | 39 | 44 | 37 | 42 | 17 | 38 | 43 | 51 |
| Incubation | 13 | 46 | 43 | 34 | 20 | 20 | 18 | 72 | 19 |
| Outcome | D | S | S | D | D | S | D | S | D |

|  | \multicolumn{9}{c}{Person} |
|---|---|---|---|---|---|---|---|---|---|
|  | 10 | 11 | 12 | 13 | 14 | 15 | 16 | 17 | 18 |
| Age | 30 | 32 | 59 | 33 | 31 | 32 | 32 | 36 | 50 |
| Incubation | 36 | 48 | 44 | 21 | 32 | 86 | 48 | 28 | 16 |
| Outcome | D | D | S | D | D | S | D | S | D |

(a) Make a scatterplot of incubation period against age, using different symbols for people who died and those who survived.

(b) Is there an overall relationship between age and incubation period? If so, describe it.

(c) More important, is there a relationship between either age or incubation period and whether the victim survived? Describe any relations that seem important here.

(d) Are there any unusual cases that may require individual investigation?

**Review 5**   Manatees are large, gentle sea creatures that live along the Florida coast. Many manatees are killed or injured by powerboats. Here are data on power boat registrations (in thousands) and the number of manatees killed by boats in Florida in the years 1977 to 1990:

| Year | Powerboat registrations (1000) | Manatees killed | Year | Powerboat registrations (1000) | Manatees killed |
|------|------|------|------|------|------|
| 1977 | 447 | 13 | 1984 | 559 | 34 |
| 1978 | 460 | 21 | 1985 | 585 | 33 |
| 1979 | 481 | 24 | 1986 | 614 | 33 |
| 1980 | 498 | 16 | 1987 | 645 | 39 |
| 1981 | 513 | 24 | 1988 | 675 | 43 |
| 1982 | 512 | 20 | 1989 | 711 | 50 |
| 1983 | 526 | 15 | 1990 | 719 | 47 |

(a) Make a scatterplot of these data. Find the least-squares regression line from these data for predicting manatee deaths from powerboat registrations and add the line to your scatterplot. Use the fitted line to predict the number of manatees killed in a year when 716,000 powerboats are registered.

(b) Here are four more years of manatee data, in the same form as the earlier data:

> 1991 716 53  1993 716 35
> 1992 716 38  1994 735 49

Add these points to your scatterplot. Florida took stronger measures to protect manatees during these years. Do you see any evidence that these measures succeeded?

(c) In (a), you predicted manatee deaths in a year with 716,000 powerboat registrations. In fact, powerboat registrations remained at 716,000 for the next three years. Compare the mean manatee deaths in these years with your prediction from either previous exercise. How accurate was your prediction?

**Review 6**  Suppose that in some far future year 2 million powerboats are registered in Florida. Use your regression line from the previous exercise to predict manatees killed. Explain why this prediction is very unreliable.

**Review 7**  Table 3.2 (page 24) gives data about education in the states. Examine the relationship between the median SAT verbal and mathematics scores as follows.

(a) You want to predict a state's SAT math score from its verbal score. Make a scatterplot. One state is an outlier. What state is this?

(b) Find the least-squares regression line and add it to your scatterplot. You learn that a state's median verbal score the following year was 455. Use your regression line to predict its median math score.

(c) What is the residual for the outlier you noted in (a)? How influential is this point? (To learn this, do the regression again with the outlier omitted. Add the new regression line to your plot and comment on the effect of removing the outlier.)

**Review 8**   Table 3.2 (page 24) gives data about education in the states. Give a careful description, using whatever tools you think are helpful, of the distribution of education spending per pupil in the states. Then give a careful description of the relationship between spending per pupil and median teachers' salaries. In both cases, be sure to include suitable numerical descriptions and to state your overall conclusions.

**Review 9**   Table 6.1 (page 71) gives the EPA city and highway gas mileages for 26 midsize cars and 22 sports utility vehicles. We began to study these data in Exercises 7.1 and 7.5. Now we can apply regression to the question of using city mileage to predict highway mileage.

Do a careful study of this question. Make a scatterplot (distinguish between the two vehicle types). Find regression lines for the desired prediction for each vehicle type separately and for all vehicles together. Are the lines similar? Is there an important difference between the ability to predict for the two vehicle types (give $r^2$ as a numerical measure, and explain its meaning)? Are any individual vehicles far from the fitted line? Are any strongly influential? What is the effect of removing these vehicles?

**Review 10**   Table 6.2 (page 76) gives the calorie and sodium content of a number of brands of each of three types of hot dogs. We began to examine the relationship between sodium and calories in Exercise 7.6. Now we can apply regression to this problem.

(a) Make a scatterplot suitable for using sodium to predict calories. Be sure to distinguish between the three types of hot dog. Give a careful description of the form, direction, and strength of the relationship. Do these differ from one type to another?

(b) As an aid to your study, add three regression lines to your scatterplot, one for each type of hot dog.

(c) Based on your findings in (a), are correlation and regression helpful for description and prediction? Can you combine all three or any two types in fitting a regression line for prediction?

**Review 11** Table I.1 (page 64) presents data on the monthly percent return on Wal-Mart stock for the 228 months between January 1973 and December 1991. The returns are in time order along the rows. We are interested in changes over time in the return on Wal-Mart stock.

(a) Make a scatterplot of monthly return against month (from 1 to 228). Add the least-squares regression line for predicting return from month number to your plot. What does your work show about the behavior of the average monthly returns during this 19-year period?

(b) The scatterplot suggests that the variability (called "volatility" in financial language) of the monthly returns decreased somewhat over time. How can you see this in the plot?

**Review 12** Here is another way to examine the behavior of Wal-Mart stock returns over time. Divide the data from Table I.1 into years (each successive group of 12 returns along the rows of the table forms a year, beginning with 1973). Make side-by-side boxplots of the 19 years' returns. How can you see from your graph changes in the median return over time? How can you see changes in volatility over time? What do you conclude about both aspects of change over time?

## SIR RONALD A. FISHER

The ideas and methods that we study as "statistics" were mostly invented in the nineteenth and twentieth centuries by people working on problems that required analysis of data. Astronomy, biology, social science, and even surveying can claim a role in the birth of statistics. But if anyone can claim to be "the father of statistics," that honor belongs to Sir Ronald A. Fisher (1890–1962).

Fisher's writings helped organize statistics as a distinct field of study whose methods apply to practical problems across many disciplines. He systematized the mathematical theory of statistics and invented many new techniques. The randomized comparative experiment is perhaps Fisher's greatest contribution.

Like many statistical pioneers, Fisher was driven by the demands of practical problems. Beginning in 1919, he worked on agricultural field experiments at Rothamsted in England. How should we arrange the planting of different crop varieties or the application of different fertilizers to get a fair comparison among them? Because fertility and other variables change as we move across a field, experimenters used elaborate checkerboard planting arrangements to obtain fair comparisons. Fisher had a better idea: "arrange the plots deliberately at random."

Part III explores statistical designs for producing data to answer specific questions like "Which crop variety has the highest mean yield?" Fisher's innovation, the deliberate use of chance in producing data, is the central theme of these lessons and one of the most important ideas in statistics.

PART **III**
**Generating Data**

LESSON **10**

# SAMPLE SURVEYS

Exploratory data analysis seeks to discover and summarize what data say by using graphs and numerical summaries. The conclusions we draw from data analysis apply to the specific data that we examine. Often we want to extend those conclusions to some larger group of individuals. If our data don't fairly represent the larger group, our conclusions from the data don't apply to the larger group.

EXAMPLE 10.1

The advice columnist Ann Landers once asked her readers, "If you had it to do over again, would you have children?" A few weeks later, her column was headlined "70% OF PARENTS SAY KIDS NOT WORTH IT." Indeed, 70% of the nearly 10,000 parents who wrote in said they would not have children if they could make the choice again.

These data are worthless as indicators of opinion among all American parents. The people who responded felt strongly enough to take the trouble to write Ann Landers. Their letters showed that many of them were angry at their children. These people don't fairly represent all parents. It is not surprising that a statistically designed opinion poll on the same issue a few months later found that 91% of parents *would* have children again. Ann Landers announced a 70% "No" result when the truth about parents was close to 90% "Yes." ◀

Both Ann Landers's write-in poll and the more careful opinion poll that contradicted her are examples of **sampling**. The idea is to study a part—a sample—in order to gain information about an entire group. Ann Landers used a *voluntary response sample.*

---

VOLUNTARY RESPONSE SAMPLE

A **voluntary response sample** consists of people who choose themselves by responding to a general appeal.

---

Voluntary response samples overrepresent people with strong opinions, most often negative opinions. If sample data are to give reliable information about a larger group of people or things, we must think carefully about how to choose the sample.

## SAMPLING DESIGN

An opinion poll wants to know what fraction of the public approves the president's performance in office. A quality engineer must estimate what fraction of the bearings rolling off an assembly line are defective. Government economists inquire about household income. In all these situations, we want to gather information about a large group of people or things. Time, cost, and inconvenience usually forbid inspecting every bearing or contacting every household. In such cases, we gather information about only part of the group in order to draw conclusions about the whole.

---

POPULATION, SAMPLE

The entire group of individuals that we want information about is called the **population.**

A **sample** is a part of the population that we actually examine in order to gather information.

---

Notice that the population is defined in terms of our desire for knowledge. If we wish to draw conclusions about all U.S. college students, that group is our population even if only local students are available for questioning. The sample is the part from which we draw conclusions about the whole. The **design** of a sample refers to the method used to choose the sample from the population. Poor sample designs can produce misleading conclusions. Voluntary response (Example 10.1) is one common type of bad sample design. Voluntary response samples display *bias,* or systematic error, in favoring some parts of the population over others.

---

BIAS

The design of a study is **biased** if it systematically favors certain outcomes.

---

## SIMPLE RANDOM SAMPLES

In a voluntary response sample, people choose whether to respond. The statistician's remedy is to allow impersonal chance to choose the sample. Choosing a sample by chance avoids bias by giving all individuals an equal chance to be chosen. Rich and poor, young and old, black and white, all have the same chance to be in the sample. The simplest way to use chance to select a sample is to place names in a hat (the population) and draw out a handful (the sample). This is the idea of *simple random sampling.*

---

SIMPLE RANDOM SAMPLE

A **simple random sample (SRS)** of size $n$ consists of $n$ individuals from the population chosen in such a way that every set of $n$ individuals has an equal chance to be the sample actually selected.

---

The idea of an SRS is to choose our sample by drawing names from a hat. Keeping that image in mind helps us understand random sampling. In practice, statistics software can choose an SRS almost instantly from a list of the individuals in the population. An SRS not only gives each individual an equal chance to be chosen (thus avoiding bias in the choice) but also gives every possible sample an equal chance to be chosen. There are other random sampling designs that give each individual, but not each sample, an equal chance. Exercise 10.10 describes one such design, called systematic random sampling.

## CAUTIONS ABOUT SAMPLE SURVEYS

Random selection eliminates bias in the choice of a sample from a list of the population. Sample surveys of large human populations, however, require much more than a good sampling design.[27] To begin, we need an accurate and complete list of the population. Because such a list is rarely available, most samples suffer from some degree of *undercoverage*. A sample survey of households, for example, will miss not only homeless people but prison inmates and students in dormitories. An opinion poll conducted by telephone will miss the 7% to 8% of American households without residential phones. The results of national sample surveys therefore have some bias if the people not covered—who most often are poor people—differ from the rest of the population.

A more serious source of bias in most sample surveys is *nonresponse,* which occurs when a selected individual cannot be contacted or refuses to cooperate. Nonresponse to sample surveys often reaches 30% or more, even with careful planning and several callbacks. Because nonresponse is higher in urban areas, most sample surveys substitute other people in the same area to avoid favoring rural areas in the final sample. If the people contacted differ from those who are rarely at home or who refuse to answer questions, some bias remains.

---

### UNDERCOVERAGE AND NONRESPONSE

**Undercoverage** occurs when some groups in the population are left out of the process of choosing the sample.

**Nonresponse** occurs when an individual chosen for the sample can't be contacted or refuses to cooperate.

---

**EXAMPLE 10.2**

Even the every-ten-years census, backed by the resources of the federal government, suffers from undercoverage and nonresponse. The census begins by mailing forms to every household in the country. The Census Bureau buys lists of addresses from private firms, then tries to fill in missing addresses. The final list is still incomplete, resulting in undercoverage. Despite special efforts to count homeless people (who can't be reached at any address), homelessness causes more undercoverage.

In 1990, about 35% of the households who were mailed census forms did not mail them back. In New York City, 47% did not return the form. That's nonresponse. The Census Bureau sends interviewers to these households. In central cities, the interviewers could not contact about one in five of the nonresponders, even after six tries.

The Census Bureau estimates that the 1990 census missed about 1.6% of the total population due to undercoverage and nonresponse. Because the undercount was greater in the poorer sections of large cities, the Census Bureau estimates that it failed to count 4.6% of blacks and 5.0% of Hispanics.[28]    ◄

The **wording of questions** is the most important influence on the answers given to a sample survey. Confusing or leading questions can introduce strong bias, and even minor changes in wording can change a survey's outcome. Leading questions are common in surveys that are intended to persuade rather than inform. Here are two examples.[29]

## EXAMPLE 10.3

When Levi Strauss & Co. asked college students to choose the most popular clothing item from a list, 90% chose Levi's 501 jeans—but they were the only jeans listed.

A survey paid for by makers of disposable diapers found that 84% of the sample opposed banning disposable diapers. Here is the actual question:

It is estimated that disposable diapers account for less than 2% of the trash in today's landfills. In contrast, beverage containers, third-class mail and yard wastes are estimated to account for about 21% of the trash in landfills. Given this, in your opinion, would it be fair to ban disposable diapers?

This question gives information on only one side of an issue, then asks an opinion. That's a sure way to bias the responses. A different question that described how long disposable diapers take to decay and how many tons they contribute to landfills each year would draw a quite different response.    ◄

Never trust the results of a sample survey until you have read the exact questions posed. The sampling design, the amount of nonresponse, and the date of the survey are also important. Good statistical design is a part, but only a part, of a trustworthy survey.

# INFERENCE ABOUT THE POPULATION

Despite the many practical difficulties in carrying out a sample survey, using chance to choose a sample does eliminate bias in the actual selection of the sample from the list of available individuals. It is unlikely, however, that results from a sample will be exactly the same as for the entire population. The monthly unemployment rate in the United States, for example, is estimated from a large government sample survey, the Current Population Survey. Two runs of the Current Population Survey would produce somewhat different unemployment rates. Properly designed samples avoid systematic bias, but their results are rarely exactly correct and they vary from sample to sample.

How accurate is a sample result like the monthly unemployment rate? We can't say for sure, because the result would be different if we took another sample. But the results of random sampling don't change haphazardly from sample to sample. Because we deliberately use chance, the results obey the laws of probability that govern chance behavior. We can say how large an error we are likely to make in drawing conclusions about the population from a sample. Results from a sample survey usually come with a margin of error that sets bounds on the size of the likely error. How to do this is part of the business of statistical inference. We will describe the reasoning in Lesson 17.

One point is worth making now: *larger samples give more accurate results than smaller samples.* By taking a very large sample, you can be confident that the sample result is very close to the truth about the population. The Current Population Survey takes a sample of 60,000 households in order to estimate the national unemployment rate very accurately. Of course, only random samples carry this guarantee. Ann Landers's voluntary response sample is worthless even though 10,000 people wrote in. Using a random sampling design and taking care to deal with practical difficulties reduce bias in a sample. The size of the sample then determines how close to the population truth the sample result is likely to fall.

## EXERCISES

10.1   A sociologist wants to know the opinions of employed adult women about government funding for day care. She obtains a list of the 520 members of a local business and professional women's club and mails a questionnaire to 100 of these women selected at random. Only 48 questionnaires are returned. What is the population in this study? What is the sample?

**10.2**    Different types of writing can sometimes be distinguished by the lengths of the words used. A student interested in this fact wants to study the lengths of words used by Tom Wolfe in his novels. She opens a Wolfe novel at random and records the lengths of each of the first 250 words on the page.

What is the population in this study? What is the sample? What is the variable measured?

**10.3**    For each of the following sampling situations, identify the population as exactly as possible. That is, say what kind of individuals the population consists of and say exactly which individuals fall in the population. If the information given is not complete, complete the description of the population in a reasonable way.

(a)  Each week, the Gallup Poll questions a sample of about 1500 adult U.S. residents to determine national opinion on a wide variety of issues.

(b)  The 1990 census tried to gather basic information from every household in the United States. But a "long form" requesting much additional information was sent to a sample of about 17% of households.

(c)  A machinery manufacturer purchases voltage regulators from a supplier. There are reports that variation in the output voltage of the regulators is affecting the performance of the finished products. To assess the quality of the supplier's production, the manufacturer sends a sample of 5 regulators from the last shipment to a laboratory for study.

**10.4**    A newspaper advertisement for *USA Today: The Television Show* once said:

Should handgun control be tougher? You call the shots in a special call-in poll tonight. If yes, call 1-900-720-6181. If no, call 1-900-720-6182.
Charge is 50 cents for the first minute.

Explain why this opinion poll is almost certainly biased.

**10.5**    The author Shere Hite undertook a study of women's attitudes toward sex and love by distributing 100,000 questionnaires through women's groups. Only 4.5% of the questionnaires were returned. Based on this sample of women, Hite wrote *Women and Love,* a best-selling book claiming that women are fed up with men. For example, 91% of the divorced women who responded said that they had initiated the divorce, and 70% of the married women said that they had committed adultery.

Explain briefly why Hite's sampling method is nearly certain to produce a strong bias. Are the sample results cited (91% and 70%) much higher or much lower than the truth about the population of all adult American women?

10.6 You are on the staff of a member of Congress who is considering a bill that would provide government-sponsored insurance for nursing home care. You report that 1128 letters have been received on the issue, of which 871 oppose the legislation. "I'm surprised that most of my constituents oppose the bill. I thought it would be quite popular," says the congresswoman. Are you convinced that a majority of the voters oppose the bill? How would you explain the statistical issue to the congresswoman?

10.7 We will illustrate the process of choosing an SRS, and the fact that the results of sampling vary when we repeat the process, by a small example. Table 3.1 (page 13) gives the percent of residents age 65 and over in each of the 50 states.

(a) Use software to choose an SRS of 5 states from the list of 50. Which states make up your sample? Find the mean percent age 65 and over for these 5 states. How does the mean compare with the mean percent for all 50 states?

(b) Repeat the process of choosing an SRS of 5 states 25 times. Record the mean percent age 65 and over in all 25 samples. Make a dotplot of these means and describe their spread and their relationship to the mean for all 50 states.

10.8 We will illustrate the process of choosing an SRS, and the fact that the results of sampling vary when we repeat the process, by a small example. Below are the scores of all 78 seventh-grade students in a rural school on a standard IQ test.[30] This is our population.

| 111 | 107 | 100 | 107 | 114 | 115 | 111 | 97 | 100 | 112 | 104 | 89 | 104 |
|-----|-----|-----|-----|-----|-----|-----|----|-----|-----|-----|----|-----|
| 102 | 91 | 114 | 114 | 103 | 106 | 105 | 113 | 109 | 108 | 113 | 130 | 128 |
| 128 | 118 | 113 | 120 | 132 | 111 | 124 | 127 | 128 | 136 | 106 | 118 | 119 |
| 123 | 124 | 126 | 116 | 127 | 119 | 97 | 86 | 102 | 110 | 120 | 103 | 115 |
| 93 | 72 | 111 | 103 | 123 | 79 | 119 | 110 | 110 | 107 | 74 | 105 | 112 |
| 105 | 110 | 107 | 103 | 77 | 98 | 90 | 96 | 112 | 112 | 114 | 93 | 106 |

(a) Describe this population both with a histogram and with numerical summaries. In particular, what is the mean IQ score in the population?

(b) Use software to choose an SRS of 20 scores from this population. Make a histogram of the scores in the sample, and give a numerical summary. Give a brief comparison of the distributions of scores in the sample and in the population.

(c) Repeat the process of choosing an SRS of size 20 four more times (five in all). Record the five histograms of your sample scores. Does it seem reasonable to you from this small trial that an SRS will usually produce a sample that is generally representative of the population?

(d) Now choose 20 more SRSs of size 20 (25 in all). Don't make histograms of these latest samples. Record the means of each of your 25 samples and make a histogram of the 25 means. One sign of *bias* would be that the distribution of the sample means was systematically on one side of the true population mean. Mark the population mean on your histogram of the 25 sample means. Is there a clear bias? How do the sample means vary with regard to the population mean?

10.9    Table I.1 (page 64) gives the percent return on Wal-Mart common stock for 228 consecutive months. These data show large variability and a number of outliers. Students of investing might ask how well holding Wal-Mart stock for some randomly chosen months would duplicate the percent return results of holding the stock for the entire period.

(a) For reference, make a histogram of the 228 observations. What are their mean and standard deviation?

(b) Use software to choose 25 SRSs of size 12 from this population. This imitates the record of 25 investors who held Wal-Mart stock for a year of randomly chosen months during the 19-year period of the data. Collect the 25 means for your samples and make a histogram of their distribution. Did the investor with the highest mean return (the luckiest of the 25) do much better than the mean of all 228 monthly returns? How much better? What about the least lucky?

(c) Choose 25 SRSs of size 48 from this population. Again collect the 25 sample mean returns, make a histogram, and report the largest and smallest mean. How do your findings for samples of size 48 compare with samples of size 12? In particular, compare the means and standard deviations for returns of the samples of both sizes with each other and with the mean and standard deviation of the 228 individual observations.

   (*Lesson:* On the average, the mean of an SRS is a reasonable estimate of the mean of a population. The mean of an SRS varies less from sample to sample when we choose larger samples. These facts will soon be very important to us.)

10.10    Sample surveys often use a *systematic random sample* to choose a sample of apartments in a large building or dwelling units in a block. An example will illustrate the idea of a systematic sample.

   Suppose that we must choose 4 addresses out of 100. Because $100/4 = 25$, we can think of the list as four lists of 25 addresses. Choose 1 of the first 25 addresses at random. The sample contains this address and the addresses 25, 50, and 75 places down the list from it. If we start at 13, for example, then the systematic random sample consists of the addresses numbered 13, 38, 63, and 88.

(a) Choose a systematic random sample of 5 of the 50 states, and give the percents of residents age 65 and over for the states in your sample. (Data in Table 3.1, page 13.)

(b) Like an SRS, a systematic random sample gives all individuals the same chance to be chosen. Explain why this is true. Then explain carefully why a systematic sample is nonetheless *not* an SRS.

10.11   The list of individuals from which a sample is actually selected is called the *sampling frame*. Ideally, the frame should list every individual in the population, but in practice this is often difficult. A frame that leaves out part of the population is a common source of undercoverage.

(a) Suppose that a sample of households in a community is selected at random from the telephone directory. What households are omitted from this frame? What types of people do you think are likely to live in these households? These people will probably be underrepresented in the sample.

(b) It is usual in telephone surveys to use random digit dialing equipment that selects the last four digits of a telephone number at random after being given the exchange (the first three digits). Which of the households you mentioned in your answer to (a) will be included in the sampling frame by random digit dialing?

10.12   A common form of nonresponse in telephone surveys is "ring-no-answer." That is, a call is made to an active number but no one answers. The Italian National Statistical Institute looked at nonresponse to a government survey of households in Italy during the periods January 1 to Easter and July 1 to August 31. All calls were made between 7 and 10 p.m., but 21.4% gave "ring-no-answer" in one period versus 41.5% "ring-no-answer" in the other period.[31] Which period do you think had the higher rate of no answers? Why? Explain why a high rate of nonresponse makes sample results less reliable.

10.13   Here are two wordings for the same question:

A. Should laws be passed to eliminate all possibilities of special interests giving huge sums of money to candidates?

B. Should laws be passed to prohibit interest groups from contributing to campaigns, or do groups have a right to contribute to the candidates they support?

One of these questions drew 40% favoring banning contributions; the other drew 80% with this opinion.[32] Which question produced the 40% and which got 80%? Explain why the results were so different.

10.14    Just before a presidential election, a national opinion polling firm increases the size of its weekly sample from the usual 1500 people to 4000 people. Why do you think the firm does this?

10.15    Should the United Nations continue to have its headquarters in the United States? A television program asked its viewers to call in with their opinions on that question. There were 186,000 callers, 67% of whom said "No." A nationwide random sample of 500 adults found that 72% answered "Yes" to the same question. Explain to someone who knows no statistics why the opinions of only 500 randomly chosen respondents are a better guide to what all Americans think than the opinions of 186,000 callers.

10.16    Sample surveys of large populations, such as opinion polls and the Current Population Survey, use random sampling designs more complex than an SRS. Samples are often taken in stages—first counties, then townships within the counties chosen, and so on. The sample at each stage may be an SRS, a systematic sample (see Exercise 10.10), or a *stratified sample*. To choose a stratified sample, divide the population into groups of similar individuals (such as urban, suburban, and rural counties) and choose a separate SRS from each group.

Table 6.1 (page 71) gives data on 26 midsize cars and 22 sports utility vehicles. An SRS of 10 of the 48 vehicles would be inefficient because it would mix the two types. It is better to choose a stratified sample of (say) 6 midsize cars and 4 sports utility vehicles. Choose such a sample and report which vehicles you got.

# DESIGNED EXPERIMENTS

A sample survey collects information about a population by selecting and measuring a sample from the population. The goal is a picture of the population, disturbed as little as possible by the act of gathering information. Sample surveys are one kind of *observational study*.

> ### OBSERVATION AND EXPERIMENT
>
> An **observational study** observes individuals and measures variables of interest but does not attempt to influence the responses.
>
> An **experiment,** on the other hand, deliberately imposes some treatment on individuals in order to observe their responses.

An observational study, even one based on a statistical sample, is a poor way to gauge the effect of an intervention. To see how nature responds to a change, we must actually impose the change. When our goal is to understand cause and effect, experiments are the only source of fully convincing data. Here is the basic vocabulary of experiments.

> EXPERIMENTAL UNITS, SUBJECTS, TREATMENT
>
> The individuals on which the experiment is done are the **experimental units.** When the units are human beings, they are called **subjects.** A specific experimental condition applied to the units is called a **treatment.**

Because the purpose of an experiment is to reveal the response of one variable to changes in other variables, the distinction between explanatory and response variables is essential. The explanatory variables in an experiment are often called **factors.** The **design** of an experiment describes the treatments and how the experimental units or subjects are assigned to the treatments. Poorly designed experiments can lead to incorrect conclusions.

## EXAMPLE 11.1

"Gastric freezing" is a clever treatment for ulcers in the upper intestine. The patient swallows a deflated balloon with tubes attached, then a refrigerated solution is pumped through the balloon for an hour. The idea is that cooling the stomach will reduce its production of acid and so relieve ulcers. An experiment reported in the *Journal of the American Medical Association* showed that gastric freezing did reduce acid production and relieve ulcer pain. The treatment was safe and easy and was widely used for several years. The design of the experiment was simple:

**Gastric freezing $\longrightarrow$ Observe pain relief**

The gastric freezing experiment was poorly designed. The patients' response may have been due to the **placebo effect.** A placebo is a dummy treatment that can have no physical effect. Many patients respond favorably to any treatment, even a placebo, presumably because of trust in the doctor and expectations of a cure. This response to a dummy treatment is the placebo effect.

A later experiment divided ulcer patients into two groups. One group was treated by gastric freezing as before. The other group received a placebo treatment in which the solution in the balloon was at body temperature rather than freezing. The results: 34% of the 82 patients in the treatment group improved, but so did 38% of the 78 patients in the placebo group. This and other properly designed experiments showed that gastric freezing was no better than a placebo, and its use was abandoned.[33] ◀

The first gastric freezing experiment gave misleading results because the effects of the explanatory variable were hopelessly mixed up with the placebo effect. We call this mixing of effects *confounding*. Unless experiments are carefully designed, we cannot see the effects of the explanatory variables because they are confounded with other variables in the environment. It is because of confounding that observational studies and poorly designed experiments usually fail to show that the explanatory variable actually *causes* changes in the response variable.

CONFOUNDING

Two variables (explanatory variables or outside variables) are **confounded** when their effects on a response variable cannot be distinguished from each other.

We can defeat confounding by *comparing* two groups of patients, as in the second gastric freezing experiment. The placebo effect and other irrelevant variables now operate on both groups. The only difference between the groups is the actual effect of gastric freezing. The group of patients who received a sham treatment is called a **control group**, because it enables us to control the effects of outside variables on the outcome. Control is the first of the three basic principles of statistical design of experiments.

PRINCIPLES OF EXPERIMENTAL DESIGN

The basic principles of statistical design of experiments are:

1. **Control** of the effects of irrelevant variables on the response, most simply by comparing several treatments.
2. **Randomization,** the use of impersonal chance to assign subjects to treatments.
3. **Replication** of the experiment on many subjects to reduce chance variation in the results.

## RANDOMIZED COMPARATIVE EXPERIMENTS

Figure 11.1 outlines the design of the experiment that discredited gastric freezing. This design uses all three of our principles. It is the simplest

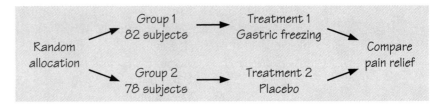

FIGURE 11.1    The design of a randomized comparative experiment that compares the pain relief provided by gastric freezing and a placebo treatment.

*randomized comparative experiment.* Impersonal chance assigns the subjects to the treatments. To do this, choose an SRS of 82 from the 160 subjects to receive gastric freezing. There are good statistical reasons to prefer roughly equal numbers of subjects in each group. As in sampling, use of chance avoids bias. Randomization produces groups of subjects that should be similar in all respects before the treatments are applied. The comparative design avoids confounding because the placebo effect and other influences affect both groups equally. The logic is:

- Similar groups of subjects, thanks to randomization.
- No differences in what the groups experience other than the treatments being compared.
- Therefore, differences in the average response of the groups (if larger than the random assignment would produce) must be *caused* by the different treatments.

The last point reminds us that deliberate use of chance allows us to use the laws of probability to verify that the differences we see are larger than chance would produce. This is why *replication* is important. We could not trust an experiment with only 3 or 4 subjects—the individual characteristics of these subjects would have too great an influence. Averaging results from about 80 subjects randomly assigned to each group reduces the influence of variation among individuals and allows trustworthy conclusions about gastric freezing.

The great advantage of randomized comparative experiments is that they can produce data that give good evidence for a cause-and-effect relationship between the explanatory and response variables. We know that in general a strong association does not imply causation. A strong association in data from a well-designed experiment does imply causation.

# CAUTIONS ABOUT EXPERIMENTATION

The logic of a randomized comparative experiment depends on our ability to treat all the subjects identically in every way except for the actual treatments being compared. Good experiments therefore require careful attention to details. The subjects in the gastric freezing experiment, for example, all got the same medical attention over the two years of the study. Moreover, the study was **double-blind**—neither the subjects themselves nor the medical personnel who worked with them knew which treatment any subject had received. The double-blind method avoids any unconscious bias by, for example, a doctor who doesn't think that "just a placebo" can relieve an ulcer.

The most serious potential weakness of experiments is **lack of realism.** The subjects or treatments or setting of an experiment may not realistically duplicate the conditions we really want to study. Here are some examples.

## EXAMPLE 11.2

A study compares two television advertisements by showing TV programs to student subjects. The students know it's "just an experiment." We can't be sure that the results apply to everyday television viewers. Many behavioral science experiments use as subjects students who know they are subjects in an experiment. That's not a realistic setting.

An industrial experiment uses a small-scale pilot production process to find the choices of catalyst concentration and temperature that maximize yield. These may not be the best choices for the operation of a full-scale plant.    ◀

Lack of realism can limit our ability to apply the conclusions of an experiment to the settings of greatest interest. Most experimenters want to generalize their conclusions to some setting wider than that of the actual experiment. Statistical analysis of the original experiment cannot tell us how far the results will generalize. Nonetheless, the randomized comparative experiment, because of its ability to give convincing evidence for causation, is one of the most important ideas in statistics.

## EXERCISES

11.1    There may be a "gender gap" in political party preference in the United States, with women more likely than men to prefer Democratic candidates.

A political scientist selects a large sample of registered voters, both men and women. She asks each voter whether they voted for the Democratic or the Republican candidate in the last congressional election. Is this study an experiment? Why or why not? What are the explanatory and response variables?

11.2    An educator wants to compare the effectiveness of computer software that teaches reading with that of a standard reading curriculum. She tests the reading ability of each student in a class of fourth graders, then divides them into two groups. One group uses the computer regularly, while the other studies a standard curriculum. At the end of the year, she retests all the students and compares the increase in reading ability in the two groups.

(a) Is this an experiment? Why or why not?

(b) What are the explanatory and response variables?

11.3    The National Halothane Study was a major investigation of the safety of anesthetics used in surgery. Records of over 850,000 operations performed in 34 major hospitals showed the following death rates for four common anesthetics:[34]

| Anesthetic | A | B | C | D |
|---|---|---|---|---|
| Death rate | 1.7% | 1.7% | 3.4% | 1.9% |

There is a clear association between the anesthetic used and the death rate of patients. Anesthetic C appears dangerous.

(a) Explain why we call the National Halothane Study an observational study rather than an experiment, even though it compared the results of using different anesthetics in actual surgery.

(b) When the study looked at other variables that are confounded with a doctor's choice of anesthetic, it found that Anesthetic C was not causing extra deaths. Suggest several variables that are mixed up with what anesthetic a patient receives.

11.4    Observational studies suggest that children who watch many hours of television get lower grades in school than those who watch less TV. Explain clearly why these studies do *not* show that watching TV *causes* poor grades. In particular, suggest some other variables that may be confounded with heavy TV viewing and may contribute to poor grades.

11.5    A manufacturer of food products uses package liners that are sealed at the top by applying heated jaws after the package is filled. The customer peels the sealed pieces apart to open the package. What effect does the temperature of the jaws have on the force needed to peel the liner? To answer this question, engineers obtain 20 pairs of pieces of package liner. They seal five pairs at each of 250° F, 275° F, 300° F, and 325° F. Then they measure the force needed to peel each seal.

(a) What are the experimental units?

(b) What are the treatments?

(c) What is the response variable?

11.6    A large study used records from Canada's national health care system to compare the effectiveness of two ways to treat prostate disease. The two treatments are traditional surgery and a new method that does not require surgery. The records described many patients whose doctors had chosen each method. The study found that patients treated by the new method were more likely to die within 8 years. We might conclude that the new method is less effective than the old.[35]

(a) Further study of the data showed that this conclusion was wrong. The extra deaths among patients treated with the new method could be explained by confounding between the choice of method and outside variables. What variables might influence a doctor's choice of surgical or nonsurgical treatment?

(b) You have 300 prostate patients who are willing to serve as subjects in an experiment to compare the two methods. Use a diagram to outline the design of a randomized comparative experiment. (When using a diagram to outline the design of an experiment, be sure to indicate the size of the treatment groups and the response variable. The diagram of the gastric freezing experiment is a model.)

11.7    Use a diagram to describe a randomized comparative experimental design for the package liner experiment of Exercise 11.5. (When using a diagram to outline the design of an experiment, be sure to indicate the size of the treatment groups and the response variable. This experiment compares more than two treatments.)

11.8    Can aspirin help prevent heart attacks? The Physicians' Health Study, a large medical experiment involving 22,000 male physicians, attempted to answer this question. One group of about 11,000 physicians took an aspirin every second day, while the rest took a placebo. After several years the study found that subjects in the aspirin group had significantly fewer heart attacks than subjects in the placebo group.[36]

(a) Identify the experimental subjects, the factor and its levels, and the response variable in the Physicians' Health Study.

(b) Use a diagram to outline a randomized comparative experimental design for the Physicians' Health Study.

11.9    A food company assesses the nutritional quality of a new "instant break-fast" product by feeding it to newly weaned male white rats. The response variable is a rat's weight gain over a 28-day period. There are 40 rats available. Outline the design of an experiment that compares the new product with a standard diet. Then label the 40 rats A, B, ..., Z, AA, ..., AN, and use software to choose an SRS of 20 rats to receive the new product. List the rats that your randomization places in each group.

11.10    Some medical researchers suspect that added calcium in the diet reduces blood pressure. You have available 40 men with high blood pressure who are willing to serve as subjects.

(a) Outline an appropriate design for the experiment, taking the placebo effect into account.

(b) The names of the subjects appear below. Do the randomization required by your design, and list the subjects to whom you will give the drug.

| | | | | |
|---|---|---|---|---|
| Alomar | Denman | Han | Liang | Rosen |
| Asihiro | Durr | Howard | Maldonado | Solomon |
| Bennett | Edwards | Hruska | Marsden | Tompkins |
| Bikalis | Farouk | Imrani | Moore | Townsend |
| Chen | Fratianna | James | O'Brian | Tullock |
| Clemente | George | Kaplan | Ogle | Underwood |
| Cranston | Green | Krushchev | Plochman | Willis |
| Curtis | Guillen | Lawless | Rodriguez | Zhang |

11.11    Does regular exercise reduce the risk of a heart attack? Here are two ways to study this question. Explain clearly why the second design will produce more trustworthy data.

1. A researcher finds 2000 men over 40 who exercise regularly and have not had heart attacks. She matches each with a similar man who does not exercise regularly, and she follows both groups for 5 years.

2. Another researcher finds 4000 men over 40 who have not had heart attacks and are willing to participate in a study. She assigns 2000 of the men to a regular program of supervised exercise. The other 2000 continue their usual habits. The researcher follows both groups for 5 years.

11.12    You read a news report of an experiment that claims to show that a meditation technique lowered the anxiety level of subjects. The experimenter interviewed the subjects and assessed their levels of anxiety. The subjects then learned how to meditate and did so regularly for a month. The experimenter reinterviewed them at the end of the month and assessed whether their anxiety levels had decreased or not.

(a) There was no control group in this experiment. Why is this a blunder? What outside variables might be confounded with the effect of meditation?

(b) The experimenter who diagnosed the effect of the treatment knew that the subjects had been meditating. Explain how this knowledge could bias the experimental conclusions.

(c) Briefly discuss a proper experimental design, with controls and blind diagnosis, to assess the effect of meditation on anxiety level.

11.13    Fizz Laboratories, a pharmaceutical company, has developed a new pain-relief medication. Sixty patients suffering from arthritis and needing pain relief are available. Each patient will be treated and asked an hour later, "About what percentage of pain relief did you experience?"

(a) Why should Fizz not simply administer the new drug and record the patients' responses?

(b) Outline the design of an experiment to compare the drug's effectiveness with that of aspirin and of a placebo.

(c) Should patients be told which drug they are receiving? How would this knowledge probably affect their reactions?

(d) If patients are not told which treatment they are receiving, the experiment is single-blind. Should this experiment be double-blind also? Explain.

11.14    It is often suggested that people will spend less on health care if their health insurance requires them to pay some part of the cost of medical treatment themselves. One experiment on this issue asked if the percent of medical costs that are paid by health insurance has an effect either on the amount of medical care that people use or on their health. The treatments were four insurance plans. Each plan paid all medical costs above a ceiling. Below the ceiling, the plans paid 100%, 75%, 50%, or 5% of costs incurred.[37]

(a) Outline the design of a randomized comparative experiment suitable for this study.

(b) Describe briefly the practical and ethical difficulties that might arise in such an experiment.

11.15    Let us examine the effects of randomization in designing an experiment. You plan to compare two methods of teaching math skills to seventh-grade students. Your subjects are 78 seventh graders. You will choose 39 students at random to be taught by Method A. The remaining 39 will be taught using Method B. The names of the students are confidential, but here are their IQ test scores:[38]

| 111 | 107 | 100 | 107 | 114 | 115 | 111 | 97 | 100 | 112 | 104 | 89 |
| 104 | 102 | 91 | 114 | 114 | 103 | 106 | 105 | 113 | 109 | 108 | 113 |
| 130 | 128 | 128 | 118 | 113 | 120 | 132 | 111 | 124 | 127 | 128 | 136 |
| 106 | 118 | 119 | 123 | 124 | 126 | 116 | 127 | 119 | 97 | 86 | 102 |
| 110 | 120 | 103 | 115 | 93 | 72 | 111 | 103 | 123 | 79 | 119 | 110 |
| 110 | 74 | 107 | 105 | 112 | 105 | 110 | 107 | 103 | 77 | 98 | 90 |
| 96 | 112 | 112 | 114 | 93 | 106 | | | | | | |

(a) Choose 39 students for the Method A group by choosing an SRS of 39 of the scores above.

(b) You would like the two groups to be similar in IQ and in all other respects—that is why you chose the Method A group at random. Find the mean IQ of the 39 students in the Method A group and also the mean IQ of the 39 students in the Method B group. Do your two groups have similar mean IQs?

(c) You might be unusually lucky (or unlucky) on one random choice. Will randomization *usually* produce two groups with similar mean IQs? To find out, repeat (a) and (b) nine more times. Make a table of the mean IQs for groups A and B on each of your 10 trials. What do you conclude?

# PART III REVIEW

Part III presents statistical ideas for designing studies to answer specific questions. The two basic types of statistically designed studies are *samples* that gather information about a specific population and *experiments* that examine the response to specific treatments. Both types of statistical design share the deliberate use of chance to choose a sample or assign subjects to treatments in an experiment. In both settings, the laws of probability describe how much chance variation is present. *"Random" in statistics does not mean "haphazard."* Random samples and randomized comparative experiments use carefully structured chance mechanisms to give predictable, rather than haphazard, results. The rest of this book explores the regular outcomes of randomized designs for producing data. Here is a review list of the most important skills you should have acquired from your study of Lessons 10 and 11.

## A. SAMPLING

1. Identify the population and the sample in a sampling situation.
2. Recognize bias due to voluntary response samples and other inferior sampling methods.

3. Select a simple random sample (SRS) from a list (frame) of a population.

4. Recognize the presence of undercoverage and nonresponse as sources of error in a sample survey. Recognize the effect of the wording of questions on the responses.

## B. EXPERIMENTS

1. Recognize whether a study is an observational study or an experiment.

2. Recognize bias due to confounding with outside variables such as the placebo effect in either an observational study or an experiment.

3. Identify the explanatory variable (factor), treatments, response variable, and experimental units or subjects in an experiment.

4. Outline the design of a randomized comparative experiment using a diagram. The diagram should show the sizes of the groups, the specific treatments, and the response variable.

5. Carry out the random assignment of subjects to treatments in a randomized comparative experiment.

6. Explain why a randomized comparative experiment can give good evidence for cause-and-effect relationships.

## REVIEW EXERCISES

**Review 1**    Before a new variety of frozen muffin is put on the market, it is subjected to extensive taste testing. People are asked to taste the new muffin and a competing brand, and to say which they prefer. (Both muffins are unidentified in the test.) Is this an observational study or an experiment? Explain your answer.

**Review 2**    For each of the following sampling situations, identify the population as exactly as possible. That is, say what kind of individuals the population consists of and say exactly which individuals fall in the population. If the information given is not complete, complete the description of the population in a reasonable way.

**(a)** A business school researcher wants to know what factors affect the survival and success of small businesses. She selects a sample of 150 eating-and-drinking establishments from those listed in the telephone directory Yellow Pages for a large city.

(b) A member of Congress wants to know whether his constituents support proposed legislation on health care. His staff reports that 228 letters have been received on the subject, of which 193 oppose the legislation.

(c) An insurance company wants to monitor the quality of its procedures for handling loss claims from its auto insurance policyholders. Each month the company selects an SRS of all auto insurance claims filed that month to examine them for accuracy and promptness.

Review 3   Some television stations take quick polls of public opinion by announcing a question on the air and asking viewers to call one of two telephone numbers to register their opinion as "Yes" or "No." Telephone companies make available "900" numbers for this purpose. Dialing a 900 number results in a small charge to your telephone bill. The first major use of call-in polling was by the ABC television network in October 1980. At the end of the first Reagan-Carter presidential election debate, ABC asked its viewers which candidate won. The call-in poll proclaimed that Reagan had won the debate by a 2 to 1 margin. But a random survey by CBS News showed only a 44% to 36% margin for Reagan, with the rest undecided. Why are call-in polls likely to be biased? Can you suggest why this bias might have favored the Republican Reagan over the Democrat Carter?

Review 4   Advice columnist Ann Landers once asked her female readers whether they would be content with affectionate treatment by men with no sex ever. Over 90,000 women wrote in, with 72% answering "Yes." Many of the letters described unfeeling treatment at the hands of men. Explain why this sample is certainly biased. What is the likely direction of the bias? That is, is that 72% probably higher or lower than the truth about the population of all adult women?

Review 5   Comment on each of the following as a potential sample survey question. Is the question clear? Is it slanted toward a desired response?

(a) Which of the following best represents your opinion on gun control?
    1. The government should confiscate our guns.
    2. We have the right to keep and bear arms.

(b) A freeze in nuclear weapons should be favored because it would begin a much-needed process to stop everyone in the world from building nuclear weapons now and reduce the possibility of nuclear war in the future. Do you agree or disagree?

(c) In view of escalating environmental degradation and incipient resource depletion, would you favor economic incentives for recycling of resource-intensive consumer goods?

**Review 6**   The Ministry of Health in the Canadian Province of Ontario wants to know whether the national health care system is achieving its goals in the province. Much information about health care comes from patient records, but that source doesn't allow us to compare people who use health services with those who don't. So the Ministry of Health conducted the Ontario Health Survey, which interviewed a random sample of 61,239 people who live in the Province of Ontario.[39]

**(a)** What is the population for this sample survey? What is the sample?

**(b)** The survey found that 76% of males and 86% of females in the sample had visited a general practitioner at least once in the past year. Do you think these estimates are close to the truth about the entire population? Why?

**Review 7**   What is the preferred treatment for breast cancer that is detected in its early stages? The most common treatment was once removal of the breast. It is now usual to remove only the tumor and nearby lymph nodes, followed by radiation. To study whether these treatments differ in their effectiveness, a medical team examines the records of 25 large hospitals and compares the survival times after surgery of all women who have had either treatment.

**(a)** What are the explanatory and response variables?

**(b)** Explain carefully why this study is not an experiment.

**(c)** Explain why the effects of outside variables will prevent this study from discovering which treatment is more effective. (The current treatment was in fact recommended after a large randomized comparative experiment.)

**Review 8**   A study of the relationship between physical fitness and leadership uses as subjects middle-aged executives who have volunteered for an exercise program. The executives are divided into a low-fitness group and a high-fitness group on the basis of a physical examination. All subjects then take a psychological test designed to measure leadership, and the results for the two groups are compared. Is this study an experiment? Explain your answer.

**Review 9**   You want to investigate the attitudes of students at your school about the school's policy on sexual harassment. You have a grant that will pay the costs of contacting about 500 students.

**(a)** Specify the exact population for your study. For example, will you include part-time students?

**(b)** Describe your sample design. Do you prefer separate samples of female and male students?

(c) Briefly discuss the practical difficulties that you anticipate. For example, how will you contact the students in your sample?

**Review 10**   Some investment advisers believe that charts of past trends in the prices of securities can help predict future prices. In an experiment to examine the effects of using charts, business students trade (hypothetically) a foreign currency at computer screens. There are 30 student subjects. Their goal is to make as much money as possible, and the best performances are rewarded with small prizes. The student traders have the price history of the foreign currency in dollars in their computers. They may or may not also be given software that charts past trends. Describe a design to test whether the chart software helps traders make more money. Label the subjects 1 to 30 and carry out the assignment of subjects to treatments called for by your design. Which subjects get the chart software?

**Review 11**   A college allows students to choose either classroom or self-paced instruction in a basic economics course. The college wants to compare the effectiveness of self-paced and regular instruction. A professor proposes giving the same final exam to all students in both versions of the course and comparing the average score of those who took the self-paced option with the average score of students in regular sections.
(a) Explain why the results of that study would not allow a trustworthy comparison of the two methods of instruction.
(b) Given 150 students who are willing to use either regular or self-paced instruction, outline an experimental design to compare the two methods of instruction.
(c) Label the subjects 1 to 150 and carry out the assignment of subjects to methods of instruction that your design calls for. List the subjects who receive self-paced instruction.

## DAVID BLACKWELL

Statistical practice rests in part on statistical theory. Statistics has been advanced not only by people concerned with practical problems, from Florence Nightingale to R. A. Fisher and John Tukey, but also by people whose first love is mathematics for its own sake. David Blackwell (1919–) is one of the major contemporary contributors to the mathematical study of statistics.

Blackwell grew up in Illinois, earned a doctorate in mathematics at the age of 22, and in 1944 joined the faculty of Howard University in Washington, D.C. "It was the ambition of every black scholar in those days to get a job at Howard University," he says. "That was the best job you could hope for." Society changed, and in 1954 Blackwell became professor of statistics at the University of California at Berkeley.

Washington, D.C., had an active statistical community, and the young mathematician Blackwell soon began to work on mathematical aspects of statistics. He explored the behavior of statistical procedures which, rather than working with a fixed sample, keep taking observations until there is enough information to reach a firm conclusion. He found insights into statistical inference by thinking of inference as a game in which nature plays against the statistician. Blackwell's work uses probability theory, the mathematics that describes chance behavior. We must travel the same route, though only a short distance. Part IV presents, in a rather informal fashion, the probabilistic ideas needed to understand the reasoning of inference.

PART **IV**

# Experience with Random Behavior

# 12

# RANDOMNESS

The reasoning of statistical inference rests on asking, "How often would this method give a correct answer if I used it very many times?" When we produce data by random sampling or randomized comparative experiments, the laws of probability answer the question "What would happen if we did this many times?"

## THE LANGUAGE OF PROBABILITY

Toss a coin, or choose an SRS. The result can't be predicted in advance because the result will vary when you toss the coin or choose the sample repeatedly. But there is still a regular pattern in the results, a pattern that only emerges after many repetitions. This remarkable fact is the basis for the idea of probability.

### EXAMPLE 12.1

When you toss a coin, there are only two possible outcomes, heads or tails. Figure 12.1 shows the results of tossing a coin 1000 times. For each number of

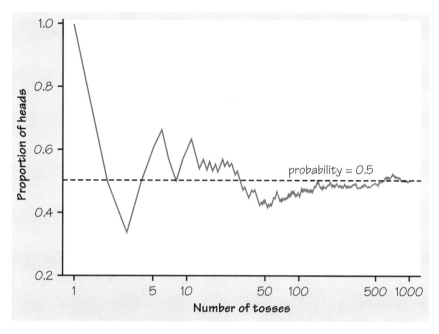

FIGURE 12.1    The proportion of tosses of a coin that give a head changes as we make more tosses. Eventually, however, the proportion approaches 0.5, the probability of a head.

tosses from 1 to 1000, we have plotted the proportion of those tosses that gave a head. The first toss was a head, so the proportion of heads starts at 1. The second toss was a tail, reducing the proportion of heads to 0.5 after two tosses. The next three tosses gave a tail followed by two heads, so the proportion of heads after five tosses is 3/5, or 0.6.

The proportion of tosses that produce heads is quite variable at first, but it settles down as we make more and more tosses. Eventually this proportion gets close to 0.5 and stays there. We say that 0.5 is the *probability* of a head. The probability 0.5 appears as a horizontal line on the graph.    ◄

"Random" in statistics is not a synonym for "haphazard" but a description of a kind of order that emerges only in the long run. We often encounter the unpredictable side of randomness in our everyday experience, but we rarely see enough repetitions of the same random phenomenon to observe the long-term regularity that probability describes. You can see that regularity emerging in Figure 12.1. In the very long run, the proportion of tosses that give a head is 0.5. This is the intuitive idea of probability. Probability 0.5 means "occurs half the time in a very large number of trials."

> RANDOMNESS AND PROBABILITY
>
> We call a phenomenon **random** if individual outcomes are uncertain but there is nonetheless a regular distribution of outcomes in a large number of repetitions.
>
> The **probability** of any outcome of a random phenomenon is the proportion of times the outcome would occur in a very long series of repetitions.

## THINKING ABOUT RANDOMNESS

That some things are random is an observed fact about the world. The outcome of a coin toss, the time between emissions of particles by a radioactive source, and the sexes of the next litter of lab rats are all random. So is the outcome of a random sample or a randomized experiment. Probability theory is the branch of mathematics that describes random behavior. Of course, we can never observe a probability exactly. We could always continue tossing the coin, for example. Mathematical probability is an idealization based on imagining what would happen in an indefinitely long series of trials.

The best way to understand randomness is to observe random behavior—not only the long-run regularity but the unpredictable results of short runs. You can do this with physical devices, as in the exercises, but computer simulations (imitations) of random behavior allow faster exploration. As you explore randomness, remember:

- You must have a long series of *independent* trials. That is, the outcome of one trial must not influence the outcome of any other. Imagine a crooked gambling house where the operator of a roulette wheel can stop it where she chooses—she can prevent the proportion of "red" from settling down to a fixed number. These trials are not independent.

- The idea of probability is empirical (based on observation). Computer simulations start with given probabilities and imitate random behavior, but we can only estimate a real-world probability by actually observing many trials.

- Nonetheless, computer simulations are very useful because we need long runs of trials. In situations such as coin tossing, the proportion of an outcome often requires several hundred trials to settle down to the probability of that outcome. The kind of physical random devices suggested in the exercises are too slow for this. Short runs give only rough estimates of a probability.

## EXERCISES

12.1    Hold a penny upright on its edge under your forefinger on a hard surface, then snap it with your other forefinger so that it spins for some time before falling. Based on 50 spins, what is the probability of heads?

12.2    You may feel that it is obvious that the probability of a head in tossing a coin is about 1/2 because the coin has two faces. Such opinions are not always correct. The previous exercise asked you to spin a penny rather than toss it—that changes the probability of a head. Now try another variation. Stand a penny on edge on a hard, flat surface. Pound the surface with your hand so that the penny falls over. What is the probability that it falls with heads upward? Make at least 50 trials to estimate the probability of a head.

12.3    When we toss a penny, experience shows that the probability (long-term proportion) of a head is close to 1/2. Suppose now that we toss the penny repeatedly until we get a head. What is the probability that the first head comes up in an odd number of tosses (1, 3, 5, and so on)? To find out, repeat this experiment 50 times, and keep a record of the number of tosses needed to get a head on each of your 50 trials.

(a) From your experiment, estimate the probability of a head on the first toss. What value should we expect this probability to have?

(b) Use your results to estimate the probability that the first head appears on an odd-numbered toss.

12.4    Toss a thumbtack on a hard surface 100 times. How many times did it land with the point up? What is the approximate probability of landing point up?

12.5    Obtain 10 identical thumbtacks and toss all 10 at once. Do this 50 times and record the number that land point up on each trial.

(a) What is the approximate probability that at least one lands point up?

(b) What is the approximate probability that more than one lands point up?

12.6    Roll a pair of dice 100 times and record the sum of the spots on the upward faces in each trial. On what proportion of these 100 trials do the dice total 8?

12.7    You read in a book on poker that the probability of being dealt three of a kind in a five-card poker hand is 1/50. Explain in simple language what this means.

12.8    A recent opinion poll showed that about 73% of married women agree that their husband does at least his fair share of household chores. Suppose that this is exactly true. Choosing a married woman at random then has probability 0.73 of getting one who agrees that her husband does his share. Use software to simulate choosing many women independently. (In most software, including *Data Desk,* the key phrase to look for is "Bernoulli trials." This is the technical term for independent trials with Yes/No outcomes. Our outcomes here are "Agree" or not.)

(a) Simulate drawing 20 women, then 80 women, then 320 women. What proportion agree in each case? We expect (but because of chance variation we can't be sure) that the proportion will be closer to 0.73 in longer runs of trials.

(b) Simulate drawing 20 women 10 times and record the percent in each trial who agree. Then simulate drawing 320 women 10 times and again record the 10 percents. Which set of 10 results is less variable? We expect the results of 320 trials to be more predictable (less variable) than the results of 20 trials. That is "long-run regularity" showing itself.

12.9    The basketball player Shaquille O'Neal makes about half of his free throws over an entire season. Use software to simulate 100 free throws shot independently by a player who has probability 0.5 of making each shot. (In most software, the key phrase to look for is "Bernoulli trials." This is the technical term for independent trials with Yes/No outcomes. Our outcomes here are "Hit" and "Miss.")

(a) What percent of the 100 shots did he hit?

(b) Examine the sequence of hits and misses. How long was the longest run of shots made? Of shots missed? (Sequences of random outcomes often show runs longer than our intuition thinks likely.)

12.10    Continue the exploration in Exercise 12.8. Your software allows you to simulate many independent Yes/No trials more quickly if all you want to save is the count of Yes outcomes. The key word "Binomial" simulates $n$ independent Bernoulli trials, each with probability $p$ of a Yes, and records just the count of Yes outcomes.

(a) Simulate 100 draws of 20 women from the population in Exercise 12.8. Record the number who say "Agree" on each draw. What is the approximate probability that out of 20 women drawn at random at least 14 agree?

**(b)** Convert the counts who agree into percents of the 20 women in each trial who agree. Make a histogram of these 100 percents. Describe the shape, center, and spread of this distribution.

**(c)** Now simulate drawing 320 women. Do this 100 times and record the percent who agree on each of the 100 draws. Make a histogram of the percents and describe the shape, center, and spread of the distribution.

**(d)** In what ways are the distributions in parts (b) and (c) alike? In what ways do they differ? (Because regularity emerges in the long run, we expect the results of drawing 320 women to be less variable than the results of drawing 20 women.)

# 13

# INTUITIVE PROBABILITY

Random phenomena are unpredictable in the short run but display regular behavior when repeated very many times. Probability uses numbers to describe this regular behavior. The idea of the probability of any event is the long-run proportion of repetitions in which that event occurs. Although the mathematics of probability theory lies behind much that we do, we need only an intuitive understanding of probability and a few basic facts.

---

PROBABILITY FACTS

- Any probability is a number between 0 and 1.
- All possible outcomes together must have probability 1.
- The probability that an event does not occur is 1 minus the probability that the event does occur.

---

*(continued on next page)*

*(continued from previous page)*

- If two events have no outcomes in common, the probability that one or the other occurs is the sum of their individual probabilities.
- If two events are independent, the probability that they both occur together is the product of their individual probabilities.

The first four facts follow from the idea of probability as "the long-run proportion of repetitions in which an event occurs."

- Any proportion is a number between 0 and 1, so any probability is also a number between 0 and 1. An event with probability 0 never occurs, and an event with probability 1 occurs on every trial.
- Because some outcome must occur on every trial, the sum of the probabilities for all possible outcomes must be exactly 1.
- If an event occurs in (say) 70% of all trials, it fails to occur in the other 30%. The probability that an event occurs and the probability that it does not occur always add to 100%, or 1.
- If one event occurs in 40% of all trials, a different event occurs in 25% of all trials, and the two can never occur together, then one or the other occurs on 65% of all trials because 40% + 25% = 65%.

The final fact—the multiplication rule for independent events—takes a bit more thought. An example will help you see why it is true.

**EXAMPLE 13.1**

Children inherit their blood types at random, with probabilities that depend on their parents' blood types. The blood types of children of the same parents are independent. That is, knowing the first child's blood type does not change the probabilities for the second child.

Each child born to parents with one particular set of blood types has probability 0.25 of having blood type O. What is the probability that the first of two children has blood type O and the second does not?

- The probability that the first child has type O blood is 0.25. That is, among many such couples, the first child has type O blood 25% of the time.

- Out of all the couples with a type O child on the first try, 75% have a child with blood type other than O on their second try. This is true because the probability of "Not type O" is $1 - 0.25$, or 0.75, and knowing the first child's type does not change this probability.
- So the proportion of couples with "O" then "Not O" for two children is 75% of the 25% who had O for the first child. That's probability $(0.25)(0.75)$.

◀

# RULES OF PROBABILITY

We sometimes want to express facts about probability in more formal language. An **event** is any collection of outcomes of a random phenomenon. For example, "I make at least 8 out of 10 basketball free throws" is an event. We use capital letters near the beginning of the alphabet to denote events. If $A$ is any event, we write its probability as $P(A)$. Here then are our probability facts in formal language.

---

PROBABILITY RULES

- The probability $P(A)$ of any event $A$ satisfies $0 \le P(A) \le 1$.
- If $S$ is the collection of all possible outcomes of a random phenomenon, then $P(S) = 1$.
- For any event $A$,

$$P(A \text{ does not occur}) = 1 - P(A)$$

- Two events $A$ and $B$ are **disjoint** if they have no outcomes in common and so can never occur simultaneously. If $A$ and $B$ are disjoint,

$$P(A \text{ or } B) = P(A) + P(B)$$

This is the **addition rule** for disjoint events.

- Two events $A$ and $B$ are **independent** if knowing that one occurs does not change the probability that the other occurs. If $A$ and $B$

---

(continued on next page)

*(continued from previous page)*

are independent,

$$P(A \text{ and } B) = P(A)P(B)$$

This is the **multiplication rule** for independent events.

## ASSIGNING PROBABILITIES: FINITE NUMBER OF OUTCOMES

Any mathematical description of probability has two parts: a description of the possible outcomes and an assignment of probabilities to events. The collection of all possible outcomes of a random phenomenon is called the **sample space.** The name "sample space" is natural in random sampling, where each possible outcome is a sample and the sample space contains all possible samples.

### EXAMPLE 13.2

A software random digit generator produces digits between 0 and 9 independently. If we produce one digit, the sample space is

$$S = \{0, 1, 2, 3, 4, 5, 6, 7, 8, 9\}$$

Careful randomization makes each outcome equally likely to be any of the 10 candidates. Because the total probability must be 1, the probability of each of the 10 outcomes must be 1/10. This assignment of probabilities to individual outcomes can be summarized in a table as follows:

| Outcome | 0 | 1 | 2 | 3 | 4 | 5 | 6 | 7 | 8 | 9 |
|---------|-----|-----|-----|-----|-----|-----|-----|-----|-----|-----|
| Probability | 0.1 | 0.1 | 0.1 | 0.1 | 0.1 | 0.1 | 0.1 | 0.1 | 0.1 | 0.1 |

We must assign probability to all events, not just to individual outcomes. The probability of an event in this example is simply the sum of the probabilities of the outcomes making up the event. For example, the probability that an odd digit is chosen is

$$P(\text{odd outcome}) = P(1) + P(3) + P(5) + P(7) + P(9) = 0.5$$

This assignment of probability satisfies all of our rules. For example, the multiplication rule for independent events says that the probability that the first digit produced is odd and the second is a 7 is

$$P(\text{first digit odd})P(\text{second digit is } 7) = (0.5)(0.1) = 0.05$$

◀

---

## PROBABILITIES IN A FINITE SAMPLE SPACE

Assign a probability to each individual outcome. These probabilities must be numbers between 0 and 1 and must have sum 1.

The probability of any event is the sum of the probabilities of the outcomes making up the event.

---

## EXAMPLE 13.3

All human blood can be typed as one of O, A, B, or AB. The distribution of the types varies a bit with race and location. Choose a black American at random. Here are the probabilities that the person you choose will have blood type O, A, or B:

| Blood type | O | A | B | AB |
|---|---|---|---|---|
| Probability | 0.49 | 0.27 | 0.20 | ? |

What is the probability of type AB blood? The probability of all possible outcomes together must be 1. The outcomes O, A, and B together have probability

$$0.49 + 0.27 + 0.20 = 0.96$$

So the probability of type AB blood must be $1 - 0.96 = 0.04$.

What is the probability that the person chosen has either type A or type B blood? No person can have both blood types, so the probability of one or the other is the sum

$$0.27 + 0.20 = 0.47$$

What is the probability that two black Americans chosen at random both have type O blood? Their blood types are independent of each other, so the multiplication rule applies:

$$(0.49)(0.49) = 0.2401$$

◀

# ASSIGNING PROBABILITIES: INTERVALS OF OUTCOMES

Suppose that we want to choose a number at random between 0 and 1, allowing *any* number between 0 and 1 as the outcome. A software random number generator will do this. The sample space is now an entire interval of numbers:

$$S = \{\text{all numbers } y \text{ such that } 0 \leq y \leq 1\}$$

How can we assign probabilities to such events as $\{0.3 \leq y \leq 0.7\}$? As in the case of selecting a random digit, we would like all possible outcomes to be equally likely. But we cannot assign probabilities to each individual value of $y$ and then sum, because there are infinitely many possible values.

Instead we use a new way of assigning probabilities directly to events— as *areas under a density curve.* Any density curve has area exactly 1 underneath it, corresponding to total probability 1.

## EXAMPLE 13.4

The random number generator will spread its output uniformly across the entire interval from 0 to 1 as we allow it to generate a long sequence of numbers. The results of many trials are represented by the uniform density curve (Figure 13.1). This density has height 1 over the interval from 0 to 1. The area under the density is 1, and the probability of any event is the area under the density and above the event in question.

As Figure 13.1(a) illustrates, the probability that the random number generator produces a number between 0.3 and 0.7 is

$$P(0.3 \leq y \leq 0.7) = 0.4$$

because the area under the density and above the interval from 0.3 to 0.7 is 0.4. The height of the density is 1 and the area of a rectangle is the product of height and length, so the probability of any interval of outcomes is just the length of the interval.

Similarly,

$$P(y \leq 0.5) = 0.5$$
$$P(y > 0.8) = 0.2$$
$$P(y \leq 0.5 \text{ or } y > 0.8) = 0.7$$

◀

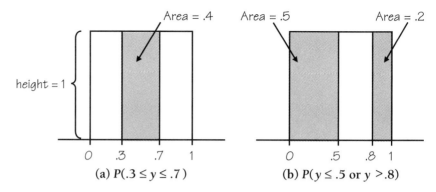

FIGURE 13.1    Probabilities as areas under a density curve. This uniform density curve distributes probability evenly between 0 and 1.

Notice that the last event consists of two nonoverlapping intervals, so the total area above the event is found by adding two areas, as illustrated by Figure 13.1(b). This assignment of probabilities obeys all of our rules for probability.    ◀

## EXERCISES

13.1    Probability is a measure of how likely an event is to occur. Match one of the probabilities that follow with each statement about an event. (The probability is usually a much more exact measure of likelihood than is the verbal statement.)

$$0, 0.01, 0.3, 0.6, 0.99, 1$$

(a) This event is impossible. It can never occur.

(b) This event is certain. It will occur on every trial of the random phenomenon.

(c) This event is very unlikely, but it will occur once in a while in a long sequence of trials.

(d) This event will occur more often than not.

13.2    Figure 13.2 displays several assignments of probabilities to the six faces of a die. We can learn which assignment is actually *accurate* for a particular die only by rolling the die many times. However, some of the assignments are not *legitimate* assignments of probability. That is, they do not obey the

Probability

| Model 1 | Model 2 | Model 3 | Model 4 |
|---------|---------|---------|---------|
| 1/3 | 1/3 | 1/7 | 1/3 |
| 0 | 1/6 | 1/7 | 1/3 |
| 1/6 | 1/6 | 1/7 | –1/6 |
| 0 | 1/6 | 1/7 | –1/6 |
| 1/6 | 1/6 | 1/7 | 1/3 |
| 1/3 | 1/6 | 1/7 | 1/3 |

FIGURE 13.2   Four assignments of probability to the faces of a die, for Exercise 13.2.

rules. Which are legitimate and which are not? In the case of the illegitimate models, explain what is wrong.

13.3    If you draw an M&M candy at random from a bag of the candies, the candy you draw will have one of six colors. The probability of drawing each color depends on the proportion of each color among all candies made.

(a) The table below gives the probability that a randomly chosen plain M&M has each color:

| Color | Brown | Red | Yellow | Green | Orange | Blue |
|-------|-------|-----|--------|-------|--------|------|
| Probability | 0.3 | 0.2 | 0.2 | 0.1 | 0.1 | ? |

What must be the probability of drawing a blue candy?

(b) The probabilities for peanut M&Ms are a bit different. Here they are:

| Color | Brown | Red | Yellow | Green | Orange | Blue |
|-------|-------|-----|--------|-------|--------|------|
| Probability | 0.2 | 0.1 | 0.2 | 0.1 | 0.1 | ? |

What is the probability that a peanut M&M chosen at random is blue?

(c) What is the probability that a plain M&M is any of red, yellow, or orange? What is the probability that a peanut M&M has one of these colors?

13.4    The probability that a randomly chosen American woman aged 20 to 24 is married is 0.31. What is the probability that she is not married?

13.5    The *New York Times* (August 21, 1989) reported a poll that interviewed a random sample of 1025 women. The married women in the sample were asked whether their husbands did their fair share of household chores. Here are the results:

| Outcome | Probability |
| --- | --- |
| Does more than his fair share | 0.12 |
| Does his fair share | 0.61 |
| Does less than his fair share | ? |

These proportions are probabilities for the random phenomenon of choosing a married woman at random and asking her opinion.

(a) What must be the probability that the woman chosen says that her husband does less than his fair share? Why?

(b) The event "I think my husband does at least his fair share" contains the first two outcomes. What is its probability?

13.6    Exactly one of Brown, Chavez, and Williams will be promoted to partner in the law firm that employs them all. Brown thinks that she has probability 0.25 of winning the promotion and that Williams has probability 0.2. What probability does Brown assign to the outcome that Chavez is the one promoted?

13.7    An American roulette wheel contains compartments numbered 1 through 36 plus 0 and 00. Of the 38 compartments, 0 and 00 are colored green, 18 of the others are red, and 18 are black. A ball is spun in the direction opposite to the wheel's motion, and bets are made on the number where the ball comes to rest. A simple wager is *red-or-black*, in which you bet that the ball will stop in, say, a red compartment. If the wheel is fair, all 38 compartments are equally likely.

(a) What is the probability of a red?

(b) In the red-or-black bet, green belongs to the house. What is the probability of green?

(c) Jim bets on red on each of two spins of the wheel. What is the probability that he will win on the first spin and lose on the second spin?

13.8    When you toss two coins, the four possible outcomes are (head, head), (head, tail), (tail, head), and (tail, tail). If the coins are balanced (probability 0.5 of a head) and are tossed independently, what are the probabilities of these four outcomes? What is the probability that exactly one of the coins shows a head?

**13.9**    Government data assign a single cause for each death that occurs in the United States. The data show that the probability is 0.42 that a randomly chosen death was due to cardiovascular (mainly heart) diseases, and 0.24 that it was due to cancer.

(a) What is the probability that a death was due either to cardiovascular disease or to cancer? What is the probability that the death was due to some other cause?

(b) Choose two Americans at random. We can assume that the causes of their eventual deaths are independent. What is the probability that both will die of cardiovascular disease?

**13.10**    Government data show that 27% of the civilian labor force have at least four years of college and that 14% of the labor force work as laborers or operators of machines or vehicles. Can you conclude that, because $(0.27)(0.14) = 0.038$, about 4% of the labor force are college-educated laborers or operators? Explain your answer.

**13.11**    In each of the following situations, describe a sample space $S$ for the indicated random phenomenon. In some cases, you have some freedom in your choice of $S$.

(a) A seed is planted in the ground. It either germinates or fails to grow.

(b) A patient with a usually fatal form of cancer is given a new treatment. The response variable is the length of time that the patient lives after treatment.

(c) A student enrolls in a statistics course and at the end of the semester receives a letter grade.

(d) A basketball player shoots two free throws.

(e) A year after knee surgery, a patient is asked to rate the amount of pain in the knee. A seven-point scale is used, with 1 corresponding to no pain and 7 corresponding to extreme discomfort.

**13.12**    Many computer systems have "random number generators" that produce numbers that are distributed uniformly between 0 and 1. Figure 13.1 graphs the density curve of the outcome $y$ of a random number generator.

(a) Check that the area under this curve is 1.

(b) What is the probability that the number $y$ is less than 0.25? (Sketch the density curve, shade the area that represents the probability, then find that area. Do this for parts (c) and (d) also.)

(c) What is the probability that $y$ lies between 0.1 and 0.9?

(d) What is the probability that $y$ is either less than 0.1 or greater than 0.9?

**(e)** The software generates two random numbers independently. What is the probability that both are less than 0.25?

13.13    Many random number generators allow users to specify the range of the random numbers to be produced. Suppose that you specify that the range is to be $0 \leq y \leq 2$. Then the density curve of the outcomes has constant height between 0 and 2, and height 0 elsewhere.

**(a)** What is the height of the density curve between 0 and 2? Draw a graph of the density curve.

**(b)** Use your graph from (a) and the fact that probability is area under the curve to find $P(y \leq 1)$.

**(c)** Find $P(0.5 < y < 1.3)$.

**(d)** Find $P(y \geq 0.8)$.

13.14    An SRS of 400 American adults is asked, "What do you think is the most serious problem facing our schools?" Suppose that in fact 30% of all adults would answer "drugs" if asked this question. The proportion $y$ of the sample who answer "drugs" will vary in repeated sampling. In fact, we can assign probabilities to values of $y$ using the normal density curve with mean 0.3 and standard deviation 0.023. Use this density curve to find the probabilities of the following events:

**(a)** At least half of the sample believes that drugs are the schools' most serious problem.

**(b)** Less than 25% of the sample believes that drugs are the most serious problem.

**(c)** The sample proportion is between 0.25 and 0.35.

## LESSON 14

# CONDITIONAL PROBABILITY*

Choose an adult American woman at random. What is the probability that the woman you choose is married? Well, there are 99,585,000 adult women in the United States, and 58,929,000 of them are married. Because "choose at random" gives an equal chance to all women, the probability we want is just the proportion of all women in the population who are married. That is,

$$P(\text{married}) = \frac{58,929,000}{99,585,000} = 0.592$$

Suppose we now learn that the woman chosen is between 18 and 24 years old. Does this information change the probability that she is married? We know that young women are less likely to be married, so we are quite sure that the probability for young women is lower than the probability for all women. In fact, there are 12,614,000 women aged 18 to 24, and 3,046,000 of these are married. So the probability that a woman is married

---

*The content of this lesson is important for an understanding of probability. However, it is not needed to understand the statistical methods in later lessons and can therefore be omitted without loss of continuity.

*given the information that she is between 18 and 24 years old is*

$$P(\text{married} \mid \text{age 18 to 24}) = \frac{3{,}046{,}000}{12{,}614{,}000} = 0.241$$

This is a *conditional probability*. That is, it gives the probability of one event (the woman chosen is married) under the condition that we know another event (she is between ages 18 and 24). You can read the bar "|" as "given the information that."

It is common sense that if we know a woman is young, we lower our guess of the probability that she is married. If we have the data, we can compute the actual conditional probability. Our task is to turn this common sense into something more general.

## DEFINING CONDITIONAL PROBABILITY

Table 14.1 gives data on the age and marital status of adult American women.[40] From these data we can find a variety of probabilities that describe a woman chosen at random. For example, the probability that we draw a married woman between 18 and 24 years old is

$$P(\text{age 18 to 24 and married}) = \frac{3{,}046}{99{,}585} = 0.031$$

This calculation used a count from the body of the table: there are 3,046,000 women who are both age 18 to 24 and married.

| **TABLE 14.1** | Women age 18 and over by age and marital status (thousands) | | | |
|---|---|---|---|---|
| | Age | | | |
| | 18 to 24 | 25 to 64 | 65 and over | Total |
| Married | 3,046 | 48,116 | 7,767 | 58,929 |
| Never married | 9,289 | 9,252 | 768 | 19,309 |
| Widowed | 19 | 2,425 | 8,636 | 11,080 |
| Divorced | 260 | 8,916 | 1,091 | 10,267 |
| Total | 12,614 | 68,709 | 18,262 | 99,585 |

If we are given no information about marital status, we find probabilities about a woman's age from the "Total" row at the bottom of the table:

| 18 to 24 | 25 to 64 | 65 and over | Total |
|----------|----------|-------------|-------|
| 12,614 | 68,709 | 18,262 | 99,585 |

For example, the probability that the woman we choose is between 18 and 24 years old is

$$P(\text{age 18 to 24}) = \frac{12,614}{99,585} = 0.127$$

If we are given some information, we look only at the women described by the information given. For example, if we are told that the woman is between 18 and 24 years old, we are confined to the "18 to 24" column in Table 14.1:

| | |
|---|---|
| Married | 3,046 |
| Never married | 9,289 |
| Widowed | 19 |
| Divorced | 260 |
| Total | 12,614 |

Thus the *conditional* probability that a woman is married, given the information that she is between 18 and 24 years of age, is

$$P(\text{married} \mid \text{age 18 to 24}) = \frac{3,046}{12,614} = 0.241$$

There is a relationship among these three probabilities. The probability that a woman is both age 18 to 24 and married is the product of the probabilities that she is age 18 to 24 and that she is married *given* that she is age 18 to 24. That is,

$$P(\text{age 18 to 24 and married}) = P(\text{age 18 to 24}) \times P(\text{married} \mid \text{age 18 to 24})$$
$$= \frac{12,614}{99,585} \times \frac{3,046}{12,614} = \frac{3,046}{99,585} \quad \text{(as before)}$$

Try to think your way through this in words before looking at the formal notation. We have just discovered the fundamental multiplication rule of probability.

MULTIPLICATION RULE

The probability that both of two events $A$ and $B$ happen together can be found by

$$P(A \text{ and } B) = P(A)P(B \mid A)$$

Here $P(B \mid A)$ is the conditional probability that $B$ occurs given the information that $A$ occurs.

If $P(A)$ and $P(A \text{ and } B)$ are given, we can rearrange the multiplication rule to produce a *definition* of the conditional probability $P(B \mid A)$ in terms of unconditional probabilities.

DEFINITION OF CONDITIONAL PROBABILITY

When $P(A) > 0$, the conditional probability of $B$ given $A$ is

$$P(B \mid A) = \frac{P(A \text{ and } B)}{P(A)}$$

## USING CONDITIONAL PROBABILITY

The definition of conditional probability reminds us that in principle all probabilities, including conditional probabilities, can be found from the assignment of probabilities to events that describes a random phenomenon. More often, however, conditional probabilities are part of the information given to us in a probability model, and the multiplication rule is used to compute $P(A \text{ and } B)$.

EXAMPLE 14.1

Slim is a professional poker player. At the moment, he wants very much to draw two diamonds in a row. As he sits at the table looking at his hand and at the upturned cards on the table, Slim sees 11 cards. Of these, 4 are diamonds.

The full deck contains 13 diamonds among its 52 cards so 9 of the 41 unseen cards are diamonds. Because the deck was carefully shuffled, each card that Slim draws is equally likely to be any of the cards that he has not seen.

To find Slim's probability of drawing two diamonds, first calculate

$$P(\text{first card diamond}) = \frac{9}{41}$$

$$P(\text{second card diamond} \mid \text{first card diamond}) = \frac{8}{40}$$

Slim finds both probabilities by counting cards. The probability that the first card drawn is a diamond is 9/41 because 9 of the 41 unseen cards are diamonds. If the first card is a diamond, that leaves 8 diamonds among the 40 remaining cards. So the *conditional* probability of another diamond is 8/40. The multiplication rule now says that

$$P(\text{both cards diamonds}) = \frac{9}{41} \times \frac{8}{40} = 0.044$$

Slim will need luck to draw his diamonds.    ◀

# INDEPENDENCE

The conditional probability $P(B \mid A)$ is generally not equal to the unconditional probability $P(B)$. That is, knowing that event $A$ occurs generally gives us some additional information about whether or not event $B$ occurs. If knowing that $A$ occurs gives no additional information about $B$, then $A$ and $B$ are *independent events*. We can define independence in terms of conditional probability.

---

INDEPENDENT EVENTS

Two events $A$ and $B$ that both have positive probability are independent if

$$P(B \mid A) = P(B)$$

That is, knowing that $A$ occurs does not change the probability that $B$ occurs.

---

## EXAMPLE 14.2

We saw that 59.2% of all adult women are married, but that only 24.1% of women aged 18 to 24 are married. That is, if we choose an adult woman at random,

$$P(\text{married}) = 0.592 \quad \text{and} \quad P(\text{married} \mid \text{age 18 to 24}) = 0.241$$

The events "a woman is married" and "a woman is age 18 to 24" are *not* independent. ◀

We now see that the multiplication rule for independent events, $P(A \text{ and } B) = P(A)P(B)$, is a special case of the general multiplication rule, $P(A \text{ and } B) = P(A)P(B \mid A)$. The multiplication rule also extends to collections of more than two events, provided that all are independent. Independence of events $A$, $B$, and $C$ means that no information about any one or any two can change the probability of the remaining events. The formal definition is a bit messy. Fortunately, independence is usually *assumed* in setting up a probability model for events that seem to be physically independent of each other. We can then use the multiplication rule freely, as in this example.

## EXAMPLE 14.3

A transatlantic telephone cable contains repeaters at regular intervals to amplify the signal. If a repeater fails it must be replaced by fishing the cable to the surface at great expense. Each repeater has probability 0.999 of functioning without failure for 10 years. Repeaters fail independently of each other. (This assumption means that there are no "common causes" such as earthquakes that would affect several repeaters at once.) Let $A_i$ be the event that the $i$th repeater operates successfully for 10 years.

The probability that two repeaters both last 10 years is

$$P(A_1 \text{ and } A_2) = P(A_1)P(A_2) = 0.999 \times 0.999 = 0.998$$

For a cable with 10 repeaters the probability of no failures in 10 years is

$$\begin{aligned}
P(A_1 \text{ and } A_2 \text{ and } \ldots \text{ and } A_{10}) &= P(A_1)P(A_2) \cdots P(A_{10}) \\
&= 0.999 \times 0.999 \times \cdots \times 0.999 \\
&= 0.999^{10} = 0.990
\end{aligned}$$

Cables with 2 or 10 repeaters are quite reliable. Unfortunately, a transatlantic cable has 300 repeaters. The probability that all 300 work for 10 years is

$$P(A_1 \text{ and } A_2 \text{ and } \ldots \text{ and } A_{300}) = 0.999^{300} = 0.741$$

There is therefore about one chance in four that the cable will have to be fished up for replacement of a repeater sometime during the next 10 years. Repeaters are in fact designed to be much more reliable than 0.999 in 10 years. Some transatlantic cables have served for more than 20 years with no failures. ◀

By combining the rules we have learned, we can compute probabilities for rather complex events. Here is an example.

## EXAMPLE 14.4

A diagnostic test for the presence of the AIDS virus has probability 0.005 of producing a false positive. That is, when a person free of the AIDS virus is tested, the test has probability 0.005 of falsely indicating that the virus is present. If the 140 employees of a medical clinic are tested and all 140 are free of AIDS, what is the probability that at least one false positive will occur?

It is reasonable to assume that the test results for different individuals are independent. The probability that the test is positive for a single person is 0.005, so the probability of a negative result is $1 - 0.005 = 0.995$. The probability of at least one false positive among the 140 people tested is therefore

$$
\begin{aligned}
P(\text{at least one positive}) &= 1 - P(\text{no positives}) \\
&= 1 - P(140 \text{ negatives}) \\
&= 1 - 0.995^{140} \\
&= 1 - 0.496 = 0.504
\end{aligned}
$$

The probability is greater than 1/2 that at least one of the 140 people will test positive for AIDS, even though no one has the virus. ◀

## EXERCISES

14.1    Choose an adult American woman at random. Table 14.1 describes the population from which we draw. Use the information in that table to answer the following questions.

(a) What is the probability that the woman chosen is over 65 years old?

(b) What is the conditional probability that the woman chosen is married, given that she is over 65 years old?

(c) How many women are *both* married and in the over 65 age group? What is the probability that the woman we choose is a married woman at least 65 years old?

(d) Verify that the three probabilities you found in (a), (b), and (c) satisfy the multiplication rule.

14.2    Choose an adult American woman at random. Table 14.1 describes the population from which we draw. Use the information in that table to find the following probabilities.

(a) The probability that the woman chosen is a widow.

(b) The conditional probability that the woman chosen is a widow, given that she is over 65 years old.

(c) The conditional probability that the woman chosen is a widow, given that she is between 25 and 64 years old.

(d) Are the events "widow" and "over 65 years old" independent? How do you know?

14.3    Choose an adult American woman at random. Table 14.1 describes the population from which we draw.

(a) What is the conditional probability that the woman chosen is 18 to 24 years old, given that she is married?

(b) Earlier, we found that $P(\text{married} \mid \text{age 18 to 24}) = 0.241$. Complete this sentence: 0.241 is the proportion of women who are _____ among those women who are _____.

(c) In (a), you found $P(\text{age 18 to 24} \mid \text{married})$. Write a sentence of the form given in (b) that describes the meaning of this result. The two conditional probabilities give us very different information.

14.4    Here are the counts (in thousands) of earned degrees in the United States in a recent year, classified by level and by the sex of the degree recipient:

|        | Bachelor's | Master's | Professional | Doctorates | Total |
|--------|-----------|----------|--------------|------------|-------|
| Female | 616       | 194      | 30           | 16         | 856   |
| Male   | 529       | 171      | 44           | 26         | 770   |
| Total  | 1145      | 365      | 74           | 42         | 1626  |

(a) If you choose a degree recipient at random, what is the probability that the person you choose is a woman?

**(b)** What is the conditional probability that you choose a woman, given that the person chosen received a professional degree?

**(c)** Are the events "choose a woman" and "choose a professional degree recipient" independent? How do you know?

14.5   The previous exercise gives the counts (in thousands) of earned degrees in the United States in a recent year. Use these data to answer the following questions.

**(a)** What is the probability that a randomly chosen degree recipient is a man?

**(b)** What is the conditional probability that the person chosen received a bachelor's degree, given that he is a man?

**(c)** Use the multiplication rule to find the probability of choosing a male bachelor's degree recipient. Check your result by finding this probability directly from the table of counts.

14.6   A general can plan a campaign to fight one major battle or three small battles. He believes that he has probability 0.6 of winning the large battle and probability 0.8 of winning each of the small battles. Victories or defeats in the small battles are independent. The general must win either the large battle or all three small battles to win the campaign. Which strategy should he choose?

14.7   An automobile manufacturer buys computer chips from a supplier. The supplier sends a shipment containing 5% defective chips. Each chip chosen from this shipment has probability 0.05 of being defective, and each automobile uses 12 chips selected independently. What is the probability that all 12 chips in a car will work properly?

14.8   Government data show that 27% of the civilian labor force have at least four years of college and that 14% of the labor force are laborers or operators of machines or vehicles. We see that $0.27 \times 0.14 = 0.038$. Can we conclude that 3.8% of the labor force are college-educated laborers or operators? Explain your answer.

14.9   A string of Christmas lights contains 20 lights. The lights are wired in series, so that if any light fails the whole string will go dark. Each light has probability 0.02 of failing during a 3-year period. The lights fail independently of each other. What is the probability that the string of lights will remain bright for 3 years?

14.10  The "random walk" theory of stock prices holds that price movements in disjoint time periods are independent of each other. Suppose that we record only whether the price is up or down each year, and that the probability that our portfolio rises in price in any one year is 0.65. (This probability is approximately correct for a portfolio containing equal dollar amounts of all common stocks listed on the New York Stock Exchange.)

(a) What is the probability that our portfolio goes up for three consecutive years?

(b) If you know that the portfolio has risen in price 2 years in a row, what probability do you assign to the event that it will go down next year?

(c) What is the probability that the portfolio's value moves in the same direction in both of the next 2 years?

14.11    The type of medical care a patient receives may vary with the age of the patient. A large study of women who had a breast lump investigated whether or not each woman received a mammogram and a biopsy when the lump was discovered. Here are some probabilities estimated by the study. The entries in the table are the probabilities that *both* of two events occur; for example, 0.321 is the probability that a patient is under 65 years of age *and* the tests were done. The four probabilities in the table have sum 1 because the table lists all possible outcomes.

|  | Tests done | Tests not done |
|---|---|---|
| Age under 65 | 0.321 | 0.124 |
| Age 65 or over | 0.365 | 0.190 |

(a) What is the probability that a patient in this study is under 65? That a patient is 65 or over?

(b) What is the probability that the tests were done for a patient? That they were not done?

(c) Are the events $A$ = {the patient was 65 or older} and $B$ = {the tests were done} independent? Were the tests omitted on older patients more or less frequently than would be the case if testing were independent of age?

14.12    Call a household prosperous if its income in 1993 exceeded $75,000. Call the household educated if the householder completed college. Select an American household at random, and let $A$ be the event that the selected household is prosperous and $B$ the event that it is educated. According to the Census Bureau, $P(A) = 0.125$, $P(B) = 0.237$, and the probability that a household is both prosperous and educated is $P(A \text{ and } B) = 0.077$.

(a) Use the definition of conditional probability to find $P(A \mid B)$. This is the proportion of educated households that are prosperous.

(b) Use the definition of conditional probability to find $P(B \mid A)$. This is the proportion of prosperous households that are educated.

14.13    Here is a table that describes all suicides committed in 1992:

|          | Male   | Female |
|----------|--------|--------|
| Firearms | 15,802 | 2,367  |
| Poison   | 3,262  | 2,233  |
| Hanging  | 3,822  | 856    |
| Other    | 1,571  | 571    |
| Total    | 24,457 | 6,027  |

(a) What is the conditional probability that a suicide used a firearm, given that it was a man? Given that it was a woman?

(b) Describe in simple language (don't use the word "probability") what your results in (a) tell you about the difference between men and women who commit suicide.

14.14    Choose a point at random in the square with sides $0 \leq x \leq 1$ and $0 \leq y \leq 1$. This means that the probability that the point falls in any region within the square is equal to the area of that region. Let $X$ be the $x$ coordinate and $Y$ the $y$ coordinate of the point chosen. Find the conditional probability $P(Y < 1/2 \mid Y > X)$. (Hint: Draw a diagram of the square and the events $Y < 1/2$ and $Y > X$.)

14.15    You have torn a tendon and are facing surgery to repair it. The surgeon explains the risks to you: infection occurs in 3% of such operations, the repair fails in 14%, and both infection and failure occur together in 1%. What percent of these operations succeed and are free from infection?

14.16    The distribution of blood types among white Americans is approximately as follows: 37% type A, 13% type B, 44% type O, and 6% type AB. Suppose that the blood types of married couples are independent and that both the husband and the wife follow this distribution.

(a) An individual with type B blood can safely receive transfusions only from persons with type B or type O blood. What is the probability that the husband of a woman with type B blood is an acceptable blood donor for her?

(b) What is the probability that in a randomly chosen couple the wife has type B blood and the husband has type A?

(c) What is the probability that one of a randomly chosen couple has type A blood and the other has type B?

(d) What is the probability that at least one of a randomly chosen couple has type O blood?

# RANDOM VARIABLES

The distribution of a variable tells us what values it takes and how often it takes each value. We can apply the same idea to describe the outcomes of a random phenomenon and their probabilities. We are most interested in quantitative outcomes such as proportions and means from random samples and randomized comparative experiments. These are quantitative variables that vary when we repeat the randomization, so we call them *random variables*.

---

RANDOM VARIABLE

A **random variable** is a variable whose value is a numerical outcome of a random phenomenon.

---

It is usual to denote random variables by capital letters near the end of the alphabet, such as $X$ or $Y$. Of course, the random variables of greatest interest to us are outcomes such as the mean $\bar{y}$ of a random sample, for which we will keep the familiar notation. The **probability distribution** of a random variable tells us the possible values of the variable and how probabilities are assigned to those values. In Lesson 13, we met two ways of assigning probabilities to outcomes. When the outcomes are values of a random variable, the two ways of assigning probabilities give us two types of random variables.

## DISCRETE RANDOM VARIABLES

The probability distribution of a random variable is most straightforward when the variable has only a finite number of possible values.

---

DISCRETE RANDOM VARIABLE

A **discrete random variable** $X$ has a finite number of possible values. The **probability distribution** of $X$ lists the values and their probabilities:

| Value of $X$ | $x_1$ | $x_2$ | $x_3$ | $\cdots$ | $x_k$ |
|---|---|---|---|---|---|
| Probability | $p_1$ | $p_2$ | $p_3$ | $\cdots$ | $p_k$ |

The probabilities $p_i$ must satisfy two requirements:

1. Every probability $p_i$ is a number between 0 and 1.
2. $p_1 + p_2 + \cdots + p_k = 1$.

Find the probability of any event by adding the probabilities $p_i$ of the particular values $x_i$ that make up the event.

---

**EXAMPLE 15.1**

The instructor of a large class gives 15% each of A's and D's, 30% each of B's and C's, and 10% F's. Choose a student at random from this class. To "choose at random" means to give every student the same chance to be chosen. The student's grade on a four-point scale (A = 4) is a random variable $X$.

The value of $X$ changes when we repeatedly choose students at random, but it is always one of 0, 1, 2, 3, or 4. Here is the distribution of $X$:

| Grade | 0 | 1 | 2 | 3 | 4 |
|-------|------|------|------|------|------|
| Probability | 0.10 | 0.15 | 0.30 | 0.30 | 0.15 |

The probability that the student got a B or better is the sum of the probabilities of an A and a B:

$$P(\text{grade is 3 or 4}) = P(X = 3) + P(X = 4)$$
$$= 0.30 + 0.15 = 0.45$$

◀

## CONTINUOUS RANDOM VARIABLES

Some random variables have values that are not isolated numbers but an entire interval of numbers. When you measure the height of a randomly chosen young woman, for example, you might get any number between about 3 feet and about 7 feet. In such settings, we use a density curve to assign probabilities.

CONTINUOUS RANDOM VARIABLE

A **continuous random variable** $X$ takes all values in an interval of numbers. The **probability distribution** of $X$ is described by a density curve. The probability of any event is the area under the density curve and above the values of $X$ that make up the event.

Any density curve gives the distribution of a continuous random variable. The density curves that are most familiar to us are the normal curves. So *normal distributions are probability distributions*. There is a close connection between a normal distribution as an idealized description for data and a normal probability distribution.

EXAMPLE 15.2

The heights of all young women closely follow the normal distribution with mean $\mu = 64.5$ inches and standard deviation $\sigma = 2.5$ inches. That is a distribution for a large set of data. Imagine choosing one young woman at random. Her height is a random variable $X$. If we repeat the random choice many times,

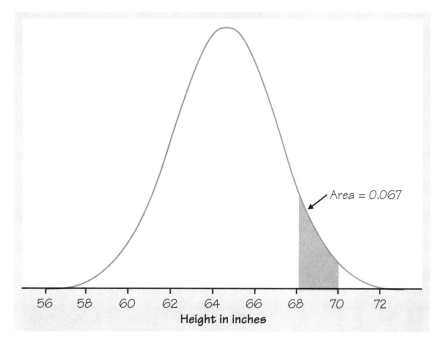

FIGURE 15.1 This normal density curve assigns probabilities when we choose a young woman at random and measure her height.

the distribution of values of $X$ is the same normal distribution that describes the population from which we are choosing.

What is the probability that a randomly chosen young woman has height between 68 and 70 inches? Figure 15.1 shows the probability as an area under a normal density curve. Calculation (be sure you recall how to do it) tells us that the probability is about 0.067. ◄

The way of thinking that this example illustrates is important in statistics. We use the probability distribution of a random variable to describe the distribution of a variable (such as height) in a population by imagining that we draw one member of the population at random.

## STATISTICAL ESTIMATION AND THE LAW OF LARGE NUMBERS

In Example 15.2, we took the mean height $\mu$ of the population of all young women to be 64.5 inches. It is more realistic to admit that we have not measured all young women and so do not know the value of their mean height $\mu$. To *estimate* $\mu$, measure an SRS of young women and use the height $\bar{y}$ of this sample as a guess for the unknown $\mu$. As we begin to use sample results to estimate population truths, we need terms to distinguish the two. Here they are.

---

PARAMETER, STATISTIC

A **parameter** is a number that describes the population. In statistical practice, the value of a parameter is not known.

A **statistic** is a number that can be computed from the sample data without making use of any unknown parameters. In practice, we often use a statistic to estimate an unknown parameter.

---

A parameter, such as the mean height $\mu$ of all young women, is in practice a fixed but unknown number. A statistic, such as the mean height $\bar{y}$ of a random sample of young women, is a random variable. It seems reasonable to use $\bar{y}$ to estimate $\mu$. An SRS should fairly represent the population, so the mean $\bar{y}$ of the sample should be somewhere near the mean $\mu$ of the population. Of course, we don't expect $\bar{y}$ to be exactly equal to $\mu$, and we realize that if we choose another SRS the luck of the draw will probably produce a different $\bar{y}$.

If $\bar{y}$ is rarely exactly right and varies from sample to sample, why is it nonetheless a reasonable estimate of the population mean $\mu$? Here is one answer: if we keep on taking larger and larger samples, the statistic $\bar{y}$ is *guaranteed* to get closer and closer to the parameter $\mu$. We have the comfort of knowing that if we can afford to keep on measuring more young women, eventually we will estimate the mean height of all young women very accurately. This remarkable fact is called the *law of large numbers*. It is remarkable because it holds for *any* population, not just for some special class such as normal distributions.

---

LAW OF LARGE NUMBERS

Draw observations at random from any population with finite mean $\mu$. As the number of observations drawn increases, the mean $\bar{y}$ of the observed values gets closer and closer to the mean $\mu$ of the population.

---

The law of large numbers can be proved mathematically starting from the basic laws of probability. The behavior of $\bar{y}$ is similar to the idea of probability. In the long run, the proportion of outcomes taking any value gets close to the probability of that value, and the average outcome gets close to the population mean. Figure 12.1 (page 147) shows how proportions approach probability in one example. Here is an example of how sample means approach the population mean.

## EXAMPLE 15.3

Figure 15.2 shows the behavior of the mean height $\bar{y}$ of $n$ women chosen at random from a population whose heights follow the $N(64.5, 2.5)$ distribution. The graph plots the values of $\bar{y}$ as we add women to our sample. The first woman drawn had height 64.21 inches, so the line starts there. The second

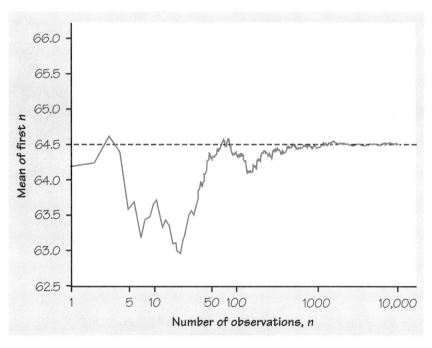

FIGURE 15.2    The law of large numbers in action. As we increase the size of our sample, the sample mean $\bar{y}$ always approaches the mean $\mu$ of the population.

had height 64.35 inches, so for $n = 2$ the mean is

$$\bar{y} = \frac{64.21 + 64.35}{2} = 64.28$$

This is the second point on the line in the graph.

At first, the graph shows that the mean of the sample changes as we take more observations. Eventually, however, the mean of the observations gets close to the population mean $\mu = 64.5$ and settles down at that value. The law of large numbers says that this *always* happens.    ◄

The law of large numbers is the foundation of such business enterprises as gambling casinos and insurance companies. The winnings (or losses) of a gambler on a few plays are uncertain—that's why gambling is exciting. In Figure 15.2, the mean of even 100 observations is not yet very close to $\mu$. It is only *in the long run* that the mean outcome is predictable. The house plays tens of thousands of times. So the house, unlike individual gamblers, can count on the long-run regularity described by the law of large numbers. The average winnings of the house on tens of thousands of plays will be very close to the mean of the distribution of winnings. Needless to say, this mean guarantees the house a profit. That's why gambling can be a business.

## EXERCISES

15.1    A study of social mobility in England looked at the social class reached by the sons of lower-class fathers. Social classes are numbered from 1 (low) to 5 (high). Take the random variable $X$ to be the class of a randomly chosen son of a father in Class 1. The study found that the distribution of $X$ is

| Son's class | 1 | 2 | 3 | 4 | 5 |
|---|---|---|---|---|---|
| Probability | 0.48 | 0.38 | 0.08 | 0.05 | 0.01 |

(a) What percent of the sons of lower-class fathers reach the highest class, Class 5?

(b) Check that this distribution satisfies the two requirements for a discrete probability distribution.

(c) What is $P(X \leq 3)$? (Be careful: the event "$X \leq 3$" includes the value 3.)

(d) What is $P(X < 3)$?

(e) Write the event "a son of a lower-class father reaches one of the two highest classes" in terms of values of $X$. What is the probability of this event?

15.2 A study of education followed a large group of fifth-grade children to see how many years of school they eventually completed. Let $X$ be the highest year of school that a randomly chosen fifth grader completes. (Students who go on to college are included in the outcome $X = 12$.) The study found this probability distribution for $X$:

| Years | 4 | 5 | 6 | 7 | 8 | 9 | 10 | 11 | 12 |
|---|---|---|---|---|---|---|---|---|---|
| Probability | 0.010 | 0.007 | 0.007 | 0.013 | 0.032 | 0.068 | 0.070 | 0.041 | 0.752 |

(a) What percent of fifth graders eventually finished twelfth grade?

(b) Check that this is a legitimate discrete probability distribution.

(c) Find $P(X \geq 6)$. (Be careful: the event "$X \geq 6$" includes the value 6.)

(d) Find $P(X > 6)$.

(e) What values of $X$ make up the event "the student completed at least one year of high school"? (High school begins with the ninth grade.) What is the probability of this event?

15.3 If a carefully made die is rolled once, it is reasonable to assign probability 1/6 to each of the six faces. What is the probability of rolling a number less than 3?

15.4 A couple plans to have three children. There are 8 possible arrangements of girls and boys. For example, GGB means the first two children are girls and the third child is a boy. All 8 arrangements are (approximately) equally likely.

(a) Write down all 8 arrangements of the sexes of three children. What is the probability of any one of these arrangements?

(b) Let $X$ be the number of girls the couple has. What is the probability that $X = 2$?

(c) Starting from your work in (a), find the distribution of $X$. That is, what values can $X$ take, and what are the probabilities for each value?

15.5 Many computer systems have "random number generators" that produce numbers that are distributed uniformly between 0 and 1. We looked at the distribution of the number $X$ produced by a random number generator in Exercise 13.12. Generate *two* random numbers between 0 and 1 and take $Y$ to be their sum. Then $Y$ is a continuous random variable that can take

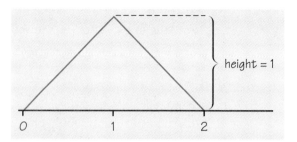

FIGURE 15.3   The density curve for the sum of two random numbers. This continuous random variable takes values between 0 and 2.

any value between 0 and 2. The density curve of $Y$ is the triangle shown in Figure 15.3.

(a) Check that the area under this curve is 1.

(b) What is the probability that $Y$ is less than 1? (Sketch the density curve, shade the area that represents the probability, then find that area. Do this for (c) also.)

(c) What is the probability that $Y$ is less than 0.5?

15.6   An opinion poll asks an SRS of 1500 adults, "Do you happen to jog?" Suppose that the population proportion who jog (a parameter) is $p = 0.15$. To estimate $p$, we use the proportion $\hat{p}$ in the sample who answer "Yes." The statistic $\hat{p}$ is a random variable that is approximately normally distributed with mean $\mu = 0.15$ and standard deviation $\sigma = 0.0092$. Find the following probabilities.

(a) $P(\hat{p} \geq 0.16)$

(b) $P(0.14 \leq \hat{p} \leq 0.16)$

State whether each boldface number in Exercises 15.7 to 15.10 is a *parameter* or a *statistic*.

15.7   The Bureau of Labor Statistics last month interviewed 60,000 members of the U.S. labor force, of whom **7.2%** were unemployed.

15.8   A carload lot of ball bearings has mean diameter **2.5003** centimeters (cm). This is within the specifications for acceptance of the lot by the purchaser. By chance, an inspector chooses 100 bearings from the lot that have mean diameter **2.5009** cm. Because this is outside the specified limits, the lot is mistakenly rejected.

15.9   A telemarketing firm in Los Angeles uses a device that dials residential telephone numbers in that city at random. Of the first 100 numbers dialed,

**48%** are unlisted. This is not surprising because **52%** of all Los Angeles residential phones are unlisted.

15.10    A researcher carries out a randomized comparative experiment with young rats to investigate the effects of a toxic compound in food. She feeds the control group a normal diet. The experimental group receives a diet with 2500 parts per million of the toxic material. After 8 weeks, the mean weight gain is **335** grams for the control group and **289** grams for the experimental group.

15.11    Figure 15.2 shows how the mean of $n$ observations behaves as we keep adding more observations to those already in hand. The first few observations are

$$64.21 \quad 64.35 \quad 65.32 \quad 63.81 \quad 60.19 \quad 64.41$$

Demonstrate that you grasp the idea of Figure 15.2: find the mean of the first one, then two, then three, . . . of these observations and plot the successive means against $n$.

15.12    It would be quite risky for you to insure the life of a 21-year-old friend for $100,000. There is a high probability that your friend will live and you will gain the few dollars you charged him in insurance premiums. But if he dies, you will have to pay $100,000. Explain carefully why selling insurance is not risky for an insurance company that insures many thousands of 21-year-old men.

15.13    One consequence of the law of large numbers is that once we have a probability distribution for a random variable, we can find its mean by simulating many outcomes and averaging them. The law of large numbers says that if we take enough outcomes, their average value is sure to approach the mean of the distribution.

I have a little bet to offer you. Toss a coin ten times. If there is no run of three or more straight heads or tails in the ten outcomes, I'll pay you $2. If there is a run of three or more, you pay me just $1. Surely you will want to take advantage of me and play this game?

Simulate enough plays of this game (the outcomes are +$1 if you win and −$2 if you lose) to estimate the mean outcome. Is it to your advantage to play?

# SAMPLING DISTRIBUTIONS

Statistical inference draws conclusions from data about some wider class of individuals. Because data are variable—if we took another sample or carried out another experiment, we would get somewhat different data—statistical inference includes a statement of how confident we can be in our conclusions. These statements use the language of probability. That is, they answer the question "What would happen if we did this many times?"

---

**EXAMPLE 16.1**

Sulfur compounds such as dimethyl sulfide (DMS) are sometimes present in wine. DMS causes "off-odors" in wine, so winemakers want to know the odor threshold, the lowest concentration of DMS that the human nose can detect. Different people have different thresholds, so we start by asking about the mean threshold $\mu$ in the population of all adults who drink wine. The number $\mu$ is a *parameter* that describes this population.

We do an experiment: we present tasters with both natural wine and the same wine spiked with DMS at different concentrations to find the lowest concentration at which they identify the spiked wine. Here are the odor thresholds (measured in micrograms of DMS per liter of wine) for 10 subjects:

$$28 \quad 28 \quad 40 \quad 33 \quad 20 \quad 31 \quad 29 \quad 27 \quad 17 \quad 21$$

The mean threshold for these subjects is $\bar{y} = 27.4$. This sample mean is a *statistic* that we use to estimate the parameter $\mu$, but it is probably not exactly equal to $\mu$. Moreover, we know that a different 10 subjects would give us a different $\bar{y}$. ◀

The law of large numbers assures us that if we measure enough subjects the statistic $\bar{y}$ will eventually get very close to the unknown parameter $\mu$. But our experiment had just 10 subjects. What can we say about $\bar{y}$ from 10 subjects as an estimate of $\mu$? We ask: "What would happen if we took many samples of 10 subjects from this population?" The answer is given by the *sampling distribution* of the sample mean $\bar{y}$.

---

SAMPLING DISTRIBUTION

The **sampling distribution** of a statistic is the distribution of values taken by the statistic in all possible samples of the same size from the same population.

---

## THE SAMPLING DISTRIBUTION OF A SAMPLE MEAN $\bar{y}$

Extensive studies have found that the DMS odor threshold of adults follows roughly a normal distribution with mean $\mu = 25$ micrograms per liter and standard deviation $\sigma = 7$ micrograms per liter. Figure 16.1 shows this population distribution. With this information, we can simulate many runs of our experiment with different subjects drawn at random from the population. Figure 16.2 is a histogram of the observed mean thresholds $\bar{y}$ for 1000 samples of 10 subjects. What can we say about the shape, center, and spread of this distribution?

- *Shape:* It looks normal! The normal probability plot in Figure 16.3 confirms that the distribution of $\bar{y}$ from many samples does have a distribution that is very close to normal.
- *Center:* The mean of the 1000 $\bar{y}$'s is 25.0726. That is, the distribution is centered very close to the population mean $\mu = 25$. This is another

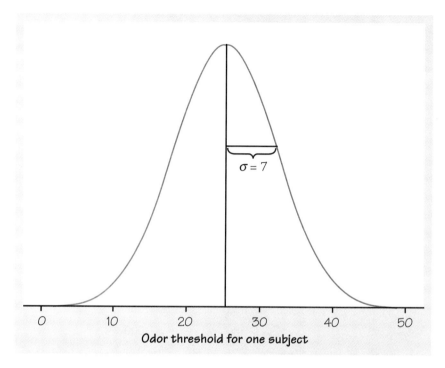

**FIGURE 16.1**    The distribution of odor threshold for DMS in wine in the population of all adults. This is also the distribution of the DMS threshold for one adult chosen at random.

version of the absence of bias in random samples. Some sample $\bar{y}$'s overestimate $\mu$ and some underestimate $\mu$, but there is no systematic deviation to one side or the other.

- **Spread:** The standard deviation of the 1000 $\bar{y}$'s is 2.1905, notably smaller than the standard deviation $\sigma = 7$ of the population of individual subjects. *Averages are less variable than individual observations.*

    Strictly speaking, the sampling distribution is the ideal pattern that would emerge if we looked at all possible samples of size 10 from our population. A distribution obtained from a fixed number of trials, like the 1000 trials in Figure 16.2, is only an approximation to the sampling distribution. We would need many simulations to understand how the sampling distribution changes when, for example, we change the number of observations $n$. One of the uses of probability theory in statistics is to obtain exact sampling distributions without simulation. Here are the facts about $\bar{y}$.

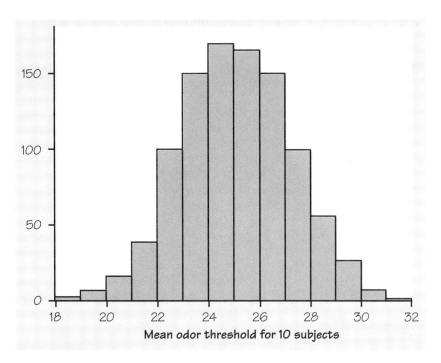

FIGURE 16.2    Histogram of 1000 sample means for SRSs of size 10 drawn from the population in Figure 16.1.

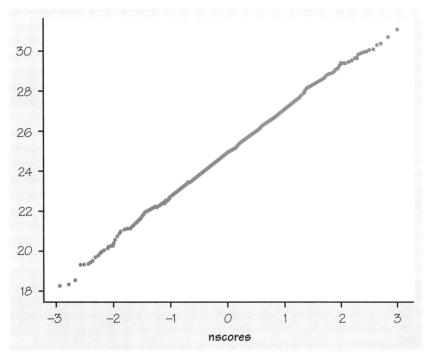

FIGURE 16.3    Normal probability plot of 1000 sample means of size 10. The distribution of these $\bar{y}$'s is very close to normal.

SAMPLING DISTRIBUTION OF A SAMPLE MEAN

Draw an SRS of size $n$ from a large[41] population that has a distribution with mean $\mu$ and standard deviation $\sigma$. The sampling distribution of the sample mean $\bar{y}$

- has mean equal to the population mean $\mu$,

$$\mu_{\bar{y}} = \mu$$

- has standard deviation equal to the population standard deviation $\sigma$ divided by the square root of the sample size,

$$\sigma_{\bar{y}} = \frac{\sigma}{\sqrt{n}}$$

- has the normal distribution $N(\mu, \sigma/\sqrt{n})$ if the population has a normal distribution $N(\mu, \sigma)$.

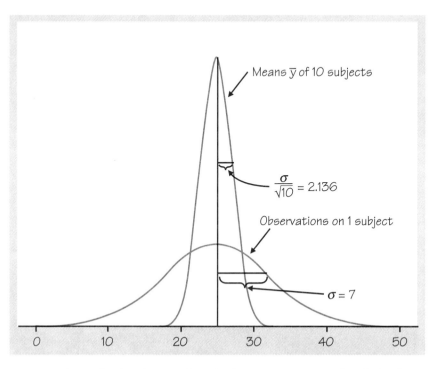

FIGURE 16.4    The distribution of single observations compared with the distribution of the mean $\bar{y}$ of 10 observations. Averages are less variable than individual observations.

We see that the standard deviation of the distribution of $\bar{y}$ is indeed smaller than the standard deviation of individual observations, but it gets smaller only at the rate $\sqrt{n}$. To cut the standard deviation of $\bar{y}$ in half, we must take four times as many observations, not just twice as many. Figure 16.4 compares the distribution of single observations (the population distribution) for odor thresholds with the sampling distribution of the means $\bar{y}$ of 10 observations.

## THE CENTRAL LIMIT THEOREM

What happens to $\bar{y}$ when the population distribution is not normal? It turns out that as the sample size increases, the distribution of $\bar{y}$ gets closer to a normal distribution. This is true no matter what shape the population distribution has, as long as the population has a finite standard deviation $\sigma$. This famous fact of probability theory is called the *central limit theorem*. It is much more useful than the fact that the distribution of $\bar{y}$ is exactly normal if the population is exactly normal.

> CENTRAL LIMIT THEOREM
>
> Draw an SRS of size $n$ from any population whatsoever with mean $\mu$ and finite standard deviation $\sigma$. When $n$ is large, the sampling distribution of the sample mean $\bar{y}$ is close to the normal distribution $N(\mu, \sigma/\sqrt{n})$ with mean $\mu$ and standard deviation $\sigma/\sqrt{n}$.

How large a sample size $n$ is needed for $\bar{y}$ to be close to normal depends on the population distribution. More observations are required if the shape of the population distribution is far from normal.

### EXAMPLE 16.2

Figure 16.5 shows the central limit theorem in action for a very nonnormal population. Figure 16.5(a) displays the density curve for the distribution of the population. The distribution is strongly right skewed, and the most probable outcomes are near 0 at one end of the range of possible values. The mean $\mu$ of this distribution is 1 and its standard deviation $\sigma$ is also 1. This particular distribution is called an *exponential distribution* from the shape of its density

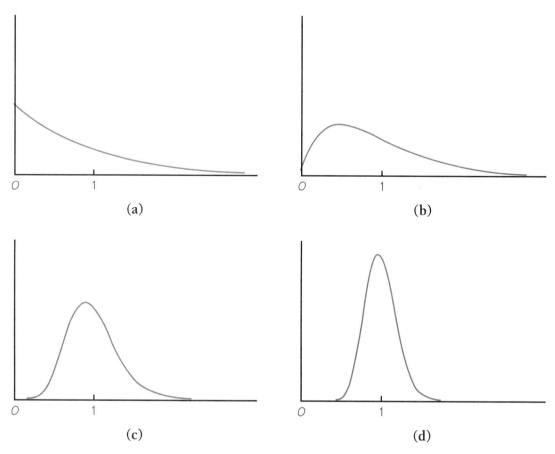

**FIGURE** 16.5    The central limit theorem in action: the distribution of sample means $\bar{y}$ from a strongly nonnormal population becomes more normal as the sample size increases. **(a)** The distribution of a single observation. **(b)** The distribution of $\bar{y}$ for 2 observations. **(c)** The distribution of $\bar{y}$ for 10 observations. **(d)** The distribution of $\bar{y}$ for 25 observations.

curve. Exponential distributions are used to describe the lifetime in service of electronic components and the time required to serve a customer or repair a machine.

Figures 16.5(b), (c), and (d) are the density curves of the sample means of 2, 10, and 25 observations from this population. As $n$ increases, the shape becomes more normal. The mean remains at $\mu = 1$ and the standard deviation decreases, taking the value $1/\sqrt{n}$. The density curve for 10 observations is still somewhat skewed to the right but already resembles a normal curve, having $\mu = 1$ and $\sigma = 1/\sqrt{10} = 0.32$. The density curve for $n = 25$ is yet more normal. The contrast between the shapes of the population distribution and of the distribution of the mean of 10 or 25 observations is striking. ◀

The central limit theorem allows us to use normal probability calculations to answer questions about sample means from many observations even when the population distribution is not normal.

**EXAMPLE 16.3**

The time that a technician requires to perform preventive maintenance on an air-conditioning unit is governed by the exponential distribution whose density curve appears in Figure 16.5(a). The mean time is $\mu = 1$ hour and the standard deviation is $\sigma = 1$ hour. Your company operates 70 of these units. What is the probability that their average maintenance time exceeds 50 minutes?

The central limit theorem says that the sample mean time $\bar{y}$ (in hours) spent working on 70 units has approximately the normal distribution with mean equal to the population mean $\mu = 1$ hour and standard deviation

$$\frac{\sigma}{\sqrt{70}} = \frac{1}{\sqrt{70}} = 0.12 \text{ hour}$$

The distribution of $\bar{y}$ is therefore approximately $N(1, 0.12)$. Figure 16.6 shows this normal curve (*solid*) and also the actual density curve of $\bar{y}$ (*dashed*).

Because 50 minutes is 50/60 of an hour, or 0.83 hour, the probability we want is $P(\bar{y} > 0.83)$. A normal distribution calculation gives this probability as 0.9222. This is the area to the right of 0.83 under the solid normal curve in Figure 16.6. The exactly correct probability is the area under the dashed density curve in the figure. It is 0.9294. The central limit theorem normal approximation is off by only about 0.007. ◄

## THE IDEA OF A SAMPLING DISTRIBUTION

Sampling distributions display the regular behavior of statistics from random samples or randomized comparative experiments. The behavior is regular in the sense that although the results of a single sample are unpredictable, there is a clear pattern when we draw many samples from the same population. That is, a sampling distribution gives a detailed answer to the question "What would happen if we did this many times?" The idea of a sampling distribution is the key to understanding statistical inference. Figure 16.7 illustrates the idea in the case of a sample mean $\bar{y}$. Keep taking random samples of size $n$ from a population. Find the mean $\bar{y}$ for each

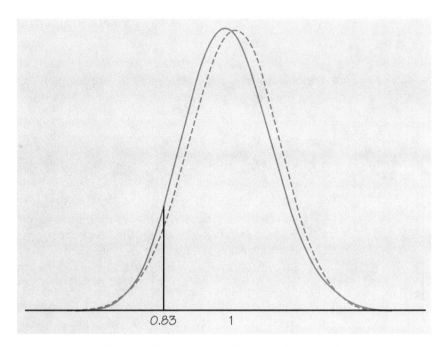

**FIGURE 16.6** The exact distribution (*dashed*) and the normal approximation from the central limit theorem (*solid*) for the average time needed to maintain an air conditioner in Example 16.3.

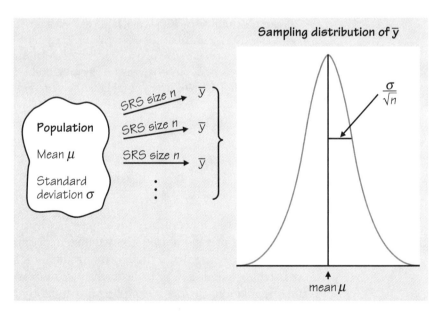

**FIGURE 16.7** The idea of a sampling distribution: choose many SRSs of size $n$ from the same population, calculate the mean $\bar{y}$ for each sample, and collect all of the $\bar{y}$'s. The distribution of the $\bar{y}$'s is the sampling distribution.

sample. Collect all the $\bar{y}$'s and display their distribution. That's the sampling distribution of $\bar{y}$. Keep this figure in mind as you go forward.

## EXERCISES

16.1   Coin tossing can illustrate the idea of a sampling distribution. The population is all outcomes (heads or tails) we would get if we tossed a coin forever. The parameter $p$ is the proportion of heads in this population. We suspect that $p$ is close to 0.5. That is, we think the coin will show about one-half heads in the long run. The sample is the outcomes of 20 tosses, and the statistic $\hat{p}$ is the proportion of heads in these 20 tosses.

   (a) Toss a coin 20 times and record the value of $\hat{p}$.

   (b) Repeat this sampling process 10 times. Make a histogram of the 10 values of $\hat{p}$. You are building up the sampling distribution of the statistic $\hat{p}$. Is the center of this distribution close to 0.5?

16.2   Let us illustrate the idea of a sampling distribution in the case of a very small sample from a very small population. The population is the scores of 20 students on an exam:

| Student | 1 | 2 | 3 | 4 | 5 | 6 | 7 | 8 | 9 | 10 |
|---------|-----|-----|-----|-----|-----|-----|-----|-----|-----|-----|
| Score | 60 | 82 | 78 | 74 | 61 | 59 | 69 | 70 | 63 | 61 |

| Student | 11 | 12 | 13 | 14 | 15 | 16 | 17 | 18 | 19 | 20 |
|---------|-----|-----|-----|-----|-----|-----|-----|-----|-----|-----|
| Score | 73 | 87 | 49 | 62 | 78 | 65 | 79 | 89 | 75 | 63 |

The parameter of interest is the mean score $\mu$ in this population. The sample is an SRS of size $n = 4$ drawn from the population.

   (a) Calculate the population mean $\mu$.

   (b) Draw an SRS of size 4 from this population. Write the four scores in your sample and calculate the mean $\bar{y}$ of the sample scores. This statistic is an estimate of $\mu$.

   (c) Repeat this process 10 times. Make a histogram of the 10 values of $\bar{y}$. You are constructing the sampling distribution of $\bar{y}$. Is the center of your histogram close to $\mu$?

   (d) Ten repetitions give a very crude approximation to the sampling distribution. Draw 90 more SRSs of size 4 (use software) to obtain 100 samples in all. Make a histogram of all the values of $\bar{y}$. Is the center close to $\mu$? Is the shape approximately normal? (Make a normal probability plot to assess normality.) This histogram is a better approximation to the sampling distribution.

16.3    Exercise 3.11 (page 26) gives the survival times of 72 guinea pigs in a medical experiment. The distribution of survival times is strongly skewed to the right. Sampling from this small population can demonstrate how averaging reduces variability and creates a more normal distribution.

(a) Use software to choose 100 SRSs of size 12 from this population. Find the mean survival time $\bar{y}$ for each of the 100 samples.

(b) Find the mean $\mu$ of the population. Make a histogram of the 100 $\bar{y}$'s and find their mean. Is the sampling distribution of $\bar{y}$ centered near $\mu$?

(c) Find the standard deviation $\sigma$ of the population. Then find the standard deviation of the 100 $\bar{y}$'s. In the long run, what do you expect to be the standard deviation of the mean from samples of size $n = 12$? How close to this value did your 100 samples come?

(d) Make normal probability plots of both the population and the 100 $\bar{y}$'s. Is the sampling distribution of the $\bar{y}$'s markedly closer to normal than the distribution of the population?

16.4    Table I.1 (page 64) gives 228 monthly percent returns on Wal-Mart stock. The distribution of returns is long-tailed. That is, the largest and smallest returns are both farther out than in a normal distribution. Sampling from this small population can demonstrate how averaging reduces variability and creates a more normal distribution.

(a) Choose 100 SRSs of size 6. Find the mean $\bar{y}$ of each sample. Make normal probability plots of the population of returns and of the 100 sample mean returns. Is the distribution of means closer to normal? In what way?

(b) Consider the distribution of the 100 sample means $\bar{y}$. What values does theory give us for the mean and standard deviation of $\bar{y}$ in samples of size $n = 6$? How close to the theoretical results are the mean and standard deviation of your 100 $\bar{y}$'s?

(c) Choose 100 SRSs of size 24 from the population of Wal-Mart returns. Find the 100 means $\bar{y}$ for these samples and repeat the analysis of (a) and (b) for these $\bar{y}$'s. We expect means of 24 observations to be closer to normal than means of 6 observations. Do your normal probability plots show this result?

16.5    Investors remember 1987 as the year stocks lost 20% of their value in a single day. For 1987 as a whole, the mean return of all common stocks on the New York Stock Exchange was $\mu = -3.5\%$. (That is, these stocks lost an average of 3.5% of their value in 1987.) The standard deviation of the returns was about $\sigma = 26\%$. The distribution of annual returns for stocks is roughly normal.

(a) What percent of stocks lost money? (That is the same as the probability that a stock chosen at random has a return less than 0.)

(b) Suppose that you held a portfolio of five stocks chosen at random from New York Stock Exchange stocks. What are the mean and standard deviation of the returns of randomly chosen portfolios of 5 stocks?

(c) What percent of such portfolios lost money? Explain the difference between this result and the result of (a).

16.6   The scores of students on the American College Testing (ACT) college entrance examination in a recent year had normal distribution with mean $\mu = 18.6$ and standard deviation $\sigma = 5.9$.

(a) What is the probability that a single student randomly chosen from all those taking the test scores 21 or higher?

(b) Now take an SRS of 50 students who took the test. What is the probability that the mean score $\bar{y}$ of these students is 21 or higher?

16.7   A laboratory weighs filters from a coal mine to measure the amount of dust in the mine atmosphere. Repeated measurements of the weight of dust on the same filter vary normally with standard deviation $\sigma = 0.08$ milligrams (mg) because the weighing is not perfectly precise. The dust on a particular filter actually weighs 123 mg. Repeated weighings will then have the normal distribution with mean 123 mg and standard deviation 0.08 mg.

(a) The laboratory reports the mean of 3 weighings. What is the distribution of this mean?

(b) What is the probability that the laboratory reports a weight of 124 mg or higher for this filter?

16.8   The number of flaws per square yard in a type of carpet material varies with mean 1.6 flaws per square yard and standard deviation 1.2 flaws per square yard. The population distribution cannot be normal, because a count takes only whole-number values. An inspector studies 200 square yards of the material, records the number of flaws found in each square yard, and calculates $\bar{y}$, the mean number of flaws per square yard inspected. Use the central limit theorem to find the approximate probability that the mean number of flaws exceeds 2 per square yard.

16.9   The weight of the eggs produced by a certain breed of hen is normally distributed with mean 65 grams (g) and standard deviation 5 g. Think of cartons of such eggs as SRSs of size 12 from the population of all eggs.

What is the probability that the weight of a carton falls between 750 g and 825 g? (Hint: Restate the problem in terms of the mean weight of the eggs in the carton.)

16.10    A company that owns and services a fleet of cars for its sales force has found that the service lifetime of disc brake pads varies from car to car according to a normal distribution with mean $\mu = 55,000$ miles and standard deviation $\sigma = 4500$ miles. The company installs a new brand of brake pads on 8 cars.

(a) If the new brand has the same lifetime distribution as the previous type, what is the distribution of the sample mean lifetime for the 8 cars?

(b) The average life of the pads on these 8 cars turns out to be $\bar{y} = 51,800$ miles. What is the probability that the sample mean lifetime is 51,800 miles or less if the lifetime distribution is unchanged? The company takes this probability as evidence that the average lifetime of the new brand of pads is less than 55,000 miles.

16.11    The level of nitrogen oxide (NOX) in the exhaust of a particular car model varies with mean 0.9 grams per mile (g/mi) and standard deviation 0.15 g/mi. A company has 125 cars of this model in its fleet. If $\bar{y}$ is the mean NOX emission level for these cars, what is the distribution of $\bar{y}$? What is the level $L$ such that the probability that $\bar{y}$ is greater than $L$ is only 0.01?

# PART IV REVIEW

Part IV concerns ideas used to study random behavior. Probability, the mathematics that describes randomness, is important in many areas of study. Here, we concentrate on informal probability as the conceptual foundation for statistical inference. Because random samples and randomized comparative experiments use chance, their results vary according to the laws of probability. Here is a review list of the most important skills you should have acquired from your study of Lessons 12 to 16.

## A. PROBABILITY

1. Recognize that some phenomena are random. Understand that the probability of an event is the proportion of times the event occurs out of very many repetitions of a random phenomenon. Use the idea of probability as long-run proportion to think about probability.

2. Use basic probability facts to detect illegitimate assignments of probability: any probability must be a number between 0 and 1, and the total probability assigned to all possible outcomes must be 1.

3. Use basic probability facts to find the probabilities of events that are formed from other events: The probability that an event does not

occur is 1 minus its probability. If two events are disjoint, the probability that one or the other occurs is the sum of their individual probabilities. If two events are independent, the probability that they both occur is the product of their probabilities.

4. Find probabilities of events in a finite sample space by adding the probabilities of their outcomes. Find probabilities of events as areas under a density curve.

5. Optional: Understand the idea of conditional probability. Find conditional probabilities for individuals chosen at random from a table of counts of possible outcomes.

6. Optional: Use the general multiplication rule to find $P(A \text{ and } B)$ from $P(A)$ and the conditional probability $P(B \mid A)$.

7. Optional: Use the multiplication rule for any number of independent events in combination with other probability rules to find the probabilities of complex events.

## B. RANDOM VARIABLES

1. Use the notation of random variables to make compact statements about random outcomes, such as $P(\bar{y} \leq 4) = 0.3$. Be able to read such statements.

2. Find probabilities of events involving the values of discrete and continuous random variables.

3. Understand how the law of large numbers describes the long-run behavior of averages of many outcomes. Understand that the average observed outcome is guaranteed to be close to the population mean *only* in the long run.

## C. SAMPLING DISTRIBUTIONS

1. Identify parameters and statistics in a sample or experiment.

2. Recognize the fact of sampling variability: a statistic will take different values when you repeat a sample or experiment.

3. Interpret a sampling distribution as describing the values taken by a statistic in all possible repetitions of a sample or experiment under the same conditions.

4. Interpret the sampling distribution of a statistic as describing the probabilities of its possible values.

5. Understand that the variability of a statistic is controlled by the size of the sample. Statistics from larger samples are less variable.

### D. SAMPLE MEANS

1. Recognize when a problem involves the mean $\bar{y}$ of a sample.
2. Find the mean and standard deviation of a sample mean $\bar{y}$ from an SRS of size $n$ when the mean $\mu$ and standard deviation $\sigma$ of the population are known.
3. Know that the standard deviation (spread) of the sampling distribution of $\bar{y}$ gets smaller at the rate $\sqrt{n}$ as the sample size $n$ gets larger.
4. Understand that $\bar{y}$ has approximately a normal distribution when the sample is large (central limit theorem). Use this normal distribution to calculate probabilities that concern $\bar{y}$.

## REVIEW EXERCISES

**Review 1**    A couple plans to have three children. What is the sample space $S$ for each of the following random phenomena?

**(a)** Record the sex (M or F) of each child in order of birth.

**(b)** Record the number of girls.

**Review 2**    Choose a student at random and record the number of dollars in bills (ignore change) that he or she is carrying. Give a reasonable sample space $S$ for this random phenomenon. (We don't know the largest amount that a student could reasonably carry, so you will have to make a choice in stating the sample space.)

**Review 3**    Which of the following are legitimate probability models for tossing three (possibly unfair) coins? Explain your answer in each case.

| Outcome | Model A | Model B | Model C | Model D |
|---------|---------|---------|---------|---------|
| H, H, H | 0.125 | 0 | 0.125 | 0.250 |
| H, H, T | 0.125 | 0.375 | 0.250 | 0.125 |
| H, T, H | 0.125 | 0 | −0.125 | 0.250 |
| H, T, T | 0.125 | 0.125 | 0.125 | 0.125 |
| T, H, H | 0.125 | 0 | 0.125 | 0.250 |
| T, H, T | 0.125 | 0.375 | 0.125 | 0.125 |
| T, T, H | 0.125 | 0 | 0.250 | 0.250 |
| T, T, T | 0.125 | 0.125 | −0.125 | 0.125 |

**Review 4**    Select a first-year college student at random and ask what his or her academic rank was in high school. Here are the probabilities, based on a large sample survey of students:

| Outcome | Top 20% | Second 20% | Third 20% | Fourth 20% | Lowest 20% |
|---|---|---|---|---|---|
| Probability | 0.41 | 0.23 | 0.29 | 0.06 | 0.01 |

(a) What is the sum of these probabilities? Why do you expect the sum to have this value?

(b) What is the probability that a randomly chosen first-year college student was not in the top 20% of his or her high school class?

(c) What is the probability that a first-year student was in the top 40% in high school?

(d) Now choose two first-year college students at random. What is the probability that both were in the top 20% of their high school classes?

**Review 5**   Choose an acre of land in Canada at random. The probability is 0.35 that it is forest and 0.03 that it is pasture.

(a) What is the probability that the acre chosen is not forested?

(b) What is the probability that it is either forest or pasture?

(c) What is the probability that a randomly chosen acre in Canada is something other than forest or pasture?

**Review 6**   Las Vegas Zeke, when asked to predict the Atlantic Coast Conference basketball champion, follows the modern practice of giving probabilistic predictions. He says, "North Carolina's probability of winning is twice Duke's. North Carolina State and Virginia each have probability 0.1 of winning, but Duke's probability is three times that. Nobody else has a chance." Has Zeke given a legitimate assignment of probabilities to the eight teams in the conference? Explain your answer.

**Review 7**   A sociologist studying social mobility in Denmark finds that the probability that the son of a lower-class father remains in the lower class is 0.46. What is the probability that the son moves to one of the higher classes?

**Review 8**   Suppose that both parents carry genes for blood types A and B. Each parent passes one of these genes to a child and is equally likely to pass either gene. The two parents pass genes independently. The child will have blood type A if both parents pass their A genes, type B if both pass their B genes, and type AB if one A and one B gene are passed. What are the probabilities that a child of these parents has type A blood? Type B? Type AB?

**Review 9**   A student makes a measurement in a chemistry laboratory and records the result in her lab report. The standard deviation of individual measurements using this instrument is $\sigma = 10$ milligrams.

(a) Suppose the student repeats the measurement 3 times and records the mean $\bar{y}$ of her 3 measurements. What is the standard deviation $\sigma_{\bar{y}}$ of the mean result?

(b) How many times must the student repeat the measurement to reduce the standard deviation of $\bar{y}$ to 5?

(c) Explain to someone who knows no statistics the advantage of reporting the average of several measurements rather than the result of a single measurement.

**Review 10**   A random sample of female college students has a mean height of **64.5** inches, which is greater than the **63**-inch mean height of all adult American women. Is each of the bold numbers a parameter or a statistic?

**Review 11**   A sample of students of high academic ability under 13 years of age was given the SAT mathematics examination, which is usually taken by high school seniors. The mean score for the females in the sample was **386**, whereas the mean score of the males was **416**. Is each of the bold numbers a parameter or a statistic?

**Review 12**   The distribution of annual returns on common stocks is roughly symmetric, but extreme observations are more frequent than in a normal distribution. Returns in successive years appear to be independent. Because the distribution is not strongly nonnormal, the mean return over even a moderate number of years is close to normal. In the long run, annual real returns on common stocks have varied with mean about 9% and standard deviation about 28%. Andrew plans to retire in 40 years and is considering investing in stocks. What is the probability (assuming that the past pattern of variation continues) that the mean annual return on common stocks over the next 40 years will exceed 15%? What is the probability that the mean return will be less than 5%?

**Review 13**   The Wechsler Adult Intelligence Scale (WAIS) is a common "IQ test" for adults. The distribution of WAIS scores for persons over 16 years of age is approximately normal with mean 100 and standard deviation 15.

(a) What is the probability that a randomly chosen individual has a WAIS score of 105 or higher?

(b) What are the mean and standard deviation of the average WAIS score $\bar{y}$ for an SRS of 60 people?

(c) What is the probability that the average WAIS score of an SRS of 60 people is 105 or higher?

(d) Would your answers to any of (a), (b), or (c) be affected if the distribution of WAIS scores in the adult population were distinctly nonnormal?

**Review 14**    A study of rush-hour traffic in San Francisco counts the number of people in each car entering a freeway at a suburban interchange. Suppose that this count has mean 1.5 and standard deviation 0.75 in the population of all cars that enter at this interchange during rush hours.

(a) Could the exact distribution of the count be normal? Why or why not?

(b) Traffic engineers estimate that the capacity of the interchange is 700 cars per hour. According to the central limit theorem, what is the approximate distribution of the mean number of persons $\bar{y}$ in 700 randomly selected cars at this interchange?

(c) What is the probability that 700 cars will carry more than 1075 people? (Hint: Restate this event in terms of the mean number of people $\bar{y}$ per car.)

**Review 15**    A maker of auto air conditioners checks a sample of 4 thermostats from each hour's production. The thermostats are set at 75° F and then placed in a chamber where the temperature rises gradually. The tester records the temperature at which the thermostat turns on the air conditioner. The mean is supposed to be $\mu = 75°$. Past experience indicates that the response temperature of properly adjusted thermostats varies with $\sigma = 0.5°$. Workers plot the mean response temperature $\bar{y}$ for each hour's sample to monitor production. If the thermostats are properly adjusted, what is the sampling distribution of $\bar{y}$? What is the probability that the observed mean is more than 0.5° away from 75°?

**Review 16**    Judy's doctor is concerned that she may suffer from hypokalemia (low potassium in the blood). There is variation both in the actual potassium level and in the blood test that measures the level. Judy's measured potassium level varies according to the normal distribution with $\mu = 3.8$ and $\sigma = 0.2$. A patient is classified as hypokalemic if the potassium level is below 3.5.

(a) If a single potassium measurement is made, what is the probability that Judy is diagnosed as hypokalemic?

(b) If measurements are made instead on 4 separate days and the mean result is compared with the criterion 3.5, what is the probability that Judy is diagnosed as hypokalemic?

**Review 17**    Children in kindergarten are sometimes given the Ravin Progressive Matrices Test (RPMT) to assess their readiness for learning.

Experience at Southwark Elementary School suggests that the RPMT scores for its kindergarten pupils have mean 13.6 and standard deviation 3.1. The distribution is close to normal. Mr. Lavin has 22 children in his kindergarten class this year. He suspects that their RPMT scores will be unusually low because the test was interrupted by a fire drill. To check this suspicion, he wants to find the level $L$ such that there is probability only 0.05 that the mean score of 22 children falls below $L$ when the usual Southwark distribution remains true. What is the value of $L$?

**Review 18 (Optional)** Choose an employed person at random. Let $A$ be the event that the person chosen is a woman, and $B$ the event that the person holds a managerial or professional job. As of 1990, $P(A) = 0.45$ and the probability of managerial and professional jobs among women is $P(B \mid A) = 0.26$. Find the probability that a randomly chosen employed person is a woman holding a managerial or professional position.

**Review 19 (Optional)** Functional Robotics Corporation buys electrical controllers from a Japanese supplier. The company's treasurer feels that there is probability 0.4 that the dollar will fall in value against the Japanese yen in the next month. The treasurer also believes that *if* the dollar falls there is probability 0.8 that the supplier will demand renegotiation of the contract. What probability has the treasurer assigned to the event that the dollar falls and the supplier demands renegotiation?

**Review 20 (Optional)** A multiple-choice quiz offers five answers for each of ten questions. Sam just guesses at all the answers. He has probability 1/5 of getting any one answer correct, and his guesses are independent. What is the probability that Sam will get all ten questions right? What is the probability that he will get all ten wrong?

**Review 21 (Optional)** You are about to visit a new neighbor. You know that the family has four children, but you do not know their age or sex. Write down all possible arrangements of girls and boys in order from youngest to oldest, such as BBGG (the two youngest are boys, the two oldest girls). The laws of genetics say that all of these arrangements are equally likely.

**(a)** What is the probability that the oldest child is a girl?

**(b)** What is the probability that the family has at least three boys?

**(c)** What is the probability that the family has at least three children of the same sex?

## JERZY NEYMAN

The most-used methods of statistical inference are confidence intervals and tests of significance. Both are products of the twentieth century. From complex and sometimes confusing origins, statistical tests took their current form in the writings of R. A. Fisher, whom we met at the beginning of Part III. Confidence intervals appeared in 1934, the brainchild of Jerzy Neyman (1894–1981).

Neyman was trained in Poland and, like Fisher, worked at an agricultural research institute. He moved to London in 1934 and in 1938 joined the University of California at Berkeley. He founded Berkeley's Statistical Laboratory and remained its head even after his official retirement as a professor in 1961. Retirement did not slow Neyman's work—he remained active until the end of his long life and almost doubled his list of publications after "retiring." Statistical problems arising from astronomy, biology, and attempts to modify the weather attracted his attention.

Neyman ranks with Fisher as a founder of modern statistical practice. In addition to introducing confidence intervals, he helped systematize the theory of sample surveys and reworked significance tests from a new point of view. Fisher, who was very argumentative, disliked Neyman's approach to tests and said so. Neyman, who wasn't shy, replied vigorously.

Tests and confidence intervals are our topic in Part V. Like most users of statistics, we will stay close to Fisher's approach to tests.

PART **V**

# Statistical Inference

# 17

# ESTIMATING WITH CONFIDENCE

To infer means to draw a conclusion. Statistical inference provides us with methods for drawing conclusions from data. We have, of course, been drawing conclusions from data all along. What is new in formal inference is that we use probability to express our confidence in our conclusions. Probability allows us to take chance variation into account and so to correct our judgment by calculation.

The methods of formal inference are based on the long-run regular behavior that probability describes. Inference is most reliable when the data are produced by a properly randomized design. *When you use statistical inference you are acting as if the data are a random sample or come from a randomized experiment*. If this is not true, your conclusions may be open to challenge.

The purpose of this lesson is to describe the reasoning used in one major type of statistical inference—*confidence intervals* that estimate the unknown value of a population parameter. The specific technique we use to present the reasoning is oversimplified, so that it is not very useful in practice. Lesson 18 shows how to modify the technique for use in the real world. There are libraries—both of books and of computer software—full of confidence intervals for use in various settings. All rely on the same reasoning. This lesson is therefore a foundation for much that follows.

## ESTIMATING A POPULATION MEAN $\mu$

Young people have a better chance of good jobs if they are good with numbers. How strong are the quantitative skills of young Americans of working age? One source of data is the National Assessment of Educational Progress (NAEP) Young Adult Literacy Assessment Survey, which is based on a nationwide random sample of households.

### EXAMPLE 17.1

The NAEP survey includes a short test of quantitative skills, covering mainly basic arithmetic and the ability to apply it to realistic problems. Scores on the test range from 0 to 500. For example, a person who scores 233 can add the amounts of two checks appearing on a bank deposit slip; someone scoring 325 can determine the price of a meal from a menu; a person scoring 375 can transform a price in cents per ounce into dollars per pound.

In a recent year, 840 men 21 to 25 years of age were in the NAEP sample. Their mean quantitative score was $\bar{y} = 272$. These 840 men are an SRS from the population of all young men. On the basis of this sample, what can we say about the mean score $\mu$ in the population of all 9.5 million young men of these ages?[42]    ◄

Because the mean score from our SRS was $\bar{y} = 272$, we guess that $\mu$ is "somewhere around 272." To make "somewhere around 272" more precise, we ask: *How would the sample mean $\bar{y}$ vary if we took many samples of 840 young men from this same population?* Recall the essential facts about the sampling distribution of $\bar{y}$:

- $\bar{y}$ has a normal distribution. (The central limit theorem tells us that the average of 840 scores has a distribution that is very close to normal.)
- The mean of this normal sampling distribution is the same as the unknown population mean $\mu$. That is, $\bar{y}$ as an estimator of $\mu$ has no bias.

- The standard deviation of $\bar{y}$ for an SRS of 840 men is $\sigma/\sqrt{840}$, where $\sigma$ is the standard deviation of individual NAEP scores among all young men.

Let us suppose that we know from long experience that the standard deviation of scores in the population of all young men is $\sigma = 60$. The standard deviation of $\bar{y}$ is then

$$\frac{\sigma}{\sqrt{n}} = \frac{60}{\sqrt{840}} \doteq 2.1$$

(It is not realistic to assume we know $\sigma$. We will see in the next lesson how to proceed when $\sigma$ is not known. For now, we are more interested in statistical reasoning than in details of realistic methods.)

If we choose many samples of size 840 and find the mean NAEP score for each sample, we get mean $\bar{y}_1$ from the first sample, $\bar{y}_2$ from the second, then $\bar{y}_3, \bar{y}_4$, and so on. If we collect all these sample means and display their distribution, we get the normal distribution with mean equal to the unknown $\mu$ and standard deviation 2.1. Inference about the unknown $\mu$ starts from this sampling distribution. Figure 17.1 displays the distribution. The different values of $\bar{y}$ appear along the axis in the figure, and the normal curve shows how probable these values are.

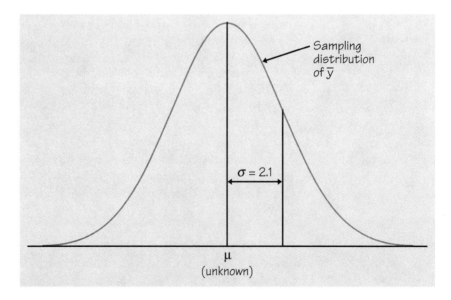

FIGURE 17.1    The sampling distribution of the mean score $\bar{y}$ of an SRS of 840 young men on the NAEP quantitative test.

## STATISTICAL CONFIDENCE

Figure 17.2 is another picture of the same sampling distribution. It illustrates the following line of thought:

- The 68–95–99.7 rule says that about 95% of the means $\bar{y}_1, \bar{y}_2, \bar{y}_3, \ldots$ will be within two standard deviations of the population mean score $\mu$. That is, 95% of the sample means $\bar{y}_i$ will be within 4.2 points of $\mu$.
- Whenever a sample has mean $\bar{y}_i$ within 4.2 points of the unknown $\mu$, then $\mu$ is within 4.2 points of the observed $\bar{y}_i$. This happens in 95% of all samples.
- So in 95% of all samples the unknown $\mu$ lies between $\bar{y} - 4.2$ and $\bar{y} + 4.2$, where $\bar{y}$ is the mean for that sample. Figure 17.3 displays this fact in picture form.

This conclusion just restates a fact about the sampling distribution of $\bar{y}$. The language of statistical inference uses this fact about what would happen in the long run to express our confidence in the results of any one sample.

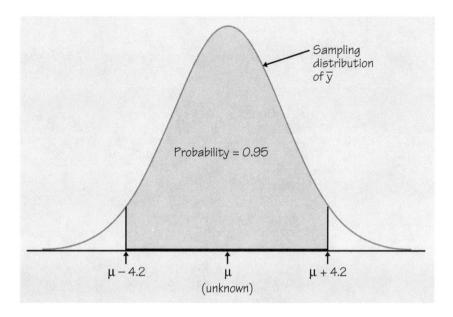

FIGURE 17.2   In 95% of all samples, $\bar{y}$ lies within $\pm 4.2$ of the unknown population mean $\mu$. So $\mu$ also lies within $\pm 4.2$ of $\bar{y}$ in those samples.

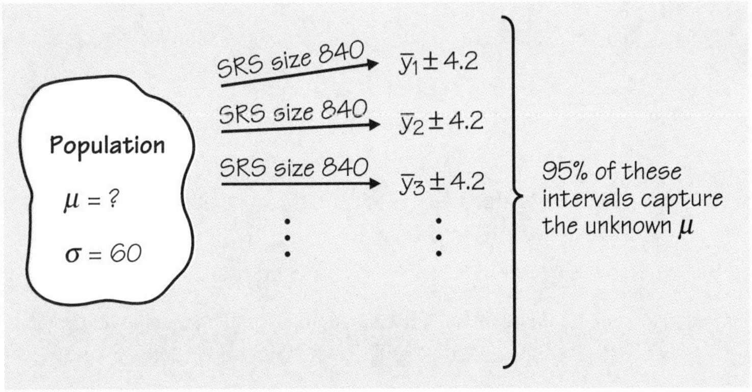

FIGURE 17.3   The meaning of "95% confidence." If we choose very many samples, 95% of the intervals $\bar{y} \pm 4.2$ will capture the population mean $\mu$.

---

**EXAMPLE 17.2**

Our sample of 840 young men gave $\bar{y} = 272$. We say that we are *95% confident* that the unknown mean NAEP quantitative score for all young men lies between

$$\bar{y} - 4.2 = 272 - 4.2 = 267.8$$

and

$$\bar{y} + 4.2 = 272 + 4.2 = 276.2$$

◀

Be sure you understand the grounds for our confidence. There are only two possibilities:

1. The interval between 267.8 and 276.2 contains the true $\mu$.

2. Our SRS was one of the few samples for which $\bar{y}$ is not within 4.2 points of the true $\mu$. Only 5% of all samples give such inaccurate results.

We cannot know whether our sample is one of the 95% for which the interval $\bar{y} \pm 4.2$ catches $\mu$, or one of the unlucky 5%. The statement that we are 95% confident that the unknown $\mu$ lies between 267.8 and 276.2 is shorthand for saying, "We got these numbers by a method that gives correct results 95% of the time."

## CONFIDENCE INTERVALS FOR THE MEAN $\mu$

The reasoning we have used is quite general. We start with the sampling distribution of the sample mean $\bar{y}$. If we standardize $\bar{y}$, we get the **one-sample $z$ statistic**

$$z = \frac{\bar{y} - \mu}{\sigma / \sqrt{n}}$$

The statistic $z$ tells us how far the observed $\bar{y}$ is from $\mu$, in units of the standard deviation of $\bar{y}$. Because $\bar{y}$ has a normal distribution, $z$ has the standard normal distribution $N(0, 1)$. This distribution is in the background of our reasoning.

Call $z^*$ the point on the standard normal distribution that catches the middle area $C$ of the total area 1 under the density curve. Figure 17.4 shows how $z^*$ and $C$ are related. If we start at the population mean $\mu$ and go out $z^*$ standard deviations of $\bar{y}$, we get an interval in which the proportion $C$ of all sample means $\bar{y}$ will fall if we take many samples. So if we start at the

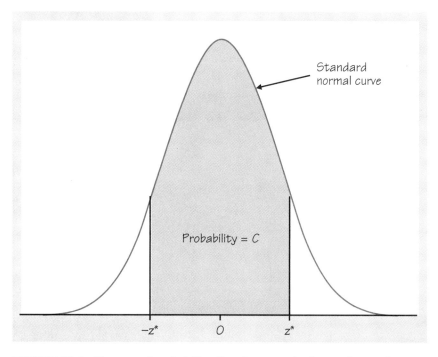

FIGURE 17.4   The central probability $C$ under a standard normal curve lies between $-z^*$ and $z^*$.

sample mean $\bar{y}$ and go out $z^*$ standard deviations, we get the interval

$$\text{from} \quad \bar{y} - z^* \frac{\sigma}{\sqrt{n}} \quad \text{to} \quad \bar{y} + z^* \frac{\sigma}{\sqrt{n}}$$

or

$$\bar{y} \pm z^* \frac{\sigma}{\sqrt{n}}$$

calculated from the sample data that captures the true mean $\mu$ in the proportion $C$ of all samples. This is a *confidence interval* for $\mu$.

---

CONFIDENCE INTERVAL

A **level $C$ confidence interval** for a parameter has two parts:

- An interval calculated from the data, usually of the form

  estimate $\pm$ margin of error

- A **confidence level** $C$, which gives the probability that the interval will capture the true parameter value in repeated samples.

---

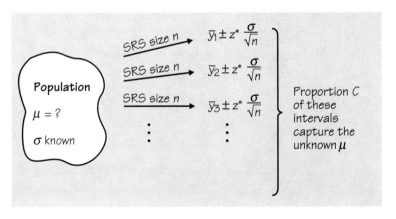

FIGURE 17.5   The meaning of a level $C$ confidence interval for a population mean $\mu$.

Figure 17.5 illustrates the meaning of confidence level $C$. Keep this picture in mind as you work with confidence intervals.

## HOW CONFIDENCE INTERVALS BEHAVE

Statistics software will calculate confidence intervals for you. Understanding what confidence means and how confidence intervals behave is therefore more important than the recipe we obtained for a specific confidence interval. The interval $\bar{y} \pm z^*\sigma/\sqrt{n}$ for the mean of a normal population helps us see the behavior of confidence intervals in general.

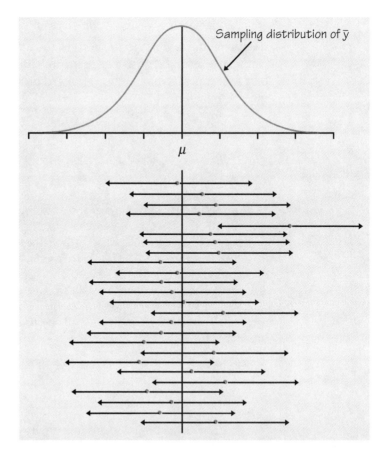

FIGURE 17.6    Twenty-five samples from the same population gave these 95% confidence intervals. In the long run, 95% of all samples give an interval that contains the population mean $\mu$.

- Confidence 95% means "We got this result using a method that captures the true parameter 95% of the time." Figure 17.6 displays the behavior of 95% confidence intervals in repeated sampling. The center of each interval is at $\bar{y}$ and therefore varies from sample to sample. The sampling distribution of $\bar{y}$ appears at the top of the figure to show the long-term pattern of this variation. The 95% confidence intervals from 25 SRSs appear below. The center $\bar{y}$ of each interval is marked by a dot. The arrows on either side of the dot span the confidence interval. All except one of these 25 intervals cover the true value of $\mu$. In a very large number of samples, 95% of the confidence intervals would contain $\mu$.

- There is a trade-off between the precision with which we estimate the unknown parameter and the confidence we have in the result. Higher confidence (larger $C$) requires a wider interval. Figure 17.4 shows that larger $C$ demands a larger $z^*$, which results in a larger margin of error.

- Increasing the sample size $n$ shortens the interval for any fixed confidence level. Because $n$ appears under a square root sign, we must take four times as many observations in order to cut the length of the interval in half.

## EXAMPLE 17.3

A pharmaceutical manufacturer analyzes a specimen from each batch of a product to verify the concentration of the active ingredient. The chemical analysis is not perfectly precise. Repeated measurements on the same specimen give slightly different results. The results of repeated measurements follow a normal distribution quite closely. The analysis procedure has no bias, so the mean $\mu$ of the population of all measurements is the true concentration in the specimen. The standard deviation of this distribution is known to be $\sigma = 0.0068$ grams per liter. The laboratory analyzes each specimen three times and reports the mean result.

Three analyses of one specimen give concentrations

$$0.8403 \quad 0.8363 \quad 0.8447$$

We want a 99% confidence interval for the true concentration $\mu$. The sample mean of these three readings is $\bar{y} = 0.8404$. The standard deviation of $\bar{y}$ is

$$\frac{\sigma}{\sqrt{n}} = \frac{0.0068}{\sqrt{3}} = 0.00393$$

For 99% confidence, we must go out $z^* = 2.576$ standard deviations (see Exercise 17.1). A 99% confidence interval for $\mu$ is therefore

$$\bar{y} \pm z^* \frac{\sigma}{\sqrt{n}} = 0.8404 \pm (2.576)(0.0039)$$

$$= 0.8404 \pm 0.0101$$

That is, we are 99% confident that

$$0.8303 < \mu < 0.8505$$

If a single measurement had given the same value, $y = 0.8404$, repeating the calculation with $n = 1$ gives the 99% confidence interval $0.8404 \pm 0.0175$. The mean of three measurements gives a smaller margin of error and therefore a shorter interval than a single measurement.

If the manufacturer is content with 90% confidence rather than 99%, we need to go out only $z^* = 1.645$ standard deviations. The three measurements would then give the 90% confidence interval $0.8404 \pm 0.0065$. Settling for only 90% confidence reduces the margin of error needed to account for the variation in $\bar{y}$ in repeated samples. ◀

## SOME CAUTIONS

Any formula for inference is correct only in specific circumstances. If statistical procedures carried warning labels like those on drugs, most inference methods would have long labels indeed. Our handy formula $\bar{y} \pm z^* \sigma/\sqrt{n}$ for estimating a normal mean comes with the following list of warnings for the user.

- The data must be an SRS from the population. We are completely safe if we actually carried out the random selection of an SRS. We are not in great danger if the data can plausibly be thought of as observations taken at random from a population. That is the case in Example 17.3, where we have in mind the population resulting from a very large number of repeated analyses of the same specimen.
- There is no correct method for inference from data haphazardly collected with bias of unknown size. Fancy formulas cannot rescue badly produced data.
- Our confidence interval requires that the sample mean $\bar{y}$ have a normal distribution. Fortunately, $\bar{y}$ is much closer to normal than the

individual observations. For samples of moderate size (say, $n \geq 15$) it is enough that the distribution of the data be roughly unimodal and symmetric.

- Because $\bar{y}$ is strongly influenced by a few extreme observations, outliers can have a large effect on the confidence interval. You should search for outliers and try to correct them or justify their removal before computing the interval. If the outliers cannot be removed, ask your statistical consultant about procedures that are not sensitive to outliers.

- You must know the standard deviation $\sigma$ of the population. This unrealistic requirement renders the interval $\bar{y} \pm z^* \sigma / \sqrt{n}$ of little use in statistical practice. We will learn in the next lesson what to do when $\sigma$ is unknown. However, if the sample is large, the sample standard deviation $s$ will be close to the unknown $\sigma$. Then $\bar{y} \pm z^* s / \sqrt{n}$ is an approximate confidence interval for $\mu$.

The most important caution concerning confidence intervals is a consequence of the first of these warnings. *The margin of error in a confidence interval covers only the effects of chance variation in repeated random samples.* Practical difficulties such as undercoverage and nonresponse in a sample survey can cause additional errors that may be larger than the random sampling error. Remember this unpleasant fact when reading the results of an opinion poll or other sample survey. The practical conduct of the survey influences the trustworthiness of its results in ways that are not included in the announced margin of error.

Every inference procedure that we will meet has its own list of warnings. Because many of the warnings are similar to those above, we will not print the full warning label each time. It is easy to state (from the mathematics of probability) conditions under which a method of inference is exactly correct. These conditions are *never* fully met in practice. For example, no population is exactly normal. Deciding when a statistical procedure should be used in practice often requires judgment assisted by exploratory analysis of the data.

## EXERCISES

17.1    Use software to verify these specific cases of the relationship between $C$ and $z^*$ shown in Figure 17.4:

(a) If the probability $C = 0.90$, then $z^* = 1.645$.

(b) If the probability $C = 0.95$, then $z^* = 1.960$.

(c) If the probability $C = 0.99$, then $z^* = 2.576$.

The bottom row of Table B (inside back cover of this book) contains the values of $z^*$ for various confidence levels $C$. You can see there, as well as from your work in this exercise, that to capture more of the probability under a normal density curve, we must go out farther from the mean.

17.2    Use software to verify the three confidence intervals given in Example 17.3. The population standard deviation is $\sigma = 0.0068$ and the three observations are 0.8403, 0.8363, and 0.8447. Find confidence intervals for the mean $\mu$ in these cases:

(a) Sample size $n = 3$ and 99% confidence.

(b) Sample size $n = 1$ and 99% confidence.

(c) Sample size $n = 3$ and 90% confidence.

17.3    Find the confidence interval for 99% confidence in Example 17.3 if the laboratory measures the concentration of each specimen 12 times. Check that your interval is half as long as the interval based on 3 measurements in Example 17.3.

17.4    A *New York Times* poll on women's issues interviewed 1025 women randomly selected from the United States, excluding Alaska and Hawaii. The poll found that 47% of the women said they do not get enough time for themselves.

(a) The poll announced a margin of error of $\pm 3$ percentage points for 95% confidence in its conclusions. What is the 95% confidence interval for the percent of all adult women who think they do not get enough time for themselves?

(b) Explain to someone who knows no statistics why we can't just say that 47% of all adult women do not get enough time for themselves.

(c) Then explain clearly what "95% confidence" means.

17.5    A student reads that a 95% confidence interval for the mean NAEP quantitative score for men of ages 21 to 25 is 267.8 to 276.2. Asked to explain the meaning of this interval, the student says, "95% of all young men have scores between 267.8 and 276.2." Is the student right? Justify your answer.

17.6    Suppose that you give the NAEP test to an SRS of 1000 people from a large population in which the scores have mean $\mu = 280$ and standard deviation $\sigma = 60$. The mean $\bar{y}$ of the 1000 scores will vary if you take repeated samples.

(a) The sampling distribution of $\bar{y}$ is approximately normal. What are its mean and standard deviation?

(b) Sketch the normal curve that describes how $\bar{y}$ varies in many samples from this population. Mark its mean and the values one, two, and three standard deviations on either side of the mean.

(c) According to the 68–95–99.7 rule, about 95% of all the values of $\bar{y}$ fall within _____ of the mean of this curve. What is the missing number? Call it $m$ for "margin of error." Shade the region from the mean minus $m$ to the mean plus $m$ on the axis of your sketch, as in Figure 17.2.

(d) Whenever $\bar{y}$ falls in the region you shaded, the true value of the population mean, $\mu = 280$, lies in the confidence interval between $\bar{y} - m$ and $\bar{y} + m$. Draw the confidence interval below your sketch for one value of $\bar{y}$ inside the shaded region and one value of $\bar{y}$ outside the shaded region. (Use Figure 17.6 as a model for the drawing.)

(e) In what percent of all samples will the true mean $\mu = 280$ be covered by the confidence interval $\bar{y} \pm m$?

17.7    Oxides of nitrogen (called NOX for short) emitted by cars and trucks are important contributors to air pollution. The amount of NOX emitted by a particular model varies from vehicle to vehicle. For one light truck model, NOX emissions vary with mean $\mu$ that is unknown and standard deviation $\sigma = 0.4$ grams per mile. You test an SRS of 50 of these trucks. The sample mean NOX level $\bar{y}$ estimates the unknown $\mu$. You will get different values of $\bar{y}$ if you repeat your sampling.

(a) The sampling distribution of $\bar{y}$ is approximately normal. What are its mean and standard deviation?

(b) Sketch the normal curve for the sampling distribution of $\bar{y}$. Mark its mean and the values one, two, and three standard deviations on either side of the mean.

(c) According to the 68–95–99.7 rule, about 95% of all values of $\bar{y}$ lie within a distance $m$ of the mean of the sampling distribution. What is $m$? Shade the region on the axis of your sketch that is within $m$ of the mean, as in Figure 17.2.

(d) Whenever $\bar{y}$ falls in the region you shaded, the unknown population mean $\mu$ lies in the confidence interval $\bar{y} \pm m$. For what percent of all possible samples does this happen?

(e) Following the style of Figure 17.6, draw the confidence intervals below your sketch for two values of $\bar{y}$, one that falls within the shaded region and one that falls outside it.

17.8    A study of the career paths of hotel general managers sent questionnaires to an SRS of 160 hotels belonging to major U.S. hotel chains. There were 114 responses. The average time these 114 general managers had spent with their current company was 11.78 years. Give a 99% confidence interval for the mean number of years general managers of major-chain hotels have spent with their current company. (Take it as known that the standard deviation of time with the company for all general managers is 3.2 years.)

17.9　Here are measurements (in millimeters) of a critical dimension on a sample of auto engine crankshafts:

224.120　224.001　224.017　223.982　223.989　223.961
223.960　224.089　223.987　223.976　223.902　223.980
224.098　224.057　223.913　223.999

The data come from a production process that is known to have standard deviation $\sigma = 0.060$ mm. The process mean is supposed to be $\mu = 224$ mm but can drift away from this target during production.

(a) We expect the distribution of the dimension to be close to normal. Make a histogram and a normal probability plot of these data and describe the shape of the distribution.

(b) Give a 95% confidence interval for the process mean at the time these crankshafts were produced.

17.10　A test for the level of potassium in the blood is not perfectly precise. Moreover, the actual level of potassium in a person's blood varies slightly from day to day. Suppose that repeated measurements for the same person on different days vary normally with $\sigma = 0.2$.

(a) Julie's potassium level is measured once. The result is $y = 3.2$. Give a 90% confidence interval for her mean potassium level.

(b) If three measurements were taken on different days and the mean result is $\bar{y} = 3.2$, what is a 90% confidence interval for Julie's mean blood potassium level?

17.11　The National Assessment of Educational Progress (NAEP) test was also given to a sample of 1077 women of ages 21 to 25 years. Their mean quantitative score was 275. Take it as known that the standard deviation of all individual scores is $\sigma = 60$.

(a) Give a 95% confidence interval for the mean score $\mu$ in the population of all young women.

(b) Give the 90% and 99% confidence intervals for $\mu$.

(c) What are the lengths of the three intervals in (a) and (b)? How does increasing the confidence level affect the length of a confidence interval?

17.12　The NAEP sample of 1077 young women had mean quantitative score $\bar{y} = 275$. Take it as known that the standard deviation of all individual scores is $\sigma = 60$.

(a) Give a 95% confidence interval for the mean score $\mu$ in the population of all young women.

**(b)** Suppose that the same result, $\bar{y} = 275$, had come from a sample of 250 women. Give the 95% confidence interval for the population mean $\mu$ in this case.

**(c)** Then suppose that a sample of 4000 women had produced the sample mean $\bar{y} = 275$, and again give the 95% confidence interval for $\mu$.

**(d)** What are the lengths of the three intervals in (a), (b), and (c)? How does increasing the sample size affect the length of a confidence interval?

17.13    A radio talk show invites listeners to enter a dispute about a proposed pay increase for city council members. "What yearly pay do you think council members should get? Call us with your number." In all, 958 people call. The mean pay they suggest is $\bar{y} = \$8740$ per year, and the standard deviation of the responses is $s = \$1125$. For a large sample such as this, $s$ is very close to the unknown population $\sigma$. The station calculates the 95% confidence interval for the mean pay $\mu$ that all citizens would propose for council members to be $8669 to $8811.

**(a)** Is the station's calculation correct?

**(b)** Does their conclusion describe the population of all the city's citizens? Explain your answer.

17.14    The *New York Times*/CBS News Poll recently asked the question, "Do you favor an amendment to the Constitution that would permit organized prayer in public schools?" Sixty-six percent of the sample answered "Yes." The article describing the poll says that it "is based on telephone interviews conducted from Sept. 13 to Sept. 18 with 1,664 adults around the United States, excluding Alaska and Hawaii. . . . the telephone numbers were formed by random digits, thus permitting access to both listed and unlisted residential numbers."

**(a)** The article gives the margin of error for 95% confidence as 3 percentage points. Make a confidence statement about the percent of all adults who favor a school prayer amendment.

**(b)** The news article goes on to say: "The theoretical errors do not take into account a margin of additional error resulting from the various practical difficulties in taking any survey of public opinion." List some of the "practical difficulties" that may cause errors in addition to the ±3% margin of error. Pay particular attention to the news article's description of the sampling method.

17.15    (**Optional**) To assess the accuracy of a laboratory scale, a standard weight known to weigh 10 grams is weighed repeatedly. The scale readings are normally distributed with unknown mean (this mean is 10 grams if the scale has no bias). The standard deviation of the scale readings is known to be 0.0002 gram.

How many measurements must be averaged to get a margin of error of $\pm 0.0001$ with 95% confidence? (Hint: The margin of error is $1.96\sigma/\sqrt{n}$. You know that $\sigma = 0.0002$ and want to find $n$ so that $1.96\sigma/\sqrt{n} = 0.0001$.)

# 18

# CONFIDENCE INTERVALS FOR A MEAN

Confidence intervals for the mean $\mu$ of a normal population are based on the sample mean $\overline{y}$. The sampling distribution of $\overline{y}$ has $\mu$ as its mean. (That is, $\overline{y}$ as an estimator of the unknown $\mu$ has no bias.) The spread of the sampling distribution depends on the sample size and also on the population standard deviation $\sigma$. In the previous lesson we made the unrealistic assumption that we knew the value of $\sigma$. In practice, we rarely know $\sigma$. We must therefore estimate $\sigma$ from the data even though we are primarily interested in $\mu$. The need to estimate $\sigma$ changes some details of confidence intervals for $\mu$, but not their interpretation.

The assumptions we make in order to do inference about a population mean are the same as in Lesson 17, except that we do not know the value of $\sigma$.

---

ASSUMPTIONS FOR INFERENCE ABOUT A MEAN

- Our data are a **simple random sample** (SRS) of size $n$ from the population. This assumption is very important.
- Observations from the population have a **normal distribution** with mean $\mu$ and standard deviation $\sigma$. In practice, it is enough that the distribution be unimodal and symmetric unless the sample is very small. Both $\mu$ and $\sigma$ are unknown parameters.

---

In this setting, the sample mean $\bar{y}$ has the normal distribution with mean $\mu$ and standard deviation $\sigma/\sqrt{n}$. Because we don't know $\sigma$, we estimate it by the sample standard deviation $s$. (See page 33 in Lesson 4 to review $s$.) We then estimate the standard deviation of $\bar{y}$ by $s/\sqrt{n}$. This quantity is called the *standard error* of the sample mean $\bar{y}$.

---

STANDARD ERROR

When the standard deviation of a statistic is estimated from the data, the result is called the **standard error** of the statistic. The standard error of the sample mean $\bar{y}$ is

$$\text{SE}_{\bar{y}} = \frac{s}{\sqrt{n}}$$

---

## THE *t* DISTRIBUTIONS

When we know the value of $\sigma$, we base confidence intervals and tests for $\mu$ on the fact that the standardized sample mean

$$z = \frac{\bar{y} - \mu}{\sigma/\sqrt{n}}$$

has the standard normal distribution $N(0, 1)$. When we do not know $\sigma$, we substitute the standard error $s/\sqrt{n}$ of $\bar{y}$ for its standard deviation $\sigma/\sqrt{n}$. The statistic that results does *not* have a normal distribution. It has a distribution that is new to us, called a $t$ *distribution*.

---

THE ONE-SAMPLE $t$ STATISTIC AND THE $t$ DISTRIBUTIONS

Draw an SRS of size $n$ from a population that has the normal distribution with mean $\mu$ and standard deviation $\sigma$. The **one-sample $t$ statistic**

$$t = \frac{\bar{y} - \mu}{s/\sqrt{n}}$$

has the **$t$ distribution** with $n - 1$ degrees of freedom.

---

The $t$ statistic has the same interpretation as any standardized statistic: it says how far $\bar{y}$ is from its mean $\mu$ in standard deviation units. There is a different $t$ distribution for each sample size. We specify a particular $t$ distribution by giving its **degrees of freedom**. The degrees of freedom for the one-sample $t$ statistic come from the sample standard deviation $s$ in the denominator of $t$. We saw in Lesson 4 (page 35) that $s$ has $n - 1$ degrees of freedom. There are other $t$ statistics with different degrees of freedom, some of which we will meet later.

Figure 18.1 compares the density curves of the standard normal distribution and the $t$ distributions with 2 and 9 degrees of freedom. The figure illustrates these facts about the $t$ distributions:

- The density curves of the $t$ distributions are similar in shape to the standard normal curve. They are symmetric about zero, unimodal, and bell-shaped.

- The spread of the $t$ distributions is a bit greater than that of the standard normal distribution. The $t$ distributions in Figure 18.1 have more probability in the tails and less in the center than does the standard normal. This is true because substituting the estimate $s$ for the fixed parameter $\sigma$ introduces more variation into the statistic.

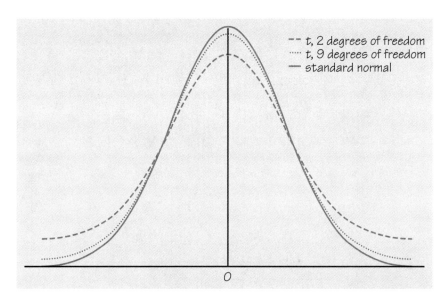

FIGURE 18.1    Density curves for the *t* distributions with 2 and 9 degrees of freedom and the standard normal distribution. All are symmetric with center 0. The *t* distributions have more probability in the tails than does the standard normal.

- As the degrees of freedom increase, the *t* density curve approaches the $N(0, 1)$ curve ever more closely. This happens because *s* estimates $\sigma$ more accurately as the sample size increases. So using *s* in place of $\sigma$ causes little extra variation when the sample is large.

## THE ONE-SAMPLE *t* CONFIDENCE INTERVALS

To give a confidence interval for the mean $\mu$ of a normal population with unknown $\sigma$, just replace the standard deviation $\sigma/\sqrt{n}$ of $\bar{y}$ by its standard error $s/\sqrt{n}$ in the *z* interval of Lesson 17. For confidence level *C*, we must use the point $t^*$ that captures the middle proportion *C* of the *t* distribution between $-t^*$ and $t^*$. Figure 18.2 shows how $t^*$ and *C* are related. There is a different value of $t^*$ for each number of degrees of freedom. Use degrees of freedom $n - 1$ to estimate $\mu$ from a sample of size *n*. Table B contains the $t^*$-values for many confidence levels *C* and many different degrees of freedom.

The one-sample *t* confidence interval is similar in both reasoning and computational detail to the *z* interval in Lesson 17. Your software will again do the calculation for you.

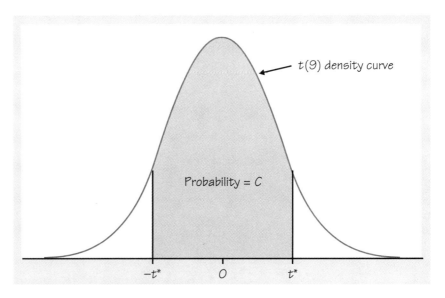

FIGURE 18.2    The central probability $C$ under a $t$ density curve lies between $-t^*$ and $t^*$. The value of $t^*$ depends on the degrees of freedom.

---

THE ONE-SAMPLE $t$ PROCEDURES

Draw an SRS of size $n$ from a population having unknown mean $\mu$. A level $C$ confidence interval for $\mu$ is

$$\bar{y} \pm t^* \frac{s}{\sqrt{n}}$$

Here $t^*$ is the point on the $t$ density curve with $n - 1$ degrees of freedom that captures the middle area $C$ under the curve between $-t^*$ and $t^*$.

This interval is exact when the population distribution is normal and is approximately correct for large $n$ in other cases.

---

**EXAMPLE 18.1**

To study the metabolism of insects, researchers fed cockroaches measured amounts of a sugar solution. After 2, 5, and 10 hours, they dissected some of the cockroaches and measured the amount of sugar in various tissues.[43] Five roaches fed the sugar D-glucose and dissected after 10 hours had the following amounts (in micrograms) of D-glucose in their hindguts:

$$55.95 \quad 68.24 \quad 52.73 \quad 21.50 \quad 23.78$$

The researchers gave a 95% confidence interval for the mean amount of D-glucose in cockroach hindguts under these conditions.

First calculate that

$$\bar{y} = 44.44 \quad \text{and} \quad s = 20.741$$

The degrees of freedom are $n - 1 = 4$. For 95% confidence, $t^* = 2.776$ (see Exercise 18.1). The confidence interval is

$$\bar{y} \pm t^* \frac{s}{\sqrt{n}} = 44.44 \pm 2.776 \frac{20.741}{\sqrt{5}}$$
$$= 44.44 \pm 25.75$$

That is, we are 95% confident that

$$18.69 < \mu < 70.19$$

The large margin of error is due to the small sample size and the rather large variation among the cockroaches, reflected in the large value of $s$.

Figure 18.3 shows the *Data Desk* output for this 95% confidence interval. The result varies slightly from our hand calculation because of differences in rounding off.  ◀

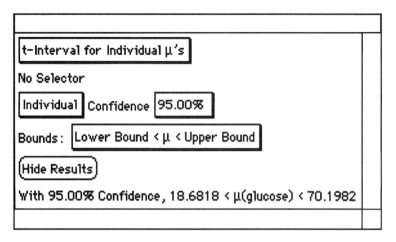

FIGURE 18.3    *Data Desk* confidence interval output.

The one-sample $t$ confidence interval has the form

$$\text{estimate} \ \pm \ t^*\text{SE}_{\text{estimate}}$$

where "SE" stands for "standard error." We will meet a number of confidence intervals that have this common form.

## SOME PRACTICAL ISSUES

The $t$ confidence interval in Example 18.1 rests on two assumptions that are reasonable but are not easy to check: random sampling and normal population distribution. We must be willing to treat the cockroaches as an SRS from a larger population if we want to draw conclusions about cockroaches in general. The roaches were chosen at random from a population grown in the laboratory for research purposes. Even though we don't actually have an SRS from the population of all cockroaches, we are willing to act as if we did. This is a matter of judgment.

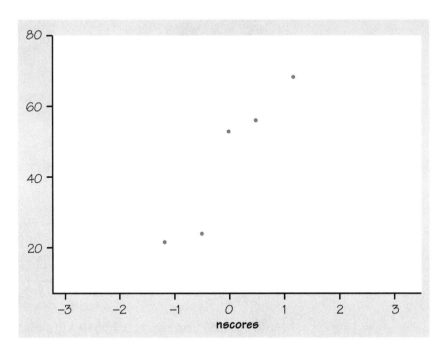

FIGURE 18.4    Normal probability plot of the 5 glucose concentrations in Example 18.1.

Because sample means are more normal than individual observations, the assumption that the population distribution is normal is less critical than the assumption of random sampling. We can't even check symmetry and unimodality effectively with only 5 observations, however. Figure 18.4 is a normal probability plot of the 5 observations. The data have a wide gap between the two smallest and the three largest observations. In observational data, this might suggest two different species of roaches. In this case we know that all five cockroaches came from a single population grown in the laboratory. The gap is just chance variation in a very small sample. There are no strong individual outliers. The researchers are willing to use the $t$ confidence interval because their experience with similar biological measurements suggests that the data should be roughly normal. This is also a matter of judgment.

## EXERCISES

18.1    Use your software to verify Example 18.1.

  (a) The point that captures the middle 0.95 probability of the $t$ distribution with 4 degrees of freedom is $t^* = 2.776$.

  (b) Enter the data and request a 95% $t$ confidence interval for $\mu$.

18.2    The scores of four roommates on the Law School Aptitude Test have mean $\bar{y} = 589$ and standard deviation $s = 37$. What is the standard error of the mean?

18.3    What point $t^*$ on a $t$ density curve satisfies each of the following conditions? Use software or Table B.

  (a) The one-sample $t$ statistic from a sample of 15 observations has probability 0.025 to the right of $t^*$.

  (b) The one-sample $t$ statistic from an SRS of 20 observations has probability 0.75 to the left of $t^*$.

  (c) The $t$ distribution with 25 degrees of freedom has probability 0.80 between $-t^*$ and $t^*$.

18.4    Poisoning by the pesticide DDT causes tremors and convulsions. In a study of DDT poisoning, researchers fed several rats a measured amount of DDT. They then made measurements on the rats' nervous systems that might explain how DDT poisoning causes tremors. One important variable was the "absolutely refractory period," the time required for a nerve to recover after a stimulus. This period varies normally. Measurements on four rats gave the data below (in milliseconds):[44]

$$1.7 \quad 1.6 \quad 1.8 \quad 1.9$$

(a) Find the mean absolutely refractory period $\bar{y}$ and the standard error of the mean.

(b) Give a 90% confidence interval for the mean absolutely refractory period for all rats of this strain when subjected to the same treatment.

18.5    The level of various substances in the blood of kidney dialysis patients is of concern because kidney failure and dialysis can lead to nutritional problems. A researcher did blood tests on several dialysis patients on six consecutive clinic visits. One variable she measured was the level of phosphate in the blood. An individual's phosphate levels tend to vary normally over time. The data on one patient, in milligrams of phosphate per deciliter of blood, are[45]

$$5.6 \quad 5.1 \quad 4.6 \quad 4.8 \quad 5.7 \quad 6.4$$

(a) Calculate the sample mean $\bar{y}$ and also its standard error.

(b) Give a 90% confidence interval for this patient's mean phosphate level.

18.6    The Acculturation Rating Scale for Mexican Americans (ARSMA) measures the extent to which Mexican Americans have adopted Anglo/English culture. During the development of ARSMA, the test was given to a group of 17 Mexicans. Their scores, from a possible range of 1.00 to 5.00, had a symmetric distribution with $\bar{y} = 1.67$ and $s = 0.25$. Because low scores should indicate a Mexican cultural orientation, these results helped to establish the validity of the test.[46]

(a) Give a 95% confidence interval for the mean ARSMA score of Mexicans.

(b) What assumptions does your confidence interval require? Which of these assumptions is most important in this case?

18.7    Here are measurements (in millimeters) of a critical dimension for 16 auto engine crankshafts:

| | | | | | |
|---|---|---|---|---|---|
| 224.120 | 224.001 | 224.017 | 223.982 | 223.989 | 223.961 |
| 223.960 | 224.089 | 223.987 | 223.976 | 223.902 | 223.980 |
| 224.098 | 224.057 | 223.913 | 223.999 | | |

(a) Check the data graphically for outliers or strong skewness that might threaten the validity of the $t$ procedures. What do you conclude?

(b) Give a 90% confidence interval for the mean dimension $\mu$ of all crankshafts produced by this process.

18.8    A study of the healing of wounds used anesthetized newts as subjects. The researchers made a razor cut in one of the newt's limbs, then measured the rate of regrowth of the skin. Here are the healing rates (in micrometers per hour) for 14 newts under one of the conditions that the researchers examined:[47]

25   13   44   45   57   42   50   36   35   38   43   31   26   48

(a) Check the data graphically for outliers or strong skewness that might threaten the validity of the $t$ procedures. What do you conclude?

(b) Give a 95% confidence interval for the mean rate of healing in this species of newt under the conditions of the experiment.

18.9    In 1798 the English scientist Henry Cavendish measured the density of the earth with great care. It is common practice to repeat careful measurements several times. Cavendish repeated his work 29 times. Here are his results (the data give the density of the earth as a multiple of the density of water):[48]

5.50   5.61   4.88   5.07   5.26   5.55   5.36   5.29   5.58   5.65
5.57   5.53   5.62   5.29   5.44   5.34   5.79   5.10   5.27   5.39
5.42   5.47   5.63   5.34   5.46   5.30   5.75   5.68   5.85

If Cavendish's measurements have no bias, the mean of all measurements he could make is the true density of the earth. Give a 99% confidence interval for the density of the earth. (Be sure to examine the data for skewness and outliers first.)

18.10   Here are the scores of 78 seventh-grade students in a rural midwestern school on a standard IQ test:[49]

111   107   100   107   114   115   111   97    100   112   104   89
104   102   91    114   114   103   106   105   113   109   108   113
130   128   128   118   113   120   132   111   124   127   128   136
106   118   119   123   124   126   116   127   119   97    86    102
110   120   103   115   93    72    111   103   123   79    119   110
110   74    107   105   112   105   110   107   103   77    98    90
96    112   112   114   93    106

(a) Assuming these students are an SRS of all students in rural midwestern schools, give a 95% confidence interval for the mean IQ score in this population.

(b) We have already examined these data and found them to be roughly normal (Example 5.3, page 53). You should nonetheless hesitate to draw conclusions about all rural midwestern seventh graders. Why?

18.11    A study of computer-assisted learning examined the learning of "Blissymbols" by children. Blissymbols are pictographs (think of Egyptian hieroglyphics) that are sometimes used to help learning-impaired children communicate. The researcher designed two computer lessons that taught the same content using the same examples. One lesson required the children to interact with the material, while in the other the children controlled only the pace of the lesson. Call these two styles "Active" and "Passive." After the lesson, the computer presented a quiz that asked the children to identify 56 Blissymbols. Here are the numbers of correct identifications by the 24 children in the Active group:[50]

29 28 24 31 15 24 27 23 20 22 23 21 24 35 21 24 44 28 17 21 21 20 28 16

(a) Make a normal probability plot of the Active data. The distribution is somewhat skewed to the right but is sufficiently close to normal to allow use of $t$ procedures.

(b) Give a 90% confidence interval for the mean number of Blissymbols identified correctly in a large population of children after the Active computer lesson.

18.12    A study of genetic influences on diabetes compared normal mice with similar mice genetically altered to remove the gene called $aP2$. Mice of both types were allowed to become obese by eating a high-fat diet. The study report gave the insulin levels in the blood of the mice (nanograms per milliliter) as $5.9 \pm 0.9$ for the normal mice and $0.75 \pm 0.2$ for the genetically altered mice. This comparison shows the effect of the $aP2$ gene. The report said, "Each value is the mean $\pm$ SEM of measurements on at least 10 mice."[51]

(a) What does "SEM" stand for?

(b) We don't know the exact number of mice, so take $n = 10$. Give a 95% confidence interval for the mean insulin level in the blood of normal mice.

(c) We do know that $n$ is "at least 10." The confidence interval you found in (b) is therefore conservative. That is, it is at least as wide as the interval you would get if you knew $n$ exactly. Explain why this is true.

# LESSON 19

# TESTING HYPOTHESES

Confidence intervals are one of the two most common types of formal statistical inference. Use them when your goal is to estimate a population parameter. The second common type of inference has a different goal: to assess the evidence provided by data about some claim concerning a population. Here is the reasoning of statistical tests in a nutshell.

EXAMPLE 19.1

I claim that I make 80% of my basketball free throws. To test my claim, you ask me to shoot 25 free throws. I make only two of the 25. "Aha!" you say. "Someone who makes 80% of his free throws would almost never make only 2 out of 25. So I don't believe your claim."

Your reasoning is based on asking what would happen if my claim were true and we repeated the sample of 25 free throws many times—I would almost never make as few as 2. This outcome is so unlikely that it gives strong evidence that my claim is not true. ◄

# THE REASONING OF STATISTICAL TESTS

The reasoning of statistical tests, like that of confidence intervals, is based on asking what would happen if we repeated the sample or experiment many times. We will again start with an unrealistic procedure in order to emphasize the reasoning. We will meet many statistical tests later. All rely on the same reasoning. This lesson is therefore a foundation for much that follows. Here is an example we will explore.

---

**EXAMPLE 19.2**

Diet colas use artificial sweeteners to avoid sugar. Colas with artificial sweeteners gradually lose their sweetness over time. Manufacturers therefore test new colas for loss of sweetness before marketing them. Trained tasters sip the cola along with drinks of standard sweetness and score the cola on a "sweetness scale" of 1 to 10. The cola is then stored for a month at high temperature to imitate the effect of four months' storage at room temperature. After a month, each taster scores the stored cola. The response variable is the difference (score before storage minus score after storage) in a taster's scores. The bigger the difference, the bigger the loss of sweetness.

Here are the sweetness losses for a new cola, as measured by 10 trained tasters:

$$2.0 \quad 0.4 \quad 0.7 \quad 2.0 \quad -0.4 \quad 2.2 \quad -1.3 \quad 1.2 \quad 1.1 \quad 2.3$$

Most are positive. That is, most tasters found a loss of sweetness. But the losses are small, and two tasters (the negative scores) thought the cola gained sweetness. *Are these data good evidence that the cola lost sweetness in storage?*    ◄

---

The mean sweetness loss for these 10 tasters is $\bar{y} = 1.02$. We will assume (unrealistically) that we know that the standard deviation for all individual tasters is $\sigma = 1$.

The reasoning is as in Example 19.1. We make a claim and ask if the data give evidence *against* it. We seek evidence that there *is* a sweetness loss, so the claim we test is that there *is not* a loss. In that case, the mean loss perceived by the population of all trained testers would be $\mu = 0$.

- If the claim that $\mu = 0$ is true, the sampling distribution of $\bar{y}$ from 10 tasters is normal with mean $\mu = 0$ and standard deviation

$$\frac{\sigma}{\sqrt{n}} = \frac{1}{\sqrt{10}} = 0.316$$

Figure 19.1 shows this sampling distribution. We can judge whether any observed $\bar{y}$ is surprising by locating it on this distribution.

- Suppose that our 10 tasters had mean loss $\bar{y} = 0.3$. It is clear from Figure 19.1 that a $\bar{y}$ this large could easily occur just by chance when the population mean is $\mu = 0$. That 10 tasters find $\bar{y} = 0.3$ is not evidence of a sweetness loss.

- In fact, our taste test produced $\bar{y} = 1.02$. That's way out on the normal curve in Figure 19.1—so far out that an observed value this large would almost never occur just by chance if the true $\mu$ were 0. This observed value is good evidence that in fact the true $\mu$ is greater than 0, that is, that the cola lost sweetness. The manufacturer must reformulate the cola and try again.

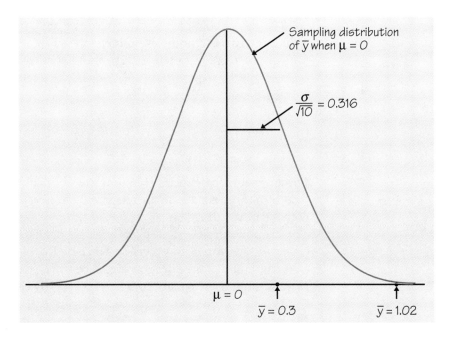

FIGURE 19.1    If a cola does not lose sweetness in storage, the mean score $\bar{y}$ for 10 tasters will have this sampling distribution. The result for one cola is $\bar{y} = 0.3$. That could easily happen just by chance. Another cola has $\bar{y} = 1.02$. This value is so far out on the normal curve that it is good evidence that the cola did lose sweetness.

## THE VOCABULARY OF STATISTICAL TESTS

A statistical test starts with a careful statement of the claims we want to compare. The claims concern a population, so we express them in terms of a population parameter. In Example 19.2, the parameter is the population mean $\mu$, the average loss in sweetness that a very large number of tasters would detect in the cola. Because the reasoning we have outlined looks for evidence *against* a claim, we start with the claim we seek evidence against, such as "no loss of sweetness." This claim is our *null hypothesis*.

---

NULL HYPOTHESIS $H_0$

The statement being tested in a statistical test is called the **null hypothesis**. The test is designed to assess the strength of the evidence against the null hypothesis. Usually the null hypothesis is a statement of "no effect" or "no difference."

---

The claim about the population that we are trying to find evidence *for* is the **alternative hypothesis**, written $H_a$. In Example 19.2, we are seeking evidence of a loss in sweetness. The null hypothesis says "no loss" on the average in a large population of tasters. The alternative hypothesis says "there is a loss." So the hypotheses are

$$H_0: \mu = 0$$
$$H_a: \mu > 0$$

The alternative hypothesis $H_a$ is **one-sided** (includes only means greater than zero) because we are interested only in deviations from the null hypothesis in one direction.

The null and alternative hypotheses are precise statements of just what claims we are testing. If we get an outcome that would be unlikely if $H_0$ were true and is in the direction suggested by $H_a$, we have evidence against $H_0$ in favor of $H_a$. We make "unlikely" precise by calculating a probability.

---

*P*-VALUE

The probability, computed assuming that $H_0$ is true, that the observed outcome would take a value as extreme or more extreme than that

---

*(continued on next page)*

*(continued from previous page)*

actually observed is called the **P-value** of the test. The smaller the P-value is, the stronger is the evidence against $H_0$ provided by the data.

## EXAMPLE 19.3

Figure 19.2 shows the P-value when 10 tasters give mean sweetness loss $\bar{y}$ = 0.3. It is the probability that, *if $\mu$ = 0 is true*, we observe a sample mean at least as large as 0.3. This probability is P = 0.1711. That is, we would observe a sweetness loss this large or larger about 17% of the time, just by the luck of the draw in choosing 10 tasters, even if the entire population of tasters would find no loss on the average. A result that would occur this often when $H_0$ is true is not good evidence against $H_0$.

In fact, the 10 tasters found $\bar{y}$ = 1.02. The P-value is again the probability of observing a $\bar{y}$ this large or larger if in fact $\mu$ = 0. This probability is P =

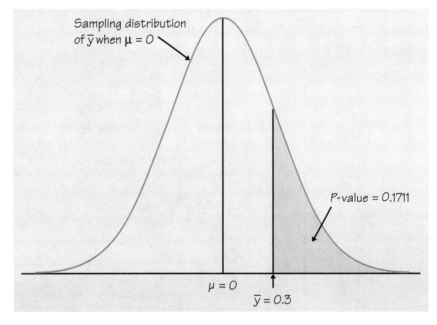

**FIGURE 19.2**   The P-value for the result $\bar{y}$ = 0.3 in the cola taste test. The P-value is the probability (when $H_0$ is true) that $\bar{y}$ takes a value as large or larger than the actually observed value.

0.0006. We would almost never get a sample sweetness loss this large if $H_0$ were true. The small $P$-value provides strong evidence against $H_0$ and in favor of the alternative $H_a$: $\mu > 0$.    ◀

In practice, software will calculate the $P$-value from the data and your statement of the null and alternative hypotheses. You must, of course, know which test to request. The test for hypotheses about the mean $\mu$ of a normal distribution with known standard deviation $\sigma$ is called the **one-sample $z$ test** because the probability calculation starts by standardizing the observed sample mean. If $\mu_0$ is the value of the population mean $\mu$ specified by the null hypothesis, the **one-sample $z$ statistic** is

$$z = \frac{\bar{y} - \mu_0}{\sigma/\sqrt{n}}$$

The statistic $z$ says how many standard deviations the observed mean $\bar{y}$ lies away from the hypothesized value $\mu_0$. Although software hides the calculation, it is usual to report $z$ as well as the $P$-value. Here is an example that illustrates the $z$ statistic as well as a two-sided alternative.

## EXAMPLE 19.4

The National Center for Health Statistics reports that the mean systolic blood pressure for males 35 to 44 years of age is 128 and that the standard deviation in this population is 15. The medical director of a large company looks at the medical records of 72 executives in this age group and finds that the mean systolic blood pressure in this sample is $\bar{y} = 126.07$. Is this evidence that the company's executives have a different mean blood pressure from the general population? As usual in this lesson, we make the unrealistic assumption that we know the population standard deviation. Assume that executives have the same $\sigma = 15$ as the general population of middle-aged males.

*Hypotheses:* The null hypothesis is "no difference" from the national mean $\mu = 128$. The alternative is **two-sided**, because the medical director did not have a particular direction in mind before examining the data. So the hypotheses about the unknown mean $\mu$ of the executive population are

$$H_0: \mu = 128$$
$$H_a: \mu \neq 128$$

*Test statistic:* The $z$ test statistic is

$$z = \frac{\bar{y} - \mu_0}{\sigma/\sqrt{n}} = \frac{126.07 - 128}{15/\sqrt{72}}$$
$$= -1.09$$

*P-value:* The P-value is always the probability of observing an outcome at least as extreme as the outcome we actually got. "At least as extreme" means at least as far out in the direction given by the alternative hypothesis. When $H_a$ is two-sided, this becomes at least as far out *in either direction* as our actual result. As always, the P-value is calculated taking the null hypothesis to be true. The P-value is the probability that

$\bar{y}$ is at least as far from 128 as the observed $\bar{y} = 126.07$

Or, in terms of standardized values, the P-value is the probability that

the $z$ statistic is at least as far from 0 as the observed $z = -1.09$

These probabilities are of course the same. Figure 19.3 illustrates the second version. Software gives $P = 0.2758$.

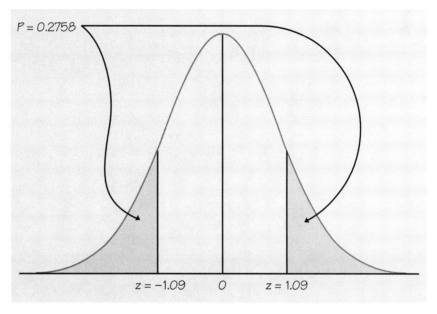

$P = 0.2758$

$z = -1.09$    $0$    $z = 1.09$

FIGURE 19.3    The P-value for the two-sided test in Example 19.4.

*Conclusion:* More than 27% of the time, an SRS of size 72 from the general male population would have a mean blood pressure at least as far from 128 as that of the executive sample. The observed $\bar{y} = 126.07$ is therefore not good evidence that executives differ from other men.     ◄

The $z$ test assumes that the 72 executives in the sample are an SRS from the population of all middle-aged male executives in the company. We should check this assumption by asking how the data were produced. If medical records are available only for executives with recent medical problems, for example, the data are of little value for our purpose. It turns out that all executives are given a free annual medical exam, and that the medical director selected 72 exam results at random.

The data in Example 19.4 do *not* establish that the mean blood pressure $\mu$ for this company's executives is 128. We sought evidence that $\mu$ differed from 128 and failed to find convincing evidence. That is all we can say. No doubt the mean blood pressure of the entire executive population is not exactly equal to 128. A large enough sample would give evidence of the difference, even if it is very small. Tests of significance assess the evidence *against* $H_0$. If the evidence is strong, we can confidently reject $H_0$ in favor of the alternative. Failing to find evidence against $H_0$ means only that the data are consistent with $H_0$, not that we have clear evidence that $H_0$ is true.

## STATISTICAL SIGNIFICANCE

Small $P$-values are evidence against $H_0$, because they say that the observed result is unlikely to occur if $H_0$ is true. A result with a small $P$-value is called **statistically significant**. That's just a way of saying that chance alone would rarely produce so extreme a result.

How small must a $P$-value be in order to persuade us? There is no fixed rule. But the level 0.05 (a result that would occur no more than 5% of the time just by chance) is a common rule of thumb. A result with a $P$-value smaller than 0.05 is "statistically significant at the 5% level." In reports of research, it is common to just write "statistically significant ($P < 0.05$)." A stronger standard for statistical evidence is significance at the 1% level ($P < 0.01$). Such a result would occur by chance no more than 1 time in 100 in many repeated samples or experiments.

## TESTS AND CONFIDENCE INTERVALS

There is a connection between tests that are significant at the 5% level and 95% confidence intervals. In fact, we can reject a null hypothesis $H_0$: $\mu = \mu_0$ against the two-sided alternative exactly when the hypothesized value $\mu_0$ lies *outside* the 95% confidence interval for $\mu$. There is a similar relationship between 99% confidence intervals and tests significant at the 1% level, and so on.

### EXAMPLE 19.5

Return to the taste test data in Example 19.2, but now ask the question "Is there good evidence that the sweetness level has changed in storage?" This question does not specify the direction of the change, so it calls for a two-sided alternative:

$$H_0: \mu = 0 \quad \text{(no change)}$$
$$H_a: \mu \neq 0 \quad \text{(either gain or loss)}$$

Our 10 tasters gave $\bar{y} = 1.02$. We are still assuming that we know $\sigma = 1$. For 99% confidence, Table B tells us that $z^* = 2.576$. The 99% $z$ confidence interval for the mean sweetness loss $\mu$ is therefore

$$\bar{y} \pm z^* \frac{\sigma}{\sqrt{n}} = 1.02 \pm 2.576 \frac{1}{\sqrt{10}}$$
$$= 1.02 \pm 0.815$$

The null hypothesis value $\mu = 0$ lies outside this interval. We therefore know that we can reject the hypothesis that $\mu = 0$ at the 1% significance level. ◀

## SOME CAUTIONS ABOUT STATISTICAL TESTS

The list of warnings for users of confidence intervals (page 215) applies to tests as well. Testing is a bit more subtle than estimation, so we offer some additional warnings.

**Statistical significance is not the same thing as practical significance.** A small $P$-value means we have good evidence that an effect is present. But the $P$-value tells us nothing about the size of the effect or its practical importance. A very large sample will find very small effects to be statistically significant. Exercise 19.16 demonstrates in detail how increasing the sample size drives down the P-value.

The remedy for attaching too much importance to statistical significance is to pay attention to the actual data as well as to the P-value. Plot your data and examine them carefully. Is the effect you are seeking visible in your plots? If not, ask yourself if the effect is large enough to be practically important. It is often wise to give a confidence interval for the parameter in which you are interested. A confidence interval actually estimates the size of an effect, rather than simply asking if it is too large to reasonably occur by chance alone.

**There is no sharp border between "significant" and "not significant,"** only increasingly strong evidence as the P-value gets smaller. There is no practical distinction between the P-values 0.049 and 0.051. It makes no sense to treat $P < 0.05$ as a universal rule for what is significant.

**Beware of multiple tests**. It is tempting (and easy when we use software) to test many different hypotheses on the same data. Remember that significance at the 5% level will occur 1 time in 20 just by chance even if all your null hypotheses are true. Running dozens of tests and finding that one or two give significant results is meaningless—we have no way to decide if anything other than chance explains the few significant results.

## EXERCISES

19.1   The Survey of Study Habits and Attitudes (SSHA) is a psychological test that measures the study habits and attitude toward school of students. Scores range from 0 to 200. The mean score for college students is about 115, and the standard deviation is about 30. A teacher suspects that older students have better attitudes toward school. She gives the SSHA to 25 students who are at least 30 years of age. Assume that scores in the population of older students are normally distributed with standard deviation $\sigma = 30$. The teacher wants to test the hypotheses

$$H_0: \mu = 115$$
$$H_a: \mu > 115$$

(a) What is the sampling distribution of the mean score $\bar{y}$ of a sample of 25 older students if the null hypothesis is true? Sketch the density curve of this distribution. (Hint: Sketch a normal curve first, then mark the axis using what you know about locating $\mu$ and $\sigma$ on a normal curve.)

(b) Suppose that the sample data give $\bar{y} = 118.6$. Mark this point on the axis of your sketch. In fact, the result was $\bar{y} = 125.7$. Mark this point on your sketch. Using your sketch, explain in simple language why one result is good evidence that the mean score of all older students is greater than 115 and why the other outcome is not.

(c) Shade the area under the curve that is the $P$-value for the sample result $\bar{y} = 118.6$.

(d) Use software or Table A to find the $P$-values for both observed values of $\bar{y}$. The two $P$-values express in numbers the comparison you made informally in (b).

19.2    The Census Bureau reports that households spend an average of 31% of their total spending on housing. A home-builders association in Cleveland believes that this average is lower in their area. They interview a sample of 40 households in the Cleveland metropolitan area to learn what percent of their spending goes toward housing. Take $\mu$ to be the mean percent of spending devoted to housing among all Cleveland households. We want to test the hypotheses

$$H_0: \mu = 31\%$$
$$H_a: \mu < 31\%$$

The population standard deviation is $\sigma = 9.6\%$.

(a) What is the sampling distribution of the mean percent $\bar{y}$ that the sample spends on housing if the null hypothesis is true? Sketch the density curve of the sampling distribution. (Hint: Sketch a normal curve first, then mark the axis using what you know about locating $\mu$ and $\sigma$ on a normal curve.)

(b) Suppose that the study finds $\bar{y} = 30.2\%$ for the 40 households in the sample. Mark this point on the axis in your sketch. Then suppose that the study result is $\bar{y} = 27.6\%$. Mark this point on your sketch. Referring to your sketch, explain in simple language why one result is good evidence that average Cleveland spending on housing is less than 31% and the other result is not.

(c) Shade the area under the curve that gives the $P$-value for the result $\bar{y} = 30.2\%$. (Note that we are looking for evidence that spending is *less* than the null hypothesis states.)

(d) Use software or Table A to find the $P$-values for both observed values of $\bar{y}$. The two $P$-values express in numbers the comparison you made informally in (b).

Each of Exercises 19.3 to 19.6 calls for a statistical test about a population mean $\mu$. State the null hypothesis $H_0$ and the alternative hypothesis $H_a$ in each case.

19.3    The diameter of a spindle in a small motor is supposed to be 5 mm. If the spindle is either too small or too large, the motor will not work properly. The manufacturer measures the diameter in a sample of motors to determine whether the mean diameter has moved away from the target.

19.4    Census Bureau data show that the mean household income in the area served by a shopping mall is $42,500 per year. A market research firm questions shoppers at the mall. The researchers suspect the mean household income of mall shoppers is higher than that of the general population.

19.5    The examinations in a large accounting class are scaled after grading so that the mean score is 50. The professor thinks that one teaching assistant is a poor teacher and suspects that his students have a lower mean score than the class as a whole. The TA's students this semester can be considered a sample from the population of all students in the course, so the professor compares their mean score with 50.

19.6    Last year, your company's service technicians took an average of 2.6 hours to respond to trouble calls from business customers who had purchased service contracts. Do this year's data show a different average response time?

19.7    Here are measurements (in millimeters) of a critical dimension on a sample of automobile engine crankshafts:

|       |       |       |       |       |       |
|-------|-------|-------|-------|-------|-------|
| 224.120 | 224.001 | 224.017 | 223.982 | 223.989 | 223.961 |
| 223.960 | 224.089 | 223.987 | 223.976 | 223.902 | 223.980 |
| 224.098 | 224.057 | 223.913 | 223.999 |       |       |

The manufacturing process is known to vary normally with standard deviation $\sigma = 0.060$ mm. The process mean is supposed to be 224 mm. Do these data give evidence that the process mean is not equal to the target value 224 mm? (State hypotheses, find the values of $z$ and $P$, and report your conclusion.)

19.8    A computer has a random number generator designed to produce random numbers that are uniformly distributed on the interval from 0 to 1. If this is true, the numbers generated come from a population with $\mu = 0.5$ and $\sigma = 0.2887$. A command to generate 100 random numbers gives outcomes with mean $\bar{y} = 0.4365$. Assume that the population $\sigma$ remains fixed. We want to test

$$H_0: \mu = 0.5$$
$$H_a: \mu \neq 0.5$$

Give the value of the $z$ statistic and its $P$-value. Is the result significant at the 5% level? At the 1% level? What do you conclude?

19.9    An agronomist examines the cellulose content of a variety of alfalfa hay. Suppose that the cellulose content in the population has standard deviation $\sigma = 8$ mg/g. A sample of 15 cuttings has mean cellulose content $\bar{y} = 145$ mg/g.

(a) Give a 90% confidence interval for the mean cellulose content in the population.

(b) A previous study claimed that the mean cellulose content was $\mu = 140$ mg/g, but the agronomist believes that the mean is higher than that figure. State $H_0$ and $H_a$ and carry out a test to see if the new data support this belief.

(c) The statistical procedures used in (a) and (b) are valid when several assumptions are met. What are these assumptions?

19.10    Market pioneers, companies that are among the first to develop a new product or service, tend to have higher market shares than latecomers to the market. What accounts for this advantage? Here is an excerpt from the conclusions of a study of a sample of 1209 manufacturers of industrial goods:

> Can patent protection explain pioneer share advantages? Only 21% of the pioneers claim a significant benefit from either a product patent or a trade secret. Though their average share is two points higher than that of pioneers without this benefit, the increase is not statistically significant ($z = 1.13$). Thus, at least in mature industrial markets, product patents and trade secrets have little connection to pioneer share advantages.[52]

Find the $P$-value for the given $z$. Then explain to someone who knows no statistics what "not statistically significant" in the study's conclusion means. Why does the author conclude that patents and trade secrets don't help, even though they contributed 2 percentage points to average market share?

19.11    A social psychologist reports that "in our sample, ethnocentrism was significantly higher ($P < 0.05$) among church attenders than among nonattenders." Explain what this means in language understandable to someone who knows no statistics. Do not use the word "significance" in your answer.

19.12    The financial aid office of a university asks a sample of students about their employment and earnings. The report states: "For academic year earnings, a significant difference ($P = 0.036$) was found between the sexes, with men earning more on the average. The difference between the earnings of black and white students was not significant ($P = 0.436$)." Explain both of these conclusions, for the effects of sex and of race on mean earnings, in language understandable to someone who knows no statistics.

19.13    When asked to explain the meaning of "the P-value was $P = 0.03$," a student says, "This means there is only probability 0.03 that the null hypothesis is true." Is this an essentially correct explanation? Explain your answer.

19.14    When asked why statistical significance appears so often in research reports, a student says, "Because saying that results are significant tells us that they cannot easily be explained by chance variation alone." Do you think that this statement is essentially correct? Explain your answer.

19.15    The cigarette industry has adopted a voluntary code requiring that models appearing in its advertising must appear to be at least 25 years old. Studies have shown, however, that consumers think many of the models are younger. Here is a quote from a study that asked whether different brands of cigarettes use models that appear to be of different ages:[53]

> The ANCOVA revealed that the brand variable is highly significant ($P < .001$), indicating that the average perceived age of the models is not equal across the 12 brands. As discussed previously, certain brands such as Lucky Strike Lights, Kool Milds, and Virginia Slims tended to have younger models . . .

ANCOVA is an advanced statistical technique, but significance and P-values have their usual meaning. Explain to someone who knows no statistics what "highly significant ($P < .001$)" means and why this is good evidence of differences among all advertisements of these brands even though the subjects saw only a sample of ads.

19.16    Let us suppose that scores on the mathematics part of the Scholastic Assessment Test (SATM) in the absence of coaching vary normally with mean $\mu = 475$ and $\sigma = 100$. Suppose also that coaching may change $\mu$ but does not change $\sigma$. An increase in the SATM score from 475 to 478 is of no importance in seeking admission to college, but this unimportant change can be statistically very significant if it occurs in a large enough sample. To see this, calculate the P-value for the test of

$$H_0: \mu = 475$$
$$H_a: \mu > 475$$

in each of the following situations:

(a) A coaching service coaches 100 students. Their SATM scores average $\bar{y} = 478$.

(b) By the next year, the service has coached 1000 students. Their SATM scores average $\bar{y} = 478$.

(c) An advertising campaign brings the number of students coached to 10,000. Their average score is still $\bar{y} = 478$.

19.17    Suppose that in the absence of special preparation Scholastic Assessment Test mathematics (SATM) scores vary normally with mean $\mu = 475$ and $\sigma = 100$. One hundred students go through a rigorous training program designed to raise their SATM scores by improving their mathematics skills. Carry out a test of

$$H_0\!: \ \mu = 475$$
$$H_a\!: \ \mu > 475$$

in each of the following situations:

(a) The students' average score is $\bar{y} = 491.4$. Is this result significant at the 5% level?

(b) The average score is $\bar{y} = 491.5$. Is this result significant at the 5% level?

The difference between the two outcomes in (a) and (b) is of no importance. Don't treat a fixed level of significance such as 5% as sacred.

19.18    Give a 99% confidence interval for the mean SATM score $\mu$ after coaching in each part of Exercise 19.16. For large samples, the confidence interval tells us, "Yes, the mean score is higher than 475 after coaching, but only by a small amount."

19.19    Which of the following questions does a test of significance answer?

(a) Is the sample or experiment properly designed?

(b) Is the observed effect due to chance?

(c) Is the observed effect important?

19.20    A researcher looking for evidence of extrasensory perception (ESP) tests 500 subjects. Four of these subjects do significantly better ($P < 0.01$) than random guessing.

(a) Is it proper to conclude that these four people have ESP? Explain your answer.

(b) What should the researcher now do to test whether any of these four subjects have ESP?

# TESTS FOR A MEAN

We wish to make an inference about the unknown mean $\mu$ of a population based on data from an SRS drawn from the population. The sample mean $\bar{y}$ is an unbiased estimator of $\mu$. (That is, the sampling distribution of $\bar{y}$ is centered at $\mu$.) If the population distribution is reasonably normal or the sample is large so that the central limit theorem takes hold, the sampling distribution of $\bar{y}$ is close to normal with mean $\mu$ and standard deviation $\sigma/\sqrt{n}$. When we know the standard deviation $\sigma$ of the population, we can base inference about $\mu$ on the **one-sample $z$ statistic**

$$z = \frac{\bar{y} - \mu}{\sigma/\sqrt{n}}$$

The $z$ statistic is just the standardized value of $\bar{y}$.

In practice, we rarely know $\sigma$. When we use the sample standard deviation $s$ to estimate $\sigma$, the $z$ statistic becomes the **one-sample $t$ statistic**

$$t = \frac{\bar{y} - \mu}{s/\sqrt{n}}$$

The *t* statistic has the same interpretation as $z$: it is the number of standard deviations from the sample mean $\bar{y}$ to the population mean $\mu$. In Lesson 18, we met the *t* confidence interval for estimating $\mu$. Now we meet the corresponding test, the *one-sample t test*.

# THE ONE-SAMPLE *t* TEST

To test the null hypothesis that the population mean $\mu$ has a specified value $\mu_0$, we ask if the sample mean $\bar{y}$ is surprisingly far from $\mu_0$. The *t* test asks this question in the standardized scale.

---

### THE ONE-SAMPLE *t* TEST

Draw an SRS of size $n$ from a population having unknown mean $\mu$. To test the hypothesis $H_0 : \mu = \mu_0$, compute the one-sample *t* statistic

$$t = \frac{\bar{y} - \mu_0}{s/\sqrt{n}}$$

Compare the observed value of $t$ with the *t* distribution with $n - 1$ degrees of freedom. The *P*-value of the test is the probability of an outcome more extreme than the observed $t$ in the direction specified by the alternative hypothesis.

The *P*-value is exact if the population distribution is normal and is approximately correct for large $n$ in other cases.

---

The one-sample *t* test statistic has the form

$$t = \frac{\text{estimate}}{\text{SE}_{\text{estimate}}}$$

where "SE" stands for "standard error." We will meet a number of test statistics that have this common form.

Let us redo a familiar example, now without the unrealistic assumption that we know the population standard deviation $\sigma$.

EXAMPLE 20.1

Cola makers test new recipes for loss of sweetness during storage. Trained tasters rate the sweetness before and after storage. Here are the sweetness losses (sweetness before storage minus sweetness after storage) found by 10 tasters for a new cola recipe:

$$2.0 \quad 0.4 \quad 0.7 \quad 2.0 \quad -0.4 \quad 2.2 \quad -1.3 \quad 1.2 \quad 1.1 \quad 2.3$$

Are these data good evidence that the cola lost sweetness?

**Data analysis:** We are willing to regard these 10 tasters as an SRS of all trained tasters. (This is a matter of judgment.) A normal probability plot (Figure 20.1) may show a slightly short right tail. Although it is difficult to assess normality from only 10 observations, there are no outliers or other strong departures from normality.

**Hypotheses:** Tasters vary in their perception of sweetness loss. So we ask the question in terms of the mean loss $\mu$ for a large population of tasters. The

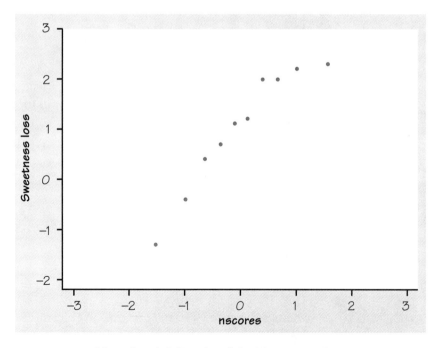

FIGURE 20.1    Normal probability plot of the 10 sweetness loss scores in Example 20.1.

hypotheses are

$$H_0 : \mu = 0$$
$$H_a : \mu > 0$$

**Test statistic:** We calculate that $\bar{y} = 1.02$ and $s = 1.196$. The one-sample $t$ statistic is

$$t = \frac{\bar{y} - \mu_0}{s/\sqrt{n}} = \frac{1.02 - 0}{1.196/\sqrt{10}}$$
$$= 2.697$$

Software will calculate $t$ from the data. It is usual to report the value of $t$ along with its $P$-value. Because $\bar{y}$ is almost 2.7 standard deviations away from 0, we know even before finding the $P$-value that this result is unlikely if the true mean is 0.

**P-value:** The $P$-value for $t = 2.697$ is the area to the right of 2.697 under the $t$ distribution curve with degrees of freedom $n - 1 = 9$. Figure 20.2 shows this area. Software tells us that $P = 0.0123$.

**Conclusion:** The result almost reaches the 1% significance level. There is quite strong evidence for a loss of sweetness. ◀

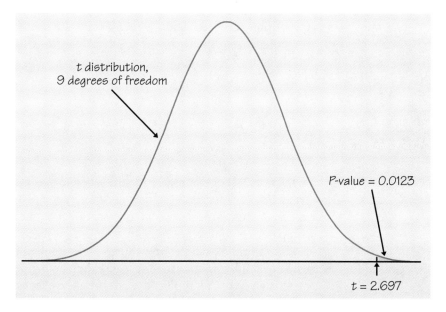

FIGURE 20.2   The $P$-value for the one-sided $t$ test in Example 20.1.

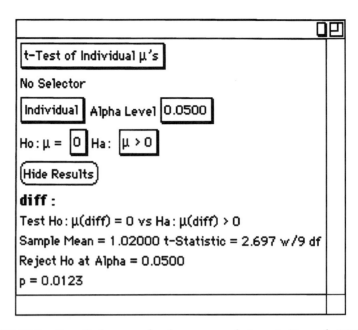

FIGURE 20.3    *Data Desk* output for the one-sample *t* test in Example 20.1.

Figure 20.3 displays the *Data Desk* output for this *t* test. The hypotheses and the values of $\bar{y}$, $t$, and $P$ all appear in the output. You can ignore the "Alpha Level" selection, which allows the user to choose a fixed level of significance if she wishes. The Greek letter $\alpha$ is the usual notation for a fixed significance level.

## PAIRED *t* PROCEDURES

In the taste test of Example 20.1, the same 10 tasters rated before-and-after sweetness. Comparative studies are more convincing than single-sample investigations. For that reason, one-sample inference is less common than comparative inference. One common design to compare two treatments makes use of one-sample procedures. In a **matched pairs design**, subjects who are similar in important respects are matched in pairs and each treatment is given to one subject in each pair. The experimenter can toss a coin to assign two treatments to the two subjects in each pair. Another situation calling for matched pairs is before-and-after observations on the same subjects, as in the taste test example.

> ### PAIRED *t* PROCEDURES
>
> To compare the mean responses to the two treatments in a matched pairs design, apply the one-sample *t* procedures to the observed differences.

The parameter $\mu$ in a paired *t* procedure is the mean difference in the responses to the two treatments within matched pairs of subjects in the entire population.

## EXAMPLE 20.2

The design of controls affects how easily people can use them. A student project investigated this effect by asking 25 right-handed students to turn a knob (with their right hands) that moved an indicator by screw action. There were two identical instruments, one with a right-hand thread (the knob turns clockwise) and the other with a left-hand thread (the knob must be turned counterclockwise). Because some people have stronger hands than others, we ask all subjects to use both instruments (in random order). This is a matched pairs design. Table 20.1 gives the times in seconds each subject took to move the indicator a fixed distance.[54]

**TABLE 20.1**    Time (seconds) to complete an experimental task

| Subject | Right thread | Left thread | Subject | Right thread | Left thread |
|---------|--------------|-------------|---------|--------------|-------------|
| 1  | 113 | 137 | 14 | 107 | 87  |
| 2  | 105 | 105 | 15 | 118 | 166 |
| 3  | 130 | 133 | 16 | 103 | 146 |
| 4  | 101 | 108 | 17 | 111 | 123 |
| 5  | 138 | 115 | 18 | 104 | 135 |
| 6  | 118 | 170 | 19 | 111 | 112 |
| 7  | 87  | 103 | 20 | 89  | 93  |
| 8  | 116 | 145 | 21 | 78  | 76  |
| 9  | 75  | 78  | 22 | 100 | 116 |
| 10 | 96  | 107 | 23 | 89  | 78  |
| 11 | 122 | 84  | 24 | 85  | 101 |
| 12 | 103 | 148 | 25 | 88  | 123 |
| 13 | 116 | 147 |    |     |     |

We suspect that right-handed people take longer (on the average) when using a left-hand thread. The means are $\bar{y}_1 = 104.12$ seconds for the right-hand thread and $\bar{y}_2 = 117.44$ seconds for the left-hand thread. Is this difference statistically significant?

**Data analysis:** Because of the matched pairs design, we analyze the differences (left − right) for the 25 subjects. A normal probability plot (Figure 20.4) shows that the 25 observations follow a normal distribution quite closely. Although the subjects were students at one university, we judge that for a study of hand use they can be considered an SRS of all young people.

**Hypotheses:** If $\mu$ is the mean difference in the population, our hypotheses are

$$H_0 : \mu = 0$$
$$H_a : \mu > 0$$

Be careful of the order. We took the "left − right" differences, so the hypothesis that left takes longer becomes $\mu > 0$.

**Test:** Figure 20.5 contains the *Data Desk* output for this paired $t$ test. There is strong evidence ($P = 0.0039$) that the left-hand thread requires more time on the average.

**Confidence interval:** How much more time does the left-hand thread require on the average? We again apply the one-sample $t$ procedure to the 25 differ-

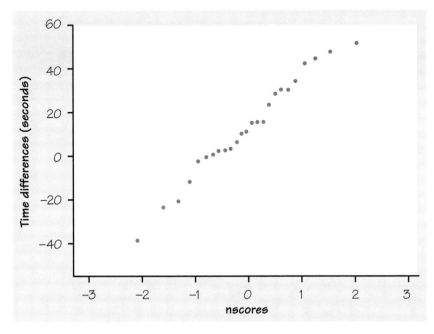

FIGURE 20.4   Normal probability plot of the time differences in Example 20.2.

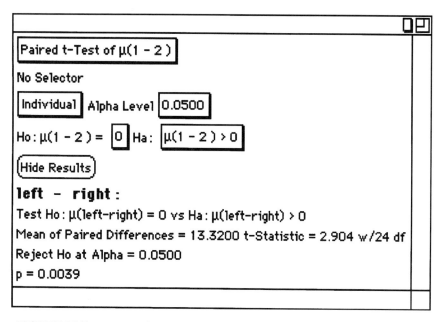

FIGURE 20.5    *Data Desk* output for the paired *t* test in Example 20.2.

ences. Software tells us that a 95% confidence interval for $\mu$ is

$$3.85 < \mu < 22.79$$

The interval is wide because the 25 subjects varied greatly, as indicated by the large standard deviation, $s = 22.936$ seconds. When there is a great deal of individual variability, it is hard to pin down the mean.    ◀

## ROBUSTNESS OF THE *t* PROCEDURES

The one-sample *t* procedures are exactly correct only when the population is normal. Real populations are never exactly normal. The usefulness of the *t* procedures in practice therefore depends on how strongly they are affected by lack of normality.

### ROBUST PROCEDURES

A confidence interval or statistical test is called **robust** if the confidence level or P-value does not change very much when the assumptions of the procedure are violated.

Like $\bar{y}$ and $s$, the $t$ procedures are strongly influenced by outliers. Fortunately, the $t$ procedures are quite robust against nonnormality of the population when there are no outliers, especially when the distribution is roughly symmetric and unimodal. Larger samples improve the accuracy of $P$-values and confidence levels from the $t$ distributions when the population is not normal. The reason for this is the central limit theorem. The $t$ statistic uses the sample mean $\bar{y}$, which becomes more nearly normal as the sample size gets larger even when the population does not have a normal distribution.

Always make a plot to check for skewness and outliers before you use the $t$ procedures for small samples. Here are practical guidelines for inference on a single mean.[55]

---

USING THE $t$ PROCEDURES

- Except in the case of small samples, the assumption that the data are an SRS from the population of interest is more important than the assumption that the population distribution is normal.
- *Sample size less than 15:* Use $t$ procedures if the data are close to normal. If the data are clearly nonnormal or if outliers are present, do not use $t$.
- *Sample size at least 15:* The $t$ procedures can be used except in the presence of outliers or strong skewness.
- *Large samples:* The $t$ procedures can be used even for clearly skewed distributions when the sample is large, roughly $n \geq 40$.

---

It is in part because they are robust that the $t$ procedures are among the most commonly used methods of inference.

## EXERCISES

20.1    Use software to verify the calculations of Example 20.1. That is, find $\bar{y}$, $s$, $t$, and the $P$-value.

20.2    Use software to verify the calculations of Example 20.2.

(a) Find the means for right-hand thread, left-hand thread, and the "left − right" differences. There is a relationship between these three numbers—what is it?

(b) Find the standard deviation of the differences and also the standard error $s/\sqrt{n}$ of their sample mean $\bar{y}$.

(c) Find the $t$ statistic from software and also by hand starting from $\bar{y}$ and its standard error. Then verify the $P$-value.

20.3    Exercise 18.4 (page 229) gives data from a study of the effects of DDT poisoning. Suppose that the mean absolutely refractory period for unpoisoned rats is known to be 1.3 milliseconds. DDT poisoning should slow nerve recovery and so increase this period. Do the data in Exercise 18.4 give good evidence for this supposition? State $H_0$ and $H_a$ and find $t$ and its $P$-value. What do you conclude from the test?

20.4    A pharmaceutical manufacturer does a chemical analysis to check the potency of products. The standard release potency for cephalothin crystals is 910. An assay of 16 lots gives the following potency data:

897    914    913    906    916    918    905    921
918    906    895    893    908    906    907    901

(a) Check the data for outliers or strong skewness that might threaten the validity of the $t$ procedures.

(b) Give a 95% confidence interval for the mean potency.

(c) Is there significant evidence at the 5% level that the mean potency is not equal to the standard release potency?

20.5    A bank wonders whether omitting the annual credit card fee for customers who charge at least $2400 in a year would increase the amount charged on its credit cards. The bank makes this offer to an SRS of 200 of its credit card customers. It then compares how much these customers charge this year with the amount that they charged last year. The mean increase is $332, and the standard deviation is $108.

(a) Is there significant evidence at the 1% level that the mean amount charged increases under the no-fee offer? State $H_0$ and $H_a$ and carry out a $t$ test. (To assess significance at the 1% level, you need only compare your $t$ with the 0.01 point of the appropriate $t$ distribution. If your $t$ lies beyond this point, it is significant.)

(b) Give a 99% confidence interval for the mean amount of the increase.

(c) The distribution of the amount charged is skewed to the right, but outliers are prevented by the credit limit that the bank enforces on each card. Use of the $t$ procedures is justified in this case even though the population distribution is not normal. Explain why.

(d) A critic points out that the customers would probably have charged more this year than last even without the new offer, because the economy is more prosperous and interest rates are lower. Briefly describe the design of an experiment to study the effect of the no-fee offer that would avoid this criticism.

20.6    The composition of the earth's atmosphere may have changed over time. To discover the nature of the atmosphere long ago, we can examine the gas in bubbles inside ancient amber. Amber is tree resin that has hardened and been trapped in rocks. The gas in bubbles within amber should be a sample of the atmosphere at the time the amber was formed. Measurements on specimens of amber from the late Cretaceous era (75 to 95 million years ago) give these percents of nitrogen:

63.4    65.0    64.4    63.3    54.8    64.5    60.8    49.1    51.0

These values are quite different from the present 78.1% of nitrogen in the atmosphere. Assume (this is not yet agreed on by experts) that these observations are an SRS from the late Cretaceous atmosphere.[56]

(a) Graph the data, and comment on skewness and outliers. The $t$ procedures will be only approximate in this case.

(b) Use a $t$ test to assess the strength of the evidence that the nitrogen content of ancient air differed from that of modern air.

(c) Give a 95% $t$ confidence interval for the mean percent of nitrogen in ancient air.

20.7    Exercise 18.9 (page 231) gives measurements of the density of the earth made by Henry Cavendish in 1798. Did Cavendish's experiment have a bias? The modern figure for the density of the earth is 5.517 times the density of water. Do a test that assesses whether the mean of Cavendish's measurements differs from 5.517. What do you conclude?

20.8    Exercise 18.10 (page 231) gives the IQ scores of 78 students and asks you to find a 95% confidence interval for a population if these students can be regarded as an SRS from that population. Use that confidence interval to answer these questions:

(a) The mean of IQ test scores is usually given as 100. Is there significant evidence at the 5% level that the population mean differs from 100? How do you know?

(b) Is there significant evidence at the 5% level that the population mean differs from 110? How do you know?

20.9    Differences of electric potential occur naturally from point to point on a body's skin. Is the natural electric field strength best for helping wounds to heal? If so, changing the field will slow healing. The research subjects are anesthetized newts. Make a razor cut in both hind limbs. Let one heal naturally (the control). Use an electrode to change the electric field in the other to half its normal value. After two hours, measure the healing rate. Here are healing rates (in micrometers per hour) for 14 newts:[57]

| Newt | Experimental limb | Control limb | Difference in healing |
|------|-------------------|--------------|-----------------------|
| 13 | 24 | 25 | −1 |
| 14 | 23 | 13 | 10 |
| 15 | 47 | 44 | 3 |
| 16 | 42 | 45 | −3 |
| 17 | 26 | 57 | −31 |
| 18 | 46 | 42 | 4 |
| 19 | 38 | 50 | −12 |
| 20 | 33 | 36 | −3 |
| 21 | 28 | 35 | −7 |
| 22 | 28 | 38 | −10 |
| 23 | 21 | 43 | −22 |
| 24 | 27 | 31 | −4 |
| 25 | 25 | 26 | −1 |
| 26 | 45 | 48 | −3 |

(a) As is usual, the paper did not report these raw data. Readers are expected to be able to interpret the summaries that the paper did report. The paper summarized the differences in the table above as "−5.71 ± 2.82" and said, "All values are expressed as means ± standard error of the mean." Show carefully where the numbers −5.71 and 2.82 come from.

(b) The researchers want to know if changing the electric field reduces the mean healing rate for all newts. State hypotheses, carry out a test, and give your conclusion. Is the result statistically significant at the 5% level? At the 1% level? (The researchers compared several field strengths and concluded that the natural strength is about right for fastest healing.)

(c) Give a 90% confidence interval for the amount by which changing the field changes the rate of healing. Then explain in a careful sentence what it means to say that you are "90% confident" of your result.

20.10    The Acculturation Rating Scale for Mexican Americans (ARSMA) measures the extent to which Mexican Americans have adopted Anglo/English culture. The ARSMA test was compared with a similar test, the Bicultural

Inventory (BI), by administering both tests to 22 Mexican Americans. Both tests have the same range of scores (1.00 to 5.00) and are scaled to have similar means for the groups used to develop them. There was a high correlation between the two scores, giving evidence that both are measuring the same characteristics. The researchers wanted to know whether the population mean scores for the two tests are the same. The differences in scores (ARSMA − BI) for the 22 subjects had $\bar{y} = 0.2519$ and $s = 0.2767$.

(a) Describe briefly how to arrange the administration of the two tests to the subjects, including randomization.

(b) Carry out a statistical test for the hypothesis that the two tests have the same population mean. Give the test statistic and the $P$-value and state your conclusion.

(c) Give a 95% confidence interval for the difference between the two population mean scores.

20.11  An agricultural field trial compares the yield of two varieties of tomatoes for commercial use. The researchers divide in half each of 10 small plots of land in different locations and plant each tomato variety on one half of each plot. After harvest, they compare the yields in pounds per plant at each location. The 10 differences (Variety A − Variety B) give $\bar{y} = 0.34$ and $s = 0.83$. Is there convincing evidence that Variety A has the higher mean yield?

(a) Describe in words what the parameter $\mu$ is in this setting.

(b) State $H_0$ and $H_a$.

(c) Find the $t$ statistic and give a $P$-value. What do you conclude?

20.12  In a study of the effectiveness of a weight-loss program, 47 subjects who were at least 20% overweight took part in the program for 10 weeks. Private weighings determined each subject's weight at the beginning of the program and 6 months after the program's end. The paired $t$ test was used to assess the significance of the average weight loss. The paper reporting the study said, "The subjects lost a significant amount of weight over time, $t(46) = 4.68$, $p < .01$." It is common to report the results of statistical tests in this abbreviated style.[58]

(a) Why was the paired $t$ test appropriate?

(b) Explain to someone who knows no statistics but is interested in weight-loss programs what the practical conclusion is.

(c) The paper follows the tradition of reporting significance only at fixed levels such as the 1% level. In fact, the results are more significant than "$p < .01$" suggests. Use software to find the $P$-value of the $t$ test.

| Day   | 1   | 2   | 3   | 4   | 5   | 6   | 7   | 8   | 9   | 10  | 11  | 12  |
|-------|-----|-----|-----|-----|-----|-----|-----|-----|-----|-----|-----|-----|
| **TABLE 20.2**  Particulate levels (grams) in two nearby locations | | | | | | | | | | | | |
| Day   | 1   | 2   | 3   | 4   | 5   | 6   | 7   | 8   | 9   | 10  | 11  | 12  |
| Rural | NA  | 67  | 42  | 33  | 46  | NA  | 43  | 54  | NA  | NA  | NA  | NA  |
| City  | 39  | 68  | 42  | 34  | 48  | 82  | 45  | NA  | NA  | 60  | 57  | NA  |
| Day   | 13  | 14  | 15  | 16  | 17  | 18  | 19  | 20  | 21  | 22  | 23  | 24  |
| Rural | 38  | 88  | 108 | 57  | 70  | 42  | 43  | 39  | NA  | 52  | 48  | 56  |
| City  | 39  | NA  | 123 | 59  | 71  | 41  | 42  | 38  | NA  | 57  | 50  | 58  |
| Day   | 25  | 26  | 27  | 28  | 29  | 30  | 31  | 32  | 33  | 34  | 35  | 36  |
| Rural | 44  | 51  | 21  | 74  | 48  | 84  | 51  | 43  | 45  | 41  | 47  | 35  |
| City  | 45  | 69  | 23  | 72  | 49  | 86  | 51  | 42  | 46  | NA  | 44  | 42  |

The remaining exercises concern a study of air pollution. Airborne particles such as dust and smoke are an important part of air pollution. To measure particulate pollution, a vacuum motor draws air through a filter for 24 hours. Weigh the filter at the beginning and end of the period. The weight gained is a measure of the concentration of particles in the air. A study of air pollution made measurements every 6 days with identical instruments in the center of a small city and at a rural location 10 miles southwest of the city. Because the prevailing winds blow from the west, we suspect that the rural readings will be generally lower than the city readings, but that the city readings can be predicted from the rural readings. Table 20.2 gives readings taken every 6 days over a 7-month period. The entry NA means that the reading for that date is not available, usually because of equipment failure.[59]

Missing data are common in field studies like this one. We think that equipment failures are not related to pollution levels. If that is true, the missing data do not introduce bias. We can work with the data that are not missing as if they are a random sample of days. We can analyze these data in different ways to answer different questions. For each of the three exercises below, do a careful descriptive analysis with graphs and summary statistics and whatever formal inference is called for. Then present and interpret your findings.

20.13   We want to assess the level of particulate pollution in the city center. Describe the distribution of city pollution levels, and estimate the mean particulate level in the city center. (All estimates should include a statistically justified margin of error.)

20.14   We want to compare the mean level of particulates in the city with the rural level on the same day. We suspect that pollution is higher in the city, and we hope that a statistical test will show that there is significant evidence to confirm this suspicion. Make a graph to check for conditions that might prevent the use of the test you plan to employ. Your graph should reflect the type of procedure that you will use. Then carry out a significance test and report your conclusion. Also estimate the mean amount by which the city particulate level exceeds the rural level on the same day.

20.15   We hope to use the rural particulate level to predict the city level on the same day. Make a graph to examine the relationship. Does the graph suggest that using the least-squares regression line for prediction will give approximately correct results over the range of values appearing in the data? Calculate the least-squares line for predicting city pollution from rural pollution. What percent of the observed variation in city pollution levels does this straight-line relationship account for? On the fourteenth date in the series, the rural reading was 88 and the city reading was not available. What do you estimate the city reading to be for that date? (In Lesson 24, we will learn how to give a margin of error for the predictions we make from the regression line.)

# PART V REVIEW

Part V introduces statistical inference. Statistical inference draws conclusions about a population on the basis of sample data and uses probability to indicate how reliable the conclusions are. A confidence interval estimates an unknown parameter. A significance test shows how strong the evidence is for some claim about a parameter.

The probabilities in both confidence intervals and tests tell us what would happen if we used the recipe for the interval or test very many times. A confidence level is the probability that the recipe for a confidence interval actually produces an interval that contains the unknown parameter. A 99% confidence interval gives a correct result 99% of the time when we use it repeatedly. A P-value is the probability that the test would produce a result at least as extreme as the observed result if the null hypothesis really were true. That is, a P-value tells us how surprising the observed outcome is. Very surprising outcomes (small P-values) are good evidence that the null hypothesis is not true. Lessons 17 and 19 discuss the reasoning of confidence intervals and tests. Lessons 18 and 20 present the t procedures for inference about the mean of a population. Here are the most important things you should be able to do after studying these lessons.

## A. CONFIDENCE INTERVALS

1. State in nontechnical language what is meant by "95% confidence" and other statements of confidence in statistical reports.

2. Use the one-sample $t$ procedure to obtain a confidence interval at a stated level of confidence for the mean $\mu$ of a population.

3. Recognize when you can safely use this confidence interval recipe and when the sample design, strong outliers, or a small sample from a skewed population makes it inaccurate.

4. Understand how the margin of error of a confidence interval changes with the sample size and the level of confidence $C$.

## B. SIGNIFICANCE TESTS

1. State the null and alternative hypotheses in a testing situation when the parameter in question is a population mean $\mu$.

2. Explain in nontechnical language the meaning of the $P$-value when you are given the numerical value of $P$ for a test.

3. Carry out both one-sided and two-sided tests about the mean $\mu$ of a population using the one-sample $t$ test.

4. Recognize when you can use the $t$ test and when the data collection design, strong outliers, or a small sample from a skewed population makes it inappropriate.

5. Assess statistical significance at standard levels such as 5% and 1% by comparing the $P$-value with the desired level of significance.

6. Recognize that significance testing does not measure the size or importance of an effect.

7. Recognize matched pairs data and use the paired $t$ procedures to obtain confidence intervals and to perform tests of significance for such data.

## REVIEW EXERCISES

**Review 1**    A shipment of machined parts has a critical dimension that is normally distributed with mean 12 centimeters and standard deviation 0.01 centimeter. The acceptance sampling team measures a random sample of 25 of these parts. What is the sampling distribution of the sample mean $\bar{y}$ of the critical dimension for these parts?

**Review 2**    The Survey of Study Habits and Attitudes (SSHA) is a psychological test that evaluates college students' motivation, study habits, and attitudes toward school. A private college gives the SSHA to a sample of 18 of its incoming first-year women students. Their scores are

| 154 | 109 | 137 | 115 | 152 | 140 | 154 | 178 | 101 |
| 103 | 126 | 126 | 137 | 165 | 165 | 129 | 200 | 148 |

The college wants to estimate the mean SSHA score of all its first-year women from these data. Give a 90% confidence interval. (Be sure to examine the data for shape and outliers first.)

**Review 3**    Gas chromatography is a sensitive technique used to measure small amounts of compounds. The response of a gas chromatograph is calibrated by repeatedly testing specimens containing a known amount of the compound to be measured. A calibration study for a specimen containing 1 nanogram (that's $10^{-9}$ gram) of a compound gave the following response readings:[60]

21.6    20.0    25.0    21.9

The response is known from experience to vary according to a normal distribution unless an outlier indicates an error in the analysis. Estimate the mean response to 1 nanogram of this substance, and give the margin of error for 90% confidence. Then explain to a chemist who knows no statistics what your margin of error means.

**Review 4**    Great white sharks are big and hungry. Here are the lengths in feet of 44 great whites:[61]

| 18.7 | 12.3 | 18.6 | 16.4 | 15.7 | 18.3 | 14.6 | 15.8 | 14.9 | 17.6 | 12.1 |
| 16.4 | 16.7 | 17.8 | 16.2 | 12.6 | 17.8 | 13.8 | 12.2 | 15.2 | 14.7 | 12.4 |
| 13.2 | 15.8 | 14.3 | 16.6 | 9.4 | 18.2 | 13.2 | 13.6 | 15.3 | 16.1 | 13.5 |
| 19.1 | 16.2 | 22.8 | 16.8 | 13.6 | 13.2 | 15.7 | 19.7 | 18.7 | 13.2 | 16.8 |

(a) Examine these data. A histogram shows some outliers. Are there any outliers by the $1.5 \times IQR$ criterion? What is the overall shape of the distribution? Is it reasonably normal?

(b) The $t$ procedures can be applied safely to these data. Why?

(c) Give a 95% confidence interval for the mean length of great white sharks. Based on this interval, is there significant evidence at the 5% level to reject the claim "Great white sharks average 20 feet in length"?

(d) It isn't clear exactly what parameter $\mu$ you estimated in (c). What information do you need to say what $\mu$ is?

**Review 5**    The embryos of brine shrimp can enter a dormant phase in which metabolic activity drops to a low level. Researchers studying this dormant phase measured the level of several compounds important to normal metabolism. They reported their results in a table, with the note, "Values are means ± SEM for three independent samples." The table entry for the compound ATP was $0.84 \pm 0.01$. Biologists reading the article must be able to decipher this.[62]

(a) What does the abbreviation SEM stand for?

(b) The researchers made three measurements of ATP, which had $\bar{y} = 0.84$. What was the sample standard deviation $s$ for these measurements?

(c) Give a 90% confidence interval for the mean ATP level in dormant brine shrimp embryos.

**Review 6**   Table 3.1 (page 13) gives the percent of the population 65 years of age and older in each of the 50 states. It does not make sense to use these data to give a confidence interval for the mean proportion over 65 in the states. Why not?

**Review 7**   The amount of dust in the atmosphere of coal mines is measured by exposing a filter in the mine and then weighing the dust collected by the filter. Repeated weighings of the same filter will vary according to a normal distribution. The values that would be obtained in many weighings form the population we are interested in. The mean $\mu$ of this population is the true weight (that is, there is no bias in the weighing). The population standard deviation describes the precision of the weighing; it is known from long experience to be $\sigma = 0.08$ milligram (mg). Each filter is weighed three times and the mean weight is reported. For one filter the three weights are

$$123.1 \text{ mg} \qquad 122.5 \text{ mg} \qquad 123.7 \text{ mg}$$

(a) Find a 95% confidence interval for $\mu$ using the known $\sigma = 0.08$ mg.

(b) Find a 95% confidence interval for $\mu$ treating $\sigma$ as unknown. How does not knowing $\sigma$ change the length of the interval?

**Review 8**   The mean area of the several thousand apartments in a new development is advertised to be 1250 square feet. A tenant group thinks that the apartments are smaller than advertised. They hire an engineer to measure a sample of apartments to test their suspicion. What are the null hypothesis $H_0$ and alternative hypothesis $H_a$?

**Review 9**   Experiments on learning in animals sometimes measure how long it takes mice to find their way through a maze. The mean time is 18 seconds for one particular maze. A researcher thinks that a loud noise will cause the mice to complete the maze faster. She measures how long each of 10 mice takes with a noise as stimulus. What are the null hypothesis $H_0$ and alternative hypothesis $H_a$?

**Review 10**  Makers of generic drugs must show that they do not differ significantly from the "reference" drug that they imitate. One aspect in which drugs might differ is their extent of absorption in the blood. Table V.1 gives data for one pair of drugs taken from 20 healthy nonsmoking male subjects.[63] This is a matched pairs design. Subjects 1 to 10 received the generic drug first, and subjects 11 to 20 received the reference drug first. In all cases, a washout period separated the two drugs so that the first has disappeared from the blood before the subject took the second. The subject numbers in the table were assigned at random to decide the order of the drugs for each subject.

| **TABLE V.1** | Absorption extent for two versions of a drug | |
|---|---|---|
| Subject | Reference drug | Generic drug |
| 15 | 4108 | 1755 |
| 3 | 2526 | 1138 |
| 9 | 2779 | 1613 |
| 13 | 3852 | 2254 |
| 12 | 1833 | 1310 |
| 8 | 2463 | 2120 |
| 18 | 2059 | 1851 |
| 20 | 1709 | 1878 |
| 17 | 1829 | 1682 |
| 2 | 2594 | 2613 |
| 4 | 2344 | 2738 |
| 16 | 1864 | 2302 |
| 6 | 1022 | 1284 |
| 10 | 2256 | 3052 |
| 5 | 938 | 1287 |
| 7 | 1339 | 1930 |
| 14 | 1262 | 1964 |
| 11 | 1438 | 2549 |
| 1 | 1735 | 3340 |
| 19 | 1020 | 3050 |

(a) Do a data analysis of the differences between the absorption measures for the generic and reference drugs. Is there any reason not to
apply $t$ procedures?

(b) Use a $t$ test to answer the key question: do the drugs differ significantly in absorption?

**Review 11**    A study of the pay of corporate CEOs (chief executive of-
ficers) examined the cash compensation, adjusted for inflation, of the
CEOs of 104 corporations over the period 1977 to 1988. Among the
data are the average annual pay increases for each of the 104 CEOs. The
mean percent increase in pay was 6.9%. The data showed great variation, with a standard deviation of 17.4%. The distribution was strongly
skewed to the right.[64]

(a) Despite the skewness of the distribution, there were no extreme outliers. Explain why we can use $t$ procedures for these data.

(b) Give a 99% confidence interval for the mean increase in pay for all
corporate CEOs. What essential condition must the data satisfy if we
are to trust your result?

**Review 12**    A manufacturer of small appliances employs a market research firm to estimate retail sales of its products by gathering information from a sample of retail stores. This month an SRS of 75 stores in the
Midwest sales region finds that these stores sold an average of 24 of the
manufacturer's hand mixers, with standard deviation 11.

(a) Give a 95% confidence interval for the mean number of mixers sold
by all stores in the region.

(b) The distribution of sales is strongly right skewed, because there are
many smaller stores and a few very large stores. The use of $t$ in (a) is
reasonably safe despite this violation of the normality assumption.
Why?

**Review 13**    A company compares two package designs for a laundry
detergent by placing bottles with both designs on the shelves of several
markets at the same price. Checkout scanner data on more than 5000
bottles bought show that more shoppers bought Design A than Design
B. The difference is statistically significant ($P = 0.02$). Can we conclude
that consumers strongly prefer Design A? Explain your answer.

**Review 14**    A study compares two groups of mothers with young children who were on welfare two years ago. One group attended a voluntary training program offered free of charge at a local vocational school
and advertised in the local news media. The other group did not choose
to attend the training program. The study finds a significant difference
($P < 0.01$) between the proportions of the mothers in the two groups

who are still on welfare. The difference is not only significant but quite large. The report says that with 95% confidence the percent of the nonattending group still on welfare is 21% ± 4% higher than that of the group who attended the program. You are on the staff of a member of Congress who is interested in the plight of welfare mothers, and who asks you about the report.

(a) Explain in simple language what "a significant difference ($P < 0.01$)" means.

(b) Explain clearly and briefly what "95% confidence" means.

(c) Is this study good evidence that requiring job training of all welfare mothers would greatly reduce the percent who remain on welfare for several years?

Review 15    A group of psychologists once measured 77 variables on a sample of schizophrenic people and a sample of people who were not schizophrenic. They compared the two samples using 77 separate significance tests. Two of these tests were significant at the 5% level. Suppose that there is in fact no difference on any of the 77 variables between people who are and people who are not schizophrenic in the adult population. Then all 77 null hypotheses are true.

(a) What is the probability that one specific test shows a difference significant at the 5% level?

(b) Why is it not surprising that 2 of the 77 tests were significant at the 5% level?

Review 16    A government report gives a 99% confidence interval for the 1994 median family income as $32,264 ± $397. This result was calculated by advanced methods from the Current Population Survey, a multistage random sample of about 60,000 households.

(a) Would a 95% confidence interval be wider or narrower? Explain your answer.

(b) Would the null hypothesis that the 1994 median family income was $30,000 be rejected at the 1% significance level in favor of the two-sided alternative?

Review 17    Statisticians prefer large samples. Describe briefly the effect of increasing the size of a sample (or the number of subjects in an experiment) on each of the following:

(a) The margin of error of a 95% confidence interval.

(b) The $P$-value of a test, when $H_0$ is false and all facts about the population remain unchanged as $n$ increases.

## JANET NORWOOD

The commissioner of labor statistics is one of the nation's most influential statisticians. As head of the Bureau of Labor Statistics, the commissioner supervises the collection and interpretation of data on employment, earnings, and many other economic and social trends.

The data collected by the Bureau of Labor Statistics are often politically sensitive, as when a report released just before an election shows rising unemployment. For this reason, the bureau must remain objective and independent of political influence. To safeguard the bureau's independence, the commissioner is appointed by the president and confirmed by the Senate for a fixed term of four years. The commissioner must have statistical skill, administrative ability, and a facility for working with both Congress and the president.

Janet Norwood served three terms as commissioner, from 1979 to 1991, under three presidents. When she retired, the *New York Times* said (December 31, 1991) that she left with "a near-legendary reputation for nonpartisanship and plaudits that include one senator's designation of her as a 'national treasure.'" Norwood says, "There have been times in the past when commissioners have been in open disagreement with the Secretary of Labor or, in some cases, with the President. We have guarded our professionalism with great care."

Comparison and prediction are the themes of the various inference methods presented in Part VI. Is this month's unemployment rate significantly higher than last month's? Has the mean weekly wage of factory workers changed? How well do earnings predict saving? The Bureau of Labor Statistics, like other users of data, needs the methods we will now meet.

PART **VI**

**Topics in Inference**

LESSON **21**

# COMPARING
# TWO MEANS

Comparing two populations or two treatments based on separate samples from each is one of the most common situations encountered in statistical practice. We call such situations **two-sample problems.** A two-sample problem can arise from a randomized comparative experiment that randomly divides subjects into two groups and exposes each group to a different treatment. Comparing random samples separately selected from two populations is also a two-sample problem. Here are some typical two-sample problems.

(**a**) A medical researcher is interested in the effect on blood pressure of added calcium in our diet. She conducts a randomized comparative experiment in which one group of subjects receives a calcium supplement and a control group gets a placebo.

(**b**) A bank wants to know which of two incentive plans will most increase the use of its credit cards. It offers each incentive to a random sample of credit card customers and compares the amount charged during the following six months.

◄

Pay careful attention to the distinction between two-sample designs and the matched pairs designs studied in Lesson 20. There is no matching of the subjects in a two-sample problem, and the two samples can be of different sizes. Inference procedures for two-sample data differ from those for matched pairs. This is a general principle in statistics: *inference methods must reflect the design that produced the data.*

## COMPARING TWO POPULATION MEANS

In Lesson 6, we saw that we can examine two-sample data graphically by comparing histograms or boxplots. Now we will apply the ideas of formal inference in this setting. When both population distributions are symmetric, and especially when they are at least approximately normal, a comparison of the mean responses in the two populations is the most common goal of inference. Here are the assumptions we will make.

---

### ASSUMPTIONS FOR COMPARING TWO MEANS

- We have **two SRSs,** from two distinct populations. The samples are **independent.** That is, one sample has no influence on the other. Matching violates independence, for example. We measure the same variable for both samples.
- Both populations are **normally distributed.** The means and standard deviations of the populations are unknown.

---

Call the variable we measure $y_1$ in the first population and $y_2$ in the second because the variable may have different distributions in the two populations. Here is the notation we will use to describe the two populations:

| Population | Variable | Mean | Standard deviation |
|:---:|:---:|:---:|:---:|
| 1 | $y_1$ | $\mu_1$ | $\sigma_1$ |
| 2 | $y_2$ | $\mu_2$ | $\sigma_2$ |

There are four unknown parameters: the two means and the two standard deviations. The subscripts remind us which population a parameter describes. We want to compare the two population means, either by giving a confidence interval for their difference, $\mu_1 - \mu_2$, or by testing the hypothesis of no difference, $H_0 : \mu_1 - \mu_2 = 0$.

We use the sample means and standard deviations to estimate the unknown parameters. Again, subscripts remind us which sample a statistic comes from. Here is the notation that describes the samples:

| Population | Sample size | Sample mean | Sample standard deviation |
|------------|-------------|-------------|---------------------------|
| 1 | $n_1$ | $\bar{y}_1$ | $s_1$ |
| 2 | $n_2$ | $\bar{y}_2$ | $s_2$ |

To do inference about the difference $\mu_1 - \mu_2$ between the means of the two populations, we start from the difference $\bar{y}_1 - \bar{y}_2$ between the means of the two samples.

## EXAMPLE 21.2

Do poultry hot dogs and beef hot dogs contain different amounts of sodium? Table 21.1 contains data on 17 brands of poultry hot dogs and 20 brands of beef hot dogs, from a study by *Consumer Reports* magazine.[65] We are willing to regard the brands tested as random samples from all brands of the two types of hot dog. The sodium content is recorded in milligrams (mg). Boxplots (Figure 21.1) suggest that poultry hot dogs have somewhat more sodium, but there is substantial overlap between the two types. We can calculate that:

| Group | Type | $n$ | $\bar{y}$ | $s$ |
|-------|------|-----|-----------|-----|
| 1 | Poultry | 17 | 459.00 | 84.739 |
| 2 | Beef | 20 | 401.15 | 102.435 |

The observed difference in mean sodium content is

$$\bar{y}_1 - \bar{y}_2 = 459.00 - 401.15 = 57.85 \text{ mg}$$

Is this difference statistically significant?

| TABLE 21.1    Calories and sodium in hot dogs | | | | | |
|---|---|---|---|---|---|
| Beef hot dogs | | Meat hot dogs | | Poultry hot dogs | |
| Calories | Sodium | Calories | Sodium | Calories | Sodium |
| 186 | 495 | 173 | 458 | 129 | 430 |
| 181 | 477 | 191 | 506 | 132 | 375 |
| 176 | 425 | 182 | 473 | 102 | 396 |
| 149 | 322 | 190 | 545 | 106 | 383 |
| 184 | 482 | 172 | 496 | 94 | 387 |
| 190 | 587 | 147 | 360 | 102 | 542 |
| 158 | 370 | 146 | 387 | 87 | 359 |
| 139 | 322 | 139 | 386 | 99 | 357 |
| 175 | 479 | 175 | 507 | 170 | 528 |
| 148 | 375 | 136 | 393 | 113 | 513 |
| 152 | 330 | 179 | 405 | 135 | 426 |
| 111 | 300 | 153 | 372 | 142 | 513 |
| 141 | 386 | 107 | 144 | 86 | 358 |
| 153 | 401 | 195 | 511 | 143 | 581 |
| 190 | 645 | 135 | 405 | 152 | 588 |
| 157 | 440 | 140 | 428 | 146 | 522 |
| 131 | 317 | 138 | 339 | 144 | 545 |
| 149 | 319 | | | | |
| 135 | 298 | | | | |
| 132 | 253 | | | | |

SOURCE: *Consumer Reports,* June 1986, pp. 366–367.

Example 21.1 fits the two-sample setting. We write hypotheses in terms of the difference between the mean sodium contents in the two populations of hot dog brands, $\mu_1 - \mu_2$. The hypotheses are

$$H_0 : \mu_1 - \mu_2 = 0$$
$$H_a : \mu_1 - \mu_2 \neq 0$$

We want to test these hypotheses and also give a confidence interval for the size of the difference, $\mu_1 - \mu_2$. Our starting point is the estimate $\bar{y}_1 - \bar{y}_2$ from our sample data.

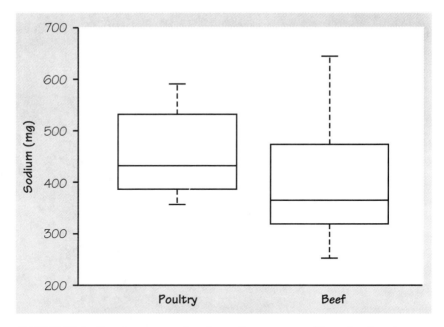

FIGURE 21.1    Boxplots comparing the sodium content of brands of poultry and beef hot dogs.

## THE TWO-SAMPLE $t$ STATISTIC

To assess the significance of the observed difference between the means of our two samples, we follow a familiar path. Whether an observed difference between two samples is surprising depends on the spread of the observations as well as on the two means. Widely different means can arise just by chance if the individual observations vary a great deal. To take variation into account, we would like to standardize the observed difference $\bar{y}_1 - \bar{y}_2$ by dividing by its standard deviation. This standard deviation is

$$\sigma_{\bar{y}_1 - \bar{y}_2} = \sqrt{\frac{\sigma_1^2}{n_1} + \frac{\sigma_2^2}{n_2}}$$

This standard deviation gets larger as either population gets more variable, that is, as $\sigma_1$ or $\sigma_2$ increases. It gets smaller as the sample sizes $n_1$ and $n_2$ increase.

Because we don't know the population standard deviations, we estimate them by the sample standard deviations from our two samples. The result is the *standard error,* or estimated standard deviation, of the difference in sample means:

$$SE_{\bar{y}_1 - \bar{y}_2} = \sqrt{\frac{s_1^2}{n_1} + \frac{s_2^2}{n_2}}$$

When we standardize the estimate by dividing it by its standard error, the result is the **two-sample *t* statistic:**

$$t = \frac{\bar{y}_1 - \bar{y}_2}{SE_{\bar{y}_1 - \bar{y}_2}}$$

The statistic *t* has the same interpretation as any *z* or *t* statistic: it says how far $\bar{y}_1 - \bar{y}_2$ is from 0 in standard deviation units.

Unfortunately, the two-sample *t* statistic does *not* have a *t* distribution. Nonetheless, statistics software uses the two-sample *t* statistic with a *t* distribution to do inference for two-sample problems. This is possible because *t* has *approximately* a *t* distribution with degrees of freedom calculated from the data by a somewhat complex formula. The approximation is very accurate, and software doesn't mind the complexity.[66]

## TWO-SAMPLE *t* PROCEDURES

Statistics software implements tests and confidence intervals based on the *t* approximation just discussed.

---

THE TWO-SAMPLE *t* PROCEDURES

Draw an SRS of size $n_1$ from a normal population with unknown mean $\mu_1$, and draw an independent SRS of size $n_2$ from another normal population with unknown mean $\mu_2$. A level *C* confidence

---

*(continued on next page)*

*(continued from previous page)*

interval for $\mu_1 - \mu_2$ given by

$$(\bar{y}_1 - \bar{y}_2) \pm t^* \text{SE}_{\bar{y}_1 - \bar{y}_2}$$

Here $t^*$ comes from a $t$ distribution with degrees of freedom $k$ found by the software.

To test the hypothesis $H_0 : \mu_1 - \mu_2 = 0$, compute the two-sample $t$ statistic

$$t = \frac{\bar{y}_1 - \bar{y}_2}{\text{SE}_{\bar{y}_1 - \bar{y}_2}}$$

and use $P$-values from the $t$ distribution with $k$ degrees of freedom.

We can now complete Example 21.2.

## EXAMPLE 21.3

***Data analysis:*** We have compared the two samples with boxplots and numerical summaries. Normal probability plots appear in Figure 21.2. The beef hot dogs show no important departures from normality. The poultry hot dogs appear to contain two groups with a gap between them. The $t$ procedures are again quite robust in the absence of outliers and strong skewness, so we can apply them to these data.

***Test:*** Figure 21.3(a) displays *Data Desk* output for testing $H_0 : \mu_1 - \mu_2 = 0$ against the two-sided alternative. We see that $t = 1.880$ with $P = 0.0685$. There is some evidence that the mean sodium contents differ, but the evidence is not significant at the 5% level. More data (measurements on additional brands of hot dogs) might produce more convincing evidence.

***Confidence interval:*** The 95% confidence interval produced by *Data Desk* appears in Figure 21.3(b). It is

$$-4.626 < \mu_1 - \mu_2 < 120.326$$

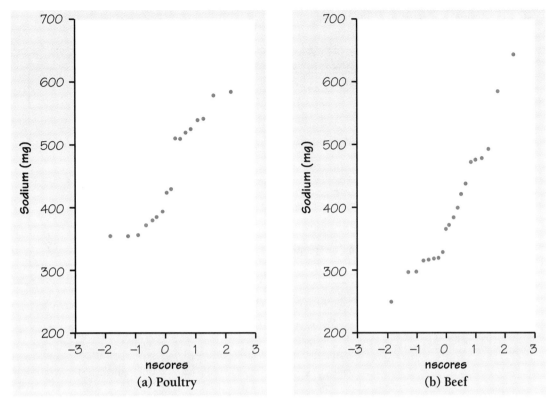

FIGURE 21.2 Normal probability plots for the sodium content of brands of poultry and beef hot dogs.

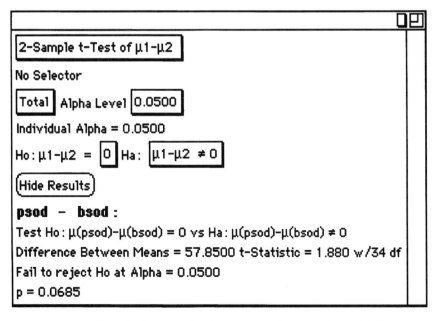

FIGURE 21.3 *Data Desk* output for the two-sample *t* procedures. (a) Test.

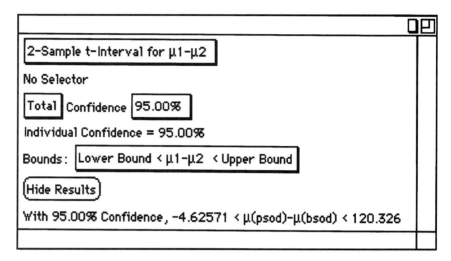

FIGURE 21.3 (*continued*)  *Data Desk* output for the two-sample *t* procedures.
(b) Confidence interval.

Because we could not reject $H_0$ at the 5% level, we expected the 95% confidence interval to include 0. The confidence interval is wide, and the test failed to find strong evidence against $H_0$, because the individual brands in each type vary so much in sodium content.    ◄

## A SHORTCUT

There is a modification of the two-sample *t* procedures that is easy to use with only a calculator that will give you $\bar{y}$ and *s* from keyed-in data. In place of the degrees of freedom found by software, just use the smaller of $n_1 - 1$ and $n_2 - 1$ as your degrees of freedom. The resulting procedures are *conservative*. That is, the actual confidence level is a bit larger than the level you ask for, and the actual *P*-value is a bit smaller than your calculations give.

**EXAMPLE 21.4**

Let us apply the conservative test to the hot dog data. The two-sample *t* statistic is unchanged: $t = 1.880$. The conservative degrees of freedom is df = 16, the smaller of $n_1 - 1 = 16$ and $n_2 - 1 = 19$. Comparing $t = 1.880$ with the entries in the df = 16 row of Table B, we find that *t* lies between the 0.025 and 0.05 tail area points. The two-sided *P*-value therefore lies between 0.05

and 0.10. Software gives the *P*-value for $t = 1.880$ with 16 degrees of free-dom as $P = 0.0784$. This is a bit larger (more conservative) than the result of Example 21.3. ◀

## THE POOLED *t* PROCEDURES

There is a special version of the two-sample *t* statistic that assumes that the two populations have the same standard deviation. This procedure in effect combines (the statistical term is "pools") the two samples to estimate the common population standard deviation. The resulting statistic is called the pooled *t* statistic. It is equal to the two-sample *t* statistic if the two sample sizes are the same, but not otherwise.

The pooled *t* statistic has the advantage that it has exactly the *t* distribution with $n_1 + n_2 - 2$ degrees of freedom *if* the two population variances really are equal. Of course, the population variances are often not equal. Moreover, the assumption of equal variances is hard to check from the data. The pooled *t* was in common use before software made it easy to use the accurate approximation to the distribution of our two-sample *t* statistic. Now it is useful only in special situations.

If the two population standard deviations are equal, the pooled estimate of the common standard deviation is

$$s_p = \sqrt{\frac{(n_1 - 1)s_1^2 + (n_2 - 1)s_2^2}{n_1 + n_2 - 2}}$$

and the **pooled *t* statistic** is

$$t = \frac{\bar{y}_1 - \bar{y}_2}{s_p\sqrt{\dfrac{1}{n_1} + \dfrac{1}{n_2}}}$$

Use the *t* distribution with $n_1 + n_2 - 2$ degrees of freedom.

## EXAMPLE 21.5

In the hot dog example, asking for the pooled procedures in software gives the two-sided test result

$$t = 1.8507 \qquad df = 35 \qquad P = 0.0727$$

and the 95% confidence interval

$$-5.608 < \mu_1 - \mu_2 < 121.308$$

There is little difference between these results and those in Example 21.3. If the results of the two methods do differ substantially, it is usually because the sample standard deviations differ greatly. The pooled method is not valid in that situation.                                    ◄

## ROBUSTNESS AGAIN

The two-sample $t$ procedures are more robust than the one-sample $t$ methods, particularly when the distributions are not symmetric. When the sizes of the two samples are equal and the two populations being compared have distributions with similar shapes, probability values from the $t$ distribution are quite accurate for a broad range of distributions when the sample sizes are as small as $n_1 = n_2 = 5$.[67] When the two population distributions have different shapes, larger samples are needed.

As a guide to practice, adapt the guidelines given on page 256 for the one-sample $t$ procedures to two-sample procedures by replacing "sample size" with the "sum of the sample sizes," $n_1 + n_2$. These guidelines err on the side of safety, especially when the two samples are of similar size. In planning a two-sample study, you should usually choose roughly equal sample sizes. The two-sample $t$ procedures are most robust against nonnormality in this case, and the conservative probability values are most accurate.

Robustness explains the great usefulness of the one-sample and two-sample $t$ procedures. Robustness (or rather, lack of robustness) also explains why we do not present methods for comparing the *spreads* of two distributions. There are methods for comparing the standard deviations of two normal populations. But these procedures are extremely sensitive to nonnormal distributions, so much so that they are of little practical value.[68]

It was once common to test equality of standard deviations as a preliminary to performing the pooled $t$ test for equality of two population means. It is better practice to check the distributions graphically, with special attention to skewness and outliers, and to use the two-sample $t$ test. This test does not require equal standard deviations.

## EXERCISES

**21.1**   Verify the calculations required in Example 21.3. Starting from the data in Table 21.1, find the means and standard deviations, the standard error $SE_{\bar{y}_1 - \bar{y}_2}$, and the two-sample $t$ statistic.

**21.2**   The Chapin Social Insight Test is a psychological test designed to measure how accurately a person appraises other people. The possible scores on the test range from 0 to 41. During the development of the Chapin test, it was given to several different groups of people. Here are the results for male and female college students majoring in the liberal arts:[69]

| Group | Sex | $n$ | $\bar{y}$ | $s$ |
|-------|------|-----|-------|------|
| 1 | Male | 133 | 25.34 | 5.05 |
| 2 | Female | 162 | 24.94 | 5.44 |

Do these data support the claim that female and male students differ in average social insight?

(a) Use the two-sample $t$ test with conservative degrees of freedom. You can use software to get the $P$-value, or you can bracket it by two values from Table B.

(b) The two sample standard deviations are almost equal. Repeat the test using the pooled-sample $t$ statistic. Compare your results with those from (a).

**21.3**   In a study of heart surgery, one issue was the effect of drugs called beta-blockers on the pulse rate of patients during surgery. The available subjects were divided at random into two groups of 30 patients each. One group received a beta-blocker; the other, a placebo. The surgical team recorded the pulse rate of each patient at a critical point during the operation. The treatment group had mean 65.2 beats per minute and standard deviation 7.8. For the control group, the mean was 70.3 and the standard deviation 8.3. The data appear roughly normal.

(a) Do beta-blockers reduce the pulse rate? State the hypotheses and calculate the two-sample $t$ statistic. Is the result significant at the 5% level? At the 1% level?

(b) Give a 99% confidence interval for the difference in mean pulse rates.

**21.4**   In a study of cereal leaf beetle damage on oats, researchers measured the number of beetle larvae per stem in small plots of oats after randomly applying one of two treatments: no pesticide, or malathion at the rate of 0.25

pound per acre. The data appear roughly normal. Here are the summary statistics:[70]

| Group | Treatment | $n$ | $\bar{y}$ | $s$ |
|-------|-----------|-----|-----------|-----|
| 1 | Control | 13 | 3.47 | 1.21 |
| 2 | Malathion | 14 | 1.36 | 0.52 |

Is there significant evidence at the 1% level that malathion reduces the mean number of larvae per stem? Be sure to state $H_0$ and $H_a$.

21.5 A business school study compared a sample of Greek firms that went bankrupt with a sample of healthy Greek businesses. One measure of a firm's financial health is the ratio of current assets to current liabilities, called $CA/CL$. For the year before bankruptcy, the study found the mean $CA/CL$ to be 1.72565 in the healthy group and 0.78640 in the group that failed. The paper reporting the study says that $t = 7.36$.[71]

You can draw a conclusion from this $t$ without calculating $P$ and even without knowing the sizes of the samples (as long as the samples are not tiny). What is your conclusion? Why don't you need more information?

21.6 The following situations require inference about a mean or means. Identify each as (1) single sample, (2) matched pairs, or (3) two samples. The procedures of Lesson 20 apply to cases (1) and (2). This lesson concerns procedures for (3).

(a) An education researcher wants to learn whether it is more effective to put questions before or after introducing a new concept in an elementary school mathematics text. He prepares two text segments that teach the concept, one with motivating questions before and the other with review questions after. He uses each text segment to teach a separate group of children. The researcher compares the scores of the groups on a test over the material.

(b) Another researcher approaches the same issue differently. She prepares text segments on two unrelated topics. Each segment comes in two versions, one with questions before and the other with questions after. The subjects are a single group of children. Each child studies both topics, one (chosen at random) with questions before and the other with questions after. The researcher compares test scores for each child on the two topics to see which topic he or she learned better.

21.7 The following situations require inference about a mean or means. Identify each as (1) single sample, (2) matched pairs, or (3) two samples. The procedures of Lesson 20 apply to cases (1) and (2). This lesson concerns procedures for (3).

(a) To check a new analytical method, a chemist obtains a reference specimen of known concentration from the National Institute of Standards and Technology. She then makes 20 measurements of the concentration of this specimen with the new method and checks for bias by comparing the mean result with the known concentration.

(b) Another chemist is checking the same new method. He has no reference specimen, but a familiar analytic method is available. He wants to know if the new and old methods agree. He takes a specimen of unknown concentration and measures the concentration 10 times with the new method and 10 times with the old method.

21.8    Ordinary corn doesn't have as much of the amino acid lysine as animals need in their feed. Plant scientists have developed varieties of corn that have increased amounts of lysine. In a test of the quality of high-lysine corn as animal feed, an experimental group of 20 one-day-old male chicks ate a ration containing the new corn. A control group of another 20 chicks received a ration that was identical except that it contained normal corn. Here are the weight gains (in grams) after 21 days:[72]

| Control | | | | Experimental | | | |
|-----|-----|-----|-----|-----|-----|-----|-----|
| 380 | 321 | 366 | 356 | 361 | 447 | 401 | 375 |
| 283 | 349 | 402 | 462 | 434 | 403 | 393 | 426 |
| 356 | 410 | 329 | 399 | 406 | 318 | 467 | 407 |
| 350 | 384 | 316 | 272 | 427 | 420 | 477 | 392 |
| 345 | 455 | 360 | 431 | 430 | 339 | 410 | 326 |

(a) Examine the data graphically. Are there outliers or strong skewness that might prevent the use of $t$ procedures?

(b) Is there good evidence that chicks fed high-lysine corn gain weight faster? Carry out a test and report your conclusions.

(c) Give a 95% confidence interval for the mean extra weight gain in chicks fed high-lysine corn.

21.9    The Survey of Study Habits and Attitudes (SSHA) is a psychological test that measures the motivation, attitude toward school, and study habits of students. Scores range from 0 to 200. A selective private college gives the SSHA to an SRS of both male and female first-year students. The data for the women are as follows:

| | | | | | | | | |
|-----|-----|-----|-----|-----|-----|-----|-----|-----|
| 154 | 109 | 137 | 115 | 152 | 140 | 154 | 178 | 101 |
| 103 | 126 | 126 | 137 | 165 | 165 | 129 | 200 | 148 |

Here are the scores of the men:

108  140  114  91  180  115  126  92  169  146
109  132  75   88  113  151  70   115 187  104

(a) Examine each sample graphically, with special attention to outliers and skewness. Is use of a *t* procedure acceptable for these data?

(b) Most studies have found that the mean SSHA score for men is lower than the mean score in a comparable group of women. Is this true for first-year students at this college? Carry out a test and give your conclusions.

(c) Give a 90% confidence interval for the mean difference between the SSHA scores of male and female first-year students at this college.

21.10   Poisoning by the pesticide DDT causes convulsions in humans and other mammals. In a randomized comparative experiment, biologists compared 6 white rats poisoned with DDT with a control group of 6 unpoisoned rats. Electrical measurements of nerve activity are the main clue to the nature of DDT poisoning. When a nerve is stimulated, its electrical response shows a sharp spike followed by a much smaller second spike. The experiment found that the second spike is larger in rats fed DDT than in normal rats. This finding was a clue about how DDT poisoning works.[73]

The researchers measured the height of the second spike as a percent of the first spike when a nerve in the rat's leg was stimulated. For the poisoned rats the results were

12.207  16.869  25.050  22.429  8.456  20.589

The control group data were

11.074  9.686  12.064  9.351  8.182  6.642

(a) Are the data reasonably normal, as far as you can judge from 6 observations?

(b) The researchers did not conjecture in advance that DDT would produce a larger spike, so they used a *t* test with a two-sided alternative. State hypotheses, give the basic descriptive statistics, carry out the test, and state your conclusion.

21.11   A study of computer-assisted learning examined the learning of "Blissymbols" by children. Blissymbols are pictographs (think of Egyptian hieroglyphics) that are sometimes used to help learning-impaired children communicate. The researcher designed two computer lessons that taught the same content using the same examples. One lesson required the children to interact with the material, while in the other the children controlled only the pace of the lesson. Call these two styles "Active" and "Passive." After the lesson, the computer presented a quiz that asked the children to identify 56 Blissymbols. Here are the numbers of correct identifications by the 24 children in the Active group:[74]

29 28 24 31 15 24 27 23 20 22 23 21 24 35 21 24 44 28 17 21 21 20 28 16

The 24 children in the Passive group had these counts of correct identifications:

16 14 17 15 26 17 12 25 21 20 18 21 20 16 18 15 26 15 13 17 21 19 15 12

Is there good evidence that active learning is superior to passive learning? State hypotheses, give a test and its $P$-value, and state your conclusion.

21.12    The Johns Hopkins Regional Talent Searches give the Scholastic Assessment Tests (intended for high school juniors and seniors) to 13-year-olds. In all, 19,883 males and 19,937 females took the tests between 1980 and 1982. The mean scores of males and females on the verbal test are nearly equal, but there is a clear difference between the sexes on the mathematics test. The reason for this difference is not understood. Here are the data summaries:[75]

| Group | $\bar{y}$ | $s$ |
|---|---|---|
| Males | 416 | 87 |
| Females | 386 | 74 |

Give a 99% confidence interval for the difference between the mean score for males and the mean score for females in the population that Johns Hopkins searches. Must SAT scores have a normal distribution in order for your confidence interval to be valid? Why?

21.13    A market research firm supplies manufacturers with estimates of the retail sales of their products from samples of retail stores. Marketing managers are prone to look at the estimate and ignore sampling error. An SRS of 75 stores this month shows mean sales of 52 units of a small appliance, with standard deviation 13 units. During the same month last year, an SRS of 53 stores gave mean sales of 49 units, with standard deviation 11 units. An increase from 49 to 52 is a rise of 6%. The marketing manager is happy, because sales are up 6%.

(a) Use the two-sample $t$ procedure to give a 95% confidence interval for the difference between this year and last year in the mean number of units sold at all retail stores.

(b) Explain in language that the manager can understand why he cannot be confident that sales rose by 6%, and that in fact sales may even have dropped.

21.14    Researchers studying the learning of speech often compare measurements made on the recorded speech of adults and children. One variable of interest is called the voice onset time (VOT). Here are the results for 6-year-old children and adults asked to pronounce the word "bees." The VOT is measured in milliseconds and can be either positive or negative.[76]

| Group | $n$ | $\bar{y}$ | $s$ |
|-------|-----|-----------|-----|
| Children | 10 | $-3.67$ | 33.89 |
| Adults | 20 | $-23.17$ | 50.74 |

(a) The researchers were investigating whether VOT distinguishes adults from children. State $H_0$ and $H_a$ and carry out a test. Give a $P$-value and report your conclusions.

(b) Give a 95% confidence interval for the difference in mean VOTs when pronouncing the word "bees." Explain why you knew from your result in (a) that this interval would contain 0 (no difference).

21.15    The researchers in the study discussed in the previous exercise looked at VOTs for adults and children pronouncing many different words. Explain why they should not do a separate test for each word and conclude that those words with a significant difference (say $P < 0.05$) distinguish children from adults. (The researchers did not make this mistake.)

21.16    Table 21.1 presents data on calories and sodium in hot dogs of three types. We are interested in their calorie content.

(a) We would like to group beef and meat together in order to compare poultry and nonpoultry hot dogs. This is reasonable if the calorie counts of beef and meat hot dogs are similar. Make boxplots and normal probability plots for beef and meat. Are the boxplots similar? The normal probability plots show similar kinds of deviation from normality. What do they show?

(b) Combine the beef and meat hot dogs to form a single group. Is there good evidence that poultry hot dogs have fewer calories on the average than nonpoultry hot dogs? (State hypotheses, examine the data, report important descriptive statistics, carry out a test, and state your conclusion.)

(c) How large is the difference between the mean calories in poultry hot dogs and other types? Give a confidence interval to answer this question.

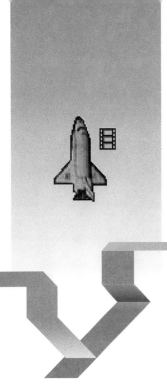

# INFERENCE FOR PROPORTIONS

Our discussion of statistical inference to this point has concerned making inferences about population *means*. Now we turn to questions about the *proportion* of some outcome in a population. Here are some examples that call for inference about population proportions.

EXAMPLE 22.1

How common is behavior that puts people at risk of AIDS? The National AIDS Behavioral Surveys interviewed a random sample of 2673 adult heterosexuals. Of these, 170 had more than one sexual partner in the past year. That's 6.36% of the sample.[77] Based on these data, what can we say about the percent of all adult heterosexuals who have multiple partners? We want to *estimate a single population proportion.* ◀

EXAMPLE 22.2

Do preschool programs for poor children make a difference in later life? A study looked at 62 children who were enrolled in a Michigan preschool in the late 1960s and at a control group of 61 similar children who were not enrolled. At 27 years of age, 61% of the preschool group and 80% of the control group had required the help of a social service agency (mainly welfare) in the previous ten years.[78] Is this significant evidence that preschool for poor children reduces later use of social services? We want to *compare two population proportions.*  ◀

## THE SAMPLING DISTRIBUTION OF A SAMPLE PROPORTION $\hat{p}$

We are interested in the unknown proportion $p$ of a population that has some characteristic. In Example 22.1, the population is adult heterosexuals, and the parameter $p$ is the proportion who have had more than one sexual partner in the past year. To estimate $p$, the National AIDS Behavioral Surveys used random dialing of telephone numbers to contact a sample of 2673 people. Of these, 170 said they had multiple sexual partners. The statistic that estimates the parameter $p$ is the *sample proportion*

$$\hat{p} = \frac{170}{2673} = 0.0636$$

In everyday language we often express proportions as percents. Statistical recipes work with proportions as decimal fractions, so 6.36% becomes 0.0636.

How good is the statistic $\hat{p}$ as an estimate of the parameter $p$? To find out, we ask, "What would happen if we took many samples?" The sampling distribution of $\hat{p}$ answers this question. Here are the facts.

SAMPLING DISTRIBUTION OF A SAMPLE PROPORTION

Choose an SRS of size $n$ from a large[79] population with population proportion $p$ having some characteristic of interest. It is usual to call whatever characteristic we are studying a "success." Let $\hat{p}$ be the

(continued on next page)

*(continued from previous page)*

**sample proportion** of successes,

$$\hat{p} = \frac{\text{count of successes in the sample}}{n}$$

Then:

- The sampling distribution of $\hat{p}$ is **approximately normal** and is closer to a normal distribution when the sample size $n$ is large.
- The **mean** of the sampling distribution is $p$.
- The **standard deviation** of the sampling distribution is

$$\sigma_{\hat{p}} = \sqrt{\frac{p(1-p)}{n}}$$

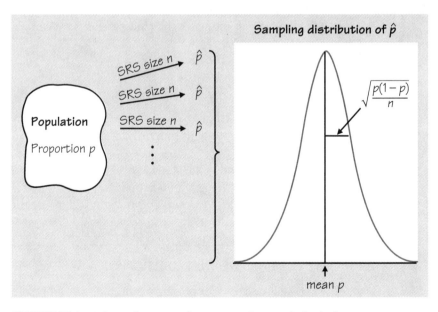

**FIGURE 22.1**    Select a large SRS from a population of which the proportion $p$ are successes. The sampling distribution of the proportion $\hat{p}$ of successes in the sample is approximately normal. The mean is $p$ and the standard deviation is $\sqrt{p(1-p)/n}$.

Figure 22.1 summarizes these facts in a form that helps you recall the big idea of a sampling distribution. The behavior of sample proportions $\hat{p}$ is similar to the behavior of sample means $\bar{y}$. The mean of the sampling distribution is the true value of the population proportion $p$, so that $\hat{p}$ estimates $p$ without bias. The standard deviation of $\hat{p}$ gets smaller as the sample size $n$ gets larger, but only at the rate $\sqrt{n}$. The sampling distribution is approximately normal. The normal approximation for the distribution of $\hat{p}$ becomes more accurate as $n$ gets larger. The approximation is least accurate when $p$ is close to 0 or close to 1. We will give guidelines to avoid using inference procedures based on the normal approximation when the sample is too small.

## TESTS FOR ONE PROPORTION

To use the sample proportion $\hat{p}$ for inference, we standardize it. If we are testing the null hypothesis $H_0 : p = p_0$ for some specified value $p_0$, the reasoning of the statistical tests asks what would happen if $H_0$ were true. So we use $p = p_0$ to standardize $\hat{p}$.

---

TEST FOR A POPULATION PROPORTION

Draw an SRS of size $n$ from a large population with unknown proportion $p$ of successes. To test the null hypothesis $H_0 : p = p_0$, use the **one-sample $z$ statistic for a proportion**:

$$z = \frac{\hat{p} - p_0}{\sqrt{\dfrac{p_0(1 - p_0)}{n}}}$$

$P$-values from the standard normal distribution $N(0, 1)$ are approximately correct when the sample size $n$ is not too small.

---

As usual, $z$ tells us how far the sample proportion $\hat{p}$ is from the hypothesized value $p_0$ in standard deviation units.

**EXAMPLE 22.3**

A coin that is balanced should come up heads half the time in the long run. The population for coin tossing contains the results of tossing the coin forever. The parameter $p$ is the probability of a head, which is the proportion of all tosses that give a head. The tosses we actually make are an SRS from this population.

The French naturalist Count Buffon (1707–1788) tossed a coin 4040 times. He got 2048 heads. The sample proportion of heads is

$$\hat{p} = \frac{2048}{4040} = 0.5069$$

Is this evidence that Buffon's coin was not balanced?

**Hypotheses:** The null hypothesis says that the coin is balanced ($p = 0.5$). The alternative hypothesis is two-sided, because we did not suspect before seeing the data that the coin favored either heads or tails. We therefore test the hypotheses

$$H_0 : p = 0.5$$
$$H_a : p \neq 0.5$$

The null hypothesis gives $p$ the value $p_0 = 0.5$.

**Test statistic:** The $z$ test statistic is

$$z = \frac{\hat{p} - p_0}{\sqrt{\dfrac{p_0(1 - p_0)}{n}}}$$

$$= \frac{0.5069 - 0.5}{\sqrt{\dfrac{(0.5)(0.5)}{4040}}} = 0.88$$

**P-value:** Because the test is two-sided, the P-value is the area under the standard normal curve more than 0.88 away from 0 in either direction. Figure 22.2 shows this area. From software or Table A we find that $P = 0.3788$.

**Conclusion:** A proportion of heads as far from one-half as Buffon's would happen 38% of the time when a balanced coin is tossed 4040 times. Buffon's result doesn't show that his coin is unbalanced.    ◀

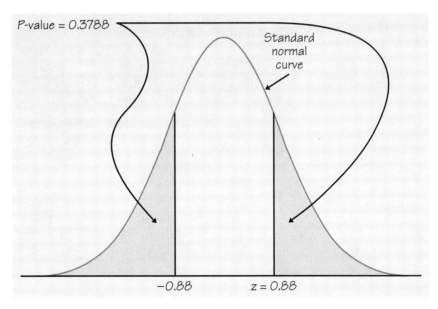

FIGURE 22.2    The *P*-value for the two-sided test of Example 22.3.

## CONFIDENCE INTERVALS FOR ONE PROPORTION

Confidence intervals for $p$ are also based on the standard normal distribution. We would like to use the interval $\hat{p} \pm z^* \sigma_{\hat{p}}$, that is,

$$\hat{p} \pm z^* \sqrt{\frac{p(1-p)}{n}}$$

We can't do that because we don't know $p$. We therefore replace $p$ by its estimate $\hat{p}$ in the margin of error. Here is the result.

---

CONFIDENCE INTERVAL FOR A POPULATION PROPORTION

Draw an SRS of size $n$ from a large population with unknown proportion $p$ of successes. An approximate level $C$ confidence interval for $p$ is

$$\hat{p} \pm z^* \sqrt{\frac{\hat{p}(1-\hat{p})}{n}}$$

---

(*continued on next page*)

*(continued from previous page)*

Here $z^*$ is the point on the standard normal density curve that captures area $C$ between $-z^*$ and $z^*$. This confidence interval is approximately correct when the sample size $n$ is not too small.

Our confidence interval has the familiar form

$$\hat{p} \pm z^* SE_{\hat{p}}$$

Notice that using the standard error does *not* result in a $t$ distribution. The reason is that we did not estimate a separate parameter $\sigma$, as we did in inference about a mean $\mu$. The population proportion $p$ is the only unknown parameter.

## EXAMPLE 22.4

The National AIDS Behavioral Surveys found that 170 of a sample of 2673 adult heterosexuals had multiple partners. That is, $\hat{p} = 0.0636$. We will act as if the sample were an SRS.

Software tells us at once that a 99% confidence interval for the proportion $p$ of all adult heterosexuals with multiple partners is

$$0.0514 < p < 0.0758$$

Here is the calculation. The $z^*$ row in Table B tells us that for 99% confidence, we must use $z^* = 2.576$. The confidence interval is

$$\hat{p} \pm z^* \sqrt{\frac{\hat{p}(1-\hat{p})}{n}} = 0.0636 \pm 2.576 \sqrt{\frac{(0.0636)(0.9364)}{2673}}$$

$$= 0.0636 \pm 0.0122$$

We are 99% confident that between 5.14% and 7.58% of all heterosexuals had multiple partners during the past year.  ◀

Recall that the margin of error we just found does not include the effects of nonresponse or other weaknesses in the conduct of the study. About 30% of the people reached by the AIDS survey refused to cooperate.

Others may not have told the truth about their sexual behavior. Although the survey was very carefully conducted, the report says that its results may underestimate the amount of risky behavior in the population.

## COMPARING TWO POPULATION PROPORTIONS

In a **two-sample problem,** we want to compare two populations or the responses to two treatments based on two independent samples. When the comparison involves the *mean* of a quantitative variable, we use the two-sample $t$ methods of Lesson 21. Now we turn to methods that compare the *proportions* of successes in two groups.

We will use notation similar to that used in our study of two-sample $t$ statistics. The groups we want to compare are Population 1 and Population 2. We have a separate SRS from each population or responses from two treatments in a randomized comparative experiment. A subscript shows which group a parameter or statistic describes. Here is our notation:

| Population | Population proportion | Sample size | Sample proportion |
|---|---|---|---|
| 1 | $p_1$ | $n_1$ | $\hat{p}_1$ |
| 2 | $p_2$ | $n_2$ | $\hat{p}_2$ |

We compare the populations by doing inference about the difference $p_1 - p_2$ between the population proportions. The statistic that estimates this difference is the difference between the two sample proportions, $\hat{p}_1 - \hat{p}_2$.

### EXAMPLE 22.5

To study the long-term effects of preschool programs for poor children, researchers have followed two groups of Michigan children since early childhood. One group of 62 attended preschool as 3- and 4-year-olds. This is a sample from Population 2, poor children who attend preschool. A control group of 61 children from the same area and similar backgrounds represents Population 1, poor children with no preschool. Thus the sample sizes are $n_1 = 61$ and $n_2 = 62$.

One response variable of interest is the need for social services as adults. In the past ten years, 38 of the preschool sample and 49 of the control sample have needed social services (mainly welfare). The sample proportions are

$$\hat{p}_1 = \frac{49}{61} = 0.803$$

$$\hat{p}_2 = \frac{38}{62} = 0.613$$

That is, about 80% of the control group uses social services, as opposed to about 61% of the preschool group.

To see if the study provides significant evidence that preschool reduces the later need for social services, we test the hypotheses

$$H_0 : p_1 - p_2 = 0$$
$$H_a : p_1 - p_2 > 0$$

To estimate how large the reduction is, we give a confidence interval for the difference, $p_1 - p_2$. Both the test and the confidence interval start from the difference of sample proportions:

$$\hat{p}_1 - \hat{p}_2 = 0.803 - 0.613 = 0.190$$

◀

When both samples are large, $\hat{p}_1 - \hat{p}_2$ has approximately a normal sampling distribution. The mean of this distribution is $p_1 - p_2$. Its standard deviation is

$$\sigma_{\hat{p}_1 - \hat{p}_2} = \sqrt{\frac{p_1(1 - p_1)}{n_1} + \frac{p_2(1 - p_2)}{n_2}}$$

## CONFIDENCE INTERVALS FOR COMPARING TWO PROPORTIONS

Figure 22.3 displays the distribution of $\hat{p}_1 - \hat{p}_2$. The standard deviation of $\hat{p}_1 - \hat{p}_2$ involves the unknown parameters $p_1$ and $p_2$. Just as in the one-sample setting, we must replace these by estimates in order to do inference. We again do this a bit differently for confidence intervals and for tests. The confidence interval is straightforward: replace the parameters $p_1$ and $p_2$ by their sample estimates $\hat{p}_1$ and $\hat{p}_2$. Here is the result.

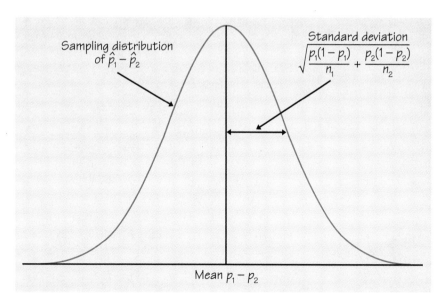

FIGURE 22.3 Select independent SRSs from two populations having proportions of successes $p_1$ and $p_2$. The proportions of successes in the two samples are $\hat{p}_1$ and $\hat{p}_2$. When the samples are large, the sampling distribution of the difference $\hat{p}_1 - \hat{p}_2$ is approximately normal.

## CONFIDENCE INTERVAL FOR THE DIFFERENCE BETWEEN TWO PROPORTIONS

Draw an SRS of size $n_1$ from a population having proportion $p_1$ of successes and draw an independent SRS of size $n_2$ from another population having proportion $p_2$ of successes. When $n_1$ and $n_2$ are large, an approximate level $C$ confidence interval for $p_1 - p_2$ is

$$(\hat{p}_1 - \hat{p}_2) \pm z^* \text{SE}_{\hat{p}_1 - \hat{p}_2}$$

In this formula the standard error of $\hat{p}_1 - \hat{p}_2$ is

$$\text{SE}_{\hat{p}_1 - \hat{p}_2} = \sqrt{\frac{\hat{p}_1(1 - \hat{p}_1)}{n_1} + \frac{\hat{p}_2(1 - \hat{p}_2)}{n_2}}$$

As usual, $z^*$ captures the middle area $C$ under the standard normal density curve.

**EXAMPLE 22.6**

Example 22.5 describes a study of the effect of preschool on later use of social services. The facts are:

| Population | Population description | Sample size | Sample proportion |
|---|---|---|---|
| 1 | Control | $n_1 = 61$ | $\hat{p}_1 = 0.803$ |
| 2 | Preschool | $n_2 = 62$ | $\hat{p}_2 = 0.613$ |

The difference $p_1 - p_2$ measures the effect of preschool in reducing the proportion of people who later need social services. Software says we can be 95% confident that

$$0.033 < p_1 - p_2 < 0.347$$

Here is the calculation in detail. The standard error is

$$\mathrm{SE}_{\hat{p}_1 - \hat{p}_2} = \sqrt{\frac{\hat{p}_1(1 - \hat{p}_1)}{n_1} + \frac{\hat{p}_2(1 - \hat{p}_2)}{n_2}}$$

$$= \sqrt{\frac{(0.803)(0.197)}{61} + \frac{(0.613)(0.387)}{62}}$$

$$= \sqrt{0.00642} = 0.0801$$

The 95% confidence interval is

$$(\hat{p}_1 - \hat{p}_2) \pm z^* \mathrm{SE}_{\hat{p}_1 - \hat{p}_2} = (0.803 - 0.613) \pm (1.960)(0.0801)$$

$$= 0.190 \pm 0.157$$

We are 95% confident that the percent needing social services is somewhere between 3.3% and 34.7% lower among people who attended preschool. The confidence interval is wide because the samples are quite small.    ◀

## TESTS FOR COMPARING TWO PROPORTIONS

A statistical test comparing two proportions asks if we have good evidence against the null hypothesis of "no difference,"

$$H_0 : p_1 - p_2 = 0$$

The alternative hypothesis says what kind of difference we expect.

**EXAMPLE 22.7**

High levels of cholesterol in the blood are associated with higher risk of heart attacks. Will using a drug to lower blood cholesterol reduce heart attacks? The Helsinki Heart Study looked at this question. Middle-aged men were assigned at random to one of two treatments: 2051 men took the drug gemfibrozil to reduce their cholesterol levels, and a control group of 2030 men took a placebo. During the next five years, 56 men in the gemfibrozil group and 84 men in the placebo group had heart attacks.

The sample proportions who had heart attacks are

$$\hat{p}_1 = \frac{56}{2051} = 0.0273 \quad \text{(gemfibrozil group)}$$

$$\hat{p}_2 = \frac{84}{2030} = 0.0414 \quad \text{(placebo group)}$$

That is, about 4.1% of the men in the placebo group had heart attacks, against only about 2.7% of the men who took the drug. Is the apparent benefit of gemfibrozil statistically significant? We hope to show that gemfibrozil reduces heart attacks, so we have a one-sided alternative:

$$H_0 : p_1 - p_2 = 0$$
$$H_a : p_1 - p_2 < 0$$

◀

Because the reasoning of statistical tests asks what would happen if $H_0$ were true, the test statistic is the estimate $\hat{p}_1 - \hat{p}_2$ standardized under the conditions stated by $H_0$. If $H_0$ is true, all the observations in both samples really come from a single population of men of whom a single unknown proportion $p$ will have a heart attack in a five-year period. So instead of estimating $p_1$ and $p_2$ separately, we pool the two samples and use the overall sample proportion to estimate the single population parameter $p$. Call this the **pooled sample proportion.** It is

$$\hat{p} = \frac{\text{count of successes in both samples combined}}{\text{count of observations in both samples combined}}$$

Use $\hat{p}$ in place of both $\hat{p}_1$ and $\hat{p}_2$ in the expression for the standard error of $\hat{p}_1 - \hat{p}_2$ to get a $z$ statistic that has the standard normal distribution when $H_0$ is true. Here is the test.

TEST FOR COMPARING TWO PROPORTIONS

To test the hypothesis

$$H_0 : p_1 - p_2 = 0$$

first find the pooled proportion $\hat{p}$ of successes in both samples combined. Then calculate the **two-sample $z$ statistic for proportions**

$$z = \frac{\hat{p}_1 - \hat{p}_2}{\sqrt{\hat{p}(1 - \hat{p})\left(\dfrac{1}{n_1} + \dfrac{1}{n_2}\right)}}$$

Use the standard normal distribution to find $P$-values.

## EXAMPLE 22.8

The pooled proportion of heart attacks for the two groups in the Helsinki Heart Study is

$$\hat{p} = \frac{\text{count of heart attacks in both samples combined}}{\text{count of subjects in both samples combined}}$$

$$= \frac{56 + 84}{2051 + 2030}$$

$$= \frac{140}{4081} = 0.0343$$

The $z$ test statistic is

$$z = \frac{\hat{p}_1 - \hat{p}_2}{\sqrt{\hat{p}(1 - \hat{p})\left(\dfrac{1}{n_1} + \dfrac{1}{n_2}\right)}}$$

$$= \frac{0.0273 - 0.0414}{\sqrt{(0.0343)(0.9657)\left(\dfrac{1}{2051} + \dfrac{1}{2030}\right)}}$$

$$= \frac{-0.0141}{0.005698} = -2.47$$

◀

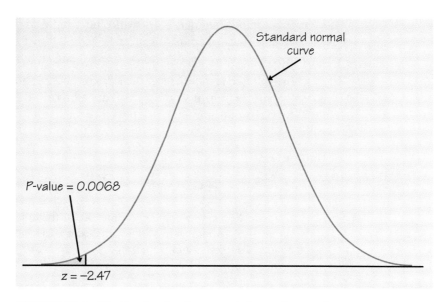

**FIGURE 22.4**   The $P$-value for the one-sided test of Example 22.8.

The one-sided $P$-value is the area under the standard normal curve to the left of $-2.47$. Figure 22.4 shows this area. The $P$-value is $P = 0.0068$. Because $P < 0.01$, the results are statistically significant at the 1% level. There is strong evidence that gemfibrozil reduced the rate of heart attacks. The large samples in the Helsinki Heart Study helped the study get highly significant results. ◀

## WHEN CAN WE USE THE $z$ PROCEDURES?

The $t$ procedures for means are exactly correct if the populations concerned are exactly normal. In practice, no population is exactly normal. We know that the $t$ procedures are robust in the sense that they give approximately correct results if the populations are unimodal and symmetric and the samples are not too small. The $z$ procedures for proportions are always just approximately correct. They are quite accurate if the samples are not too small. Here are rules of thumb for applying the $z$ procedures in practice.

- **One sample:** For a test of $H_0 : p = p_0$, the sample size $n$ should be so large that both $np_0$ and $n(1 - p_0)$ are 10 or more. For a confidence interval, $n$ should be so large that both the count of successes $n\hat{p}$ and the count of failures $n(1 - \hat{p})$ are 10 or more.

- *Two-sample confidence interval:* The sample sizes $n_1$ and $n_2$ should be so large that the counts of successes and failures in the two samples, $n_1\hat{p}_1$, $n_1(1 - \hat{p}_1)$, $n_2\hat{p}_2$, and $n_2(1 - \hat{p}_2)$, are all 5 or more.
- *Two-sample tests:* The sample sizes $n_1$ and $n_2$ should be so large that $n_1\hat{p}$, $n_1(1-\hat{p})$, $n_2\hat{p}$, and $n_2(1-\hat{p})$ are all 5 or more. Here $\hat{p}$ is the pooled sample proportion.

## EXAMPLE 22.9

Example 22.6 gave a confidence interval for the difference between two proportions. To check that this approximate confidence interval is safe, look at the counts of successes and failures in the two samples. The smallest of these four quantities is

$$n_1(1 - \hat{p}_1) = (61)(1 - 0.803) = 12$$

This is larger than 5, so the interval will be accurate.    ◀

## EXERCISES

22.1    The Internal Revenue Service plans to examine an SRS of individual federal income tax returns from each state. One variable of interest is the proportion of returns claiming itemized deductions. The total number of tax returns in a state varies from almost 14 million in California to fewer than 215,000 in Wyoming.

(a) Will the sampling variability of the sample proportion change from state to state if an SRS of 2000 tax returns is selected in each state? Explain your answer.

(b) Will the sampling variability of the sample proportion change from state to state if an SRS of 1% of all tax returns is selected in each state? Explain your answer.

22.2    While he was a prisoner of the Germans during World War II, the British statistician John Kerrich tossed a coin 10,000 times. He got 5067 heads. Take Kerrich's tosses to be an SRS from the population of all possible tosses of his coin. If the coin is perfectly balanced, $p = 0.5$. Is there reason to think that Kerrich's coin was not balanced?

22.3    How likely are patients who file complaints with a health maintenance organization (HMO) to leave the HMO? In one recent year, 639 of the more than 400,000 members of a large New England HMO filed complaints. Fifty-four of the complainers left the HMO voluntarily. (That is, they were not forced to leave by a move or a job change.)[80] Consider the year's complainers as an SRS of all patients who will complain in the future. Give a 90% confidence interval for the proportion of complainers who voluntarily leave the HMO.

22.4    The National AIDS Behavioral Surveys (Example 22.1) also interviewed a sample of adults in the cities where AIDS is most common. The sample included 803 heterosexuals who reported having more than one sexual partner in the past year. We can consider this an SRS of size 803 from the population of all heterosexuals in high-risk cities who have multiple partners. These people risk infection with the AIDS virus. Yet 304 of the respondents said they never use condoms.

(a) Is this strong evidence that more than one-third of this population never use condoms?

(b) Give a 95% confidence interval for the proportion of this population who never use condoms.

22.5    In a recent year, 73% of first-year college students responding to a national survey identified "being very well-off financially" as an important personal goal. A state university finds that 132 of an SRS of 200 of its first-year students say that this goal is important.

(a) Give a 95% confidence interval for the proportion of all first-year students at the university who would identify being well-off as an important personal goal.

(b) Is there good evidence that the proportion of first-year students at this university who think being very well-off is important differs from the national value, 73%? (Be sure to state hypotheses, give the P-value, and state your conclusion.)

(c) Check that you could safely use the methods of this lesson in both (a) and (b).

22.6    The Gallup Poll asked a sample of 1785 adults, "Did you, yourself, happen to attend church or synagogue in the last 7 days?" Of the respondents, 750 said "Yes." Suppose (it is not quite true) that Gallup's sample was an SRS of all American adults.

(a) Give a 99% confidence interval for the proportion of all adults who attended church or synagogue during the week preceding the poll.

(b) Do the results provide good evidence that less than half of the population attended church or synagogue?

22.7    A study of the survival of small businesses chose an SRS from the telephone directory's Yellow Pages listings of food-and-drink businesses in 12 counties in central Indiana. For various reasons, the study got no response from 45% of the businesses chosen. Interviews were completed with 148 businesses. Three years later, 22 of these businesses had failed.[81]

(a) Give a 95% confidence interval for the percent of all small businesses in this class that fail within three years.

(b) The authors hope that their findings describe the population of all small businesses. What about the study makes this unlikely?

22.8    One-sample procedures for proportions, like those for means, are used to analyze data from matched pairs designs. Here is an example.

Each of 50 subjects tastes two unmarked cups of coffee and says which he or she prefers. One cup in each pair contains instant coffee; the other, fresh-brewed coffee. Thirty-one of the subjects prefer the fresh-brewed coffee. Take $p$ to be the proportion of the population who would prefer fresh-brewed coffee in a blind tasting.

(a) Test the claim that a majority of people prefer the taste of fresh-brewed coffee. State hypotheses and report the $z$ statistic and its $P$-value. Is your result significant at the 5% level? What is your practical conclusion?

(b) Find a 90% confidence interval for $p$.

(c) When you do an experiment like this, in what order should you present the two cups of coffee to the subjects?

22.9    A study of "adverse symptoms" in users of over-the-counter pain relief medications assigned subjects at random to one of two common pain relievers, acetaminophen and ibuprofen. (Both of these pain relievers are sold under various brand names, sometimes combined with other ingredients.) In all, 650 subjects took acetaminophen, and 44 experienced some adverse symptom. Of the 347 subjects who took ibuprofen, 49 had an adverse symptom. How strong is the evidence that the two pain relievers differ in the proportion of people who experience an adverse symptom?

(a) State hypotheses and check that you can use the $z$ test.

(b) Find the $P$-value of the test and give your conclusion.

22.10    The 1958 Detroit Area Study was an important investigation of the influence of religion on everyday life. The sample "was basically a simple random sample of the population of the metropolitan area" of Detroit, Michigan. Of the 656 respondents, 267 were white Protestants and 230 were white Catholics.[82]

(a) The study took place at the height of the cold war. One question asked if the right of free speech included the right to make speeches in favor of communism. Of the 267 white Protestants, 104 said "Yes," while 75 of the 230 white Catholics said "Yes." Give a 95% confidence interval for the difference between the proportion of Protestants who agreed that communist speeches are protected and the proportion of Catholics who held this opinion. Be sure to check that it is safe to use the $z$ confidence interval.

**(b)** Another question asked whether the government was doing enough in areas such as housing, unemployment, and education; 161 of the Protestants and 136 of the Catholics said "No." Is there evidence that white Protestants and white Catholics differed on this issue?

22.11 Exercise 22.3 describes a study of whether patients who file complaints leave a health maintenance organization (HMO). We want to know whether complainers are more likely to leave than patients who do not file complaints. In the year of the study, 639 patients filed complaints, and 54 of these patients left the HMO voluntarily. For comparison, the HMO chose an SRS of 743 patients who had not filed complaints. Twenty-two of these patients left voluntarily.

**(a)** How much higher is the proportion of complainers who leave? Give a 90% confidence interval.

**(b)** Check that you can safely use the methods of this lesson.

22.12 The drug AZT was the first drug that seemed effective in delaying the onset of AIDS. Evidence for AZT's effectiveness came from a large randomized comparative experiment. The subjects were 1300 volunteers who were infected with HIV, the virus that causes AIDS, but did not yet have AIDS. The study assigned 435 of the subjects at random to take 500 milligrams of AZT each day, and another 435 to take a placebo. (The others were assigned to a third treatment, a higher dose of AZT. We will compare only two groups.) At the end of the study, 38 of the placebo subjects and 17 of the AZT subjects had developed AIDS. We want to test the claim that taking AZT lowers the proportion of infected people who will develop AIDS in a given period of time.

**(a)** State hypotheses, and check that you can safely use the $z$ procedures.

**(b)** How significant is the evidence that AZT is effective?

**(c)** The experiment was double-blind. Explain what this means.

(*Comment:* Medical experiments on treatments for AIDS and other fatal diseases raise hard ethical questions. Some people argue that because AIDS is always fatal, infected people should get any drug that has any hope of helping them. The counterargument is that we will then never find out which drugs really work. The placebo patients in this study were given AZT as soon as the results were in.)

22.13     The study of small-business failures described in Exercise 22.7 looked at 148 food-and-drink businesses in central Indiana. Of these, 106 were headed by men and 42 were headed by women. During a three-year period, 15 of the men's businesses and 7 of the women's businesses failed. Is there a significant difference between the failure rates of businesses headed by men and of those headed by women?

(*Comment:* This study did not select two separate samples. A single sample of businesses was divided after the fact into those headed by men and those headed by women. The two-sample $z$ procedures for comparing proportions are valid in such situations. This is an important fact about these methods.)

22.14     Nonresponse to sample surveys may differ with the season of the year. In Italy, for example, many people leave town during the summer. The Italian National Statistical Institute called random samples of telephone numbers between 7 p.m. and 10 p.m. at several seasons of the year. Here are the results for two seasons:[83]

| Dates | Number of calls | No answer | Total nonresponse |
|-------|-----------------|-----------|-------------------|
| Jan. 1 to Apr. 13 | 1558 | 333 | 491 |
| July 1 to Aug. 31 | 2075 | 861 | 1174 |

(a) How much higher is the proportion of "no answers" in July and August compared with the early part of the year? Give a 99% confidence interval.

(b) The difference between the proportions of "no answers" is so large that it is clearly statistically significant. How can you tell from your work in (a) that the difference is significant at the 1% level?

(c) Use the information given to find the counts of calls which had nonresponse for some reason other than "no answer." Do the rates of nonresponse due to other causes also differ significantly for the two seasons?

22.15     The National Assessment of Educational Progress (NAEP) Young Adult Literacy Assessment Survey interviewed a random sample of 1917 people 21 to 25 years old. The sample contained 840 men, of whom 775 were fully employed. There were 1077 women, and 680 of them were fully employed.[84]

(a) Use a 99% confidence interval to describe the difference between the proportions of young men and young women who are fully employed. Is the difference statistically significant at the 1% significance level?

(b) The mean and standard deviation of scores on the NAEP's test of quantitative skills were $\bar{y}_1 = 272.40$ and $s_1 = 59.2$ for the men in the sample. For the women, the results were $\bar{y}_2 = 274.73$ and $s_2 = 57.5$. Is the difference between the mean scores for men and women significant at the 1% level?

22.16    A television news program conducts a call-in poll about a proposed city ban on handgun ownership. Of the 2372 calls, 1921 oppose the ban. The station, following recommended practice, makes a confidence statement: "81% of the Channel 13 Pulse Poll sample opposed the ban. We can be 95% confident that the true proportion of citizens opposing a handgun ban is within 1.6% of the sample result." Is this conclusion justified?

22.17    Never forget that even small effects can be statistically significant if the samples are large. To illustrate this fact, return to the study of 148 small businesses in Exercise 22.13.

(a) Find the proportions of failures for businesses headed by women and businesses headed by men. These sample proportions are quite close to each other. Give the $P$-value for the $z$ test of the hypothesis that the same proportion of women's and men's businesses fail. (Use the two-sided alternative.) The test is very far from being significant.

(b) Now suppose that the same sample proportions came from a sample 30 times as large. That is, 210 out of 1260 businesses headed by women and 450 out of 3180 businesses headed by men fail. Verify that the proportions of failures are exactly the same as in (a). Repeat the $z$ test for the new data, and show that it is now significant at the 5% level.

(c) It is wise to use a confidence interval to estimate the size of an effect, rather than just giving a $P$-value. Give 95% confidence intervals for the difference between the proportions of women's and men's businesses that fail for the settings of both (a) and (b). What is the effect of larger samples on the confidence interval?

22.18    In which of the following situations can you safely use the methods of this lesson for inference about the population proportion $p$? Explain your answers.

(a) Glenn wonders what proportion of the students at his school think that tuition is too high. He interviews an SRS of 50 of the 2400 students at his college. Thirty-eight of those interviewed think tuition is too high.

(b) In the National AIDS Behavioral Surveys sample of 2673 adult heterosexuals, 0.2% (that's 0.002 as a decimal fraction) had both received a blood transfusion and had a sexual partner from a group at high risk of AIDS. We want to estimate the proportion $p$ in the population who share these two risk factors.

(c) You toss a coin 10 times in order to test the hypothesis $H_0 : p = 0.5$ that the coin is balanced.

(d) A college president says, "99% of the alumni support my firing of Coach Boggs." You contact an SRS of 200 of the college's 15,000 living alumni to test the hypothesis $H_0 : p = 0.99$.

22.19    **(Optional)** A national opinion poll found that 44% of all American adults agree that parents should be given vouchers good for education at any public or private school of their choice. The result was based on a small sample. How large an SRS is required to obtain a margin of error of 0.03 (that is, $\pm 3\%$) in a 95% confidence interval? (Hint: The margin of error is $z^* \sqrt{\hat{p}(1 - \hat{p})/n}$. Use the poll result to approximate $\hat{p}$ and use $z^* = 1.96$ for 95% confidence. Then find $n$ to obtain margin of error 0.03.)

# 23

# TWO-WAY TABLES

The two-sample $z$ procedures of Lesson 22 allow us to compare the proportions of successes in two groups, either two populations or two treatment groups in an experiment. What if we want to compare more than two groups? We need a new statistical test. The new test starts by presenting the data in a new way, as a **two-way table.** Two-way tables have more general uses than comparing the proportions of successes in several groups—we use them to describe relationships between any two categorical variables. We will start with the problem of comparing several proportions.

## EXAMPLE 23.1

Chronic users of cocaine need the drug to feel pleasure. Perhaps giving them a medication that fights depression will help them stay off cocaine. A three-year study compared an antidepressant called desipramine with lithium (a standard treatment for cocaine addiction) and a placebo. The subjects were 72 chronic users of cocaine who wanted to break their drug habit. Twenty-four of the subjects were randomly assigned to each treatment. Here are the counts and proportions of the subjects who remained free of cocaine use during the study:[85]

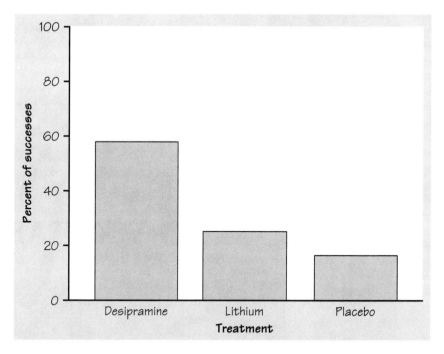

FIGURE 23.1    Bar chart comparing the success rates of three treatments for cocaine addiction, from Example 23.1.

| Group | Treatment | Subjects | Drug free | Proportion |
|---|---|---|---|---|
| 1 | Desipramine | 24 | 14 | 0.583 |
| 2 | Lithium | 24 | 6 | 0.250 |
| 3 | Placebo | 24 | 4 | 0.167 |

The sample proportions of subjects who stayed off cocaine are quite different. The bar chart in Figure 23.1 compares the results visually. Are these data good evidence that the proportions of successes for the three treatments differ in the population of all cocaine users?    ◀

## THE PROBLEM OF MULTIPLE COMPARISONS

Call the population proportions of successes in the three groups $p_1$, $p_2$, and $p_3$. We again use a subscript to remind us which group a parameter or statistic describes. To compare these three population proportions, we might use the two-sample $z$ procedures several times:

- Test $H_0 : p_1 - p_2 = 0$ to see if the success rate of desipramine differs from that of lithium.
- Test $H_0 : p_1 - p_3 = 0$ to see if desipramine differs from a placebo.
- Test $H_0 : p_2 - p_3 = 0$ to see if lithium differs from a placebo.

The weakness of doing three tests is that we get three $P$-values, one for each test alone. That doesn't tell us how likely it is that *three* sample proportions are spread apart as far as these are. It may be that 0.167 and 0.583 are significantly different if we look at just two groups, but not significantly different if we know that they are the smallest and largest proportions in three groups. As we look at more groups, we expect the gap between the smallest and largest sample proportion to get larger. (Think of comparing the tallest and shortest person in larger and larger groups of people.) We can't safely compare many parameters by doing tests or confidence intervals for two parameters at a time.

We want an overall test of the null hypothesis that there are *no differences* among the proportions of successes for addicts given the three treatments:

$$H_0 : p_1 = p_2 = p_3$$

The alternative hypothesis is that there *is* some difference, that not all three proportions are equal:

$$H_a : \text{ not all of } p_1, p_2, \text{ and } p_3 \text{ are equal}$$

The alternative hypothesis is no longer one-sided or two-sided. It is "many-sided," because it allows any relationship other than "all three equal." For example, $H_a$ includes the situation in which $p_2 = p_3$ but $p_1$ has a different value.

## TWO-WAY TABLES

The first step in the overall test for comparing several proportions is to arrange the data in a **two-way table** that gives counts for both failures and successes. Here is the two-way table for the cocaine addiction data:

|  | Drug-free | |
| | No | Yes |
| Desipramine | 10 | 14 |
| Lithium | 18 | 6 |
| Placebo | 20 | 4 |

We call this a $3 \times 2$ table because it has 3 rows and 2 columns. A table with $r$ rows and $c$ columns is an $r \times c$ **table.** The table shows the relationship between two categorical variables. The explanatory variable is the treatment (one of three drugs). The response variable is success (remaining free of cocaine) or failure. The two-way table gives the counts for all 6 combinations of values of these variables. Each of the 6 counts occupies a **cell** of the table.

## THE CHI-SQUARE STATISTIC

Our $z$ and $t$ tests all worked like this: compare a statistic (such as $\hat{p}$) with the value we would expect it to have if the null hypothesis were true (such as a specified value $p_0$). Put the difference between observed and expected on a standard scale, and ask whether it is so large that it would rarely occur just by chance. The test for

$$H_0 : p_1 = p_2 = p_3$$

is a more elaborate version of the same idea. Here it is.

---

CHI-SQUARE STATISTIC

To test the null hypothesis that there is no difference between two or more population proportions:

- Arrange the observed counts in a two-way table.
- Find the counts that would be expected in each cell of the table if $H_0$ were true.
- The **chi-square statistic** is a measure of how far the observed counts in a two-way table are from the expected counts. The

---

(continued on next page)

*(continued from previous page)*

formula for the statistic is

$$\chi^2 = \sum \frac{(\text{observed count} - \text{expected count})^2}{\text{expected count}}$$

The sum is over all $r \times c$ cells in the table.

Think of the chi-square statistic $\chi^2$ as a measure of the distance of the observed counts from the expected counts. Like any distance, it is always zero or positive, and it is zero only when the observed counts are exactly equal to the expected counts. Large values of $\chi^2$ are evidence against $H_0$ because they say that the observed counts are far from what we would expect if $H_0$ were true. Although the alternative hypothesis $H_a$ is many-sided, the chi-square test is one-sided because any violation of $H_0$ tends to produce a large value of $\chi^2$. Small values of $\chi^2$ are not evidence against $H_0$.

## EXAMPLE 23.2

Figure 23.2 displays *Data Desk* output for our two-way table. Each cell of the output table has several entries, indexed at the lower left. The counts form the basic table. The sums of the counts in each row and each column appear as "totals" at the margins of the table. The second entry in each cell is the cell count as a percent of its row total. We compared the three percents of success in Example 23.1. The third entry is the expected count for each cell.

At the bottom of the window we find the chi-square statistic and its P-value:

$$\chi^2 = 10.500 \quad P = 0.0052$$

The very small P-value gives strong reason to conclude that there *are* differences among the responses to three treatments.    ◄

Here is the idea behind the expected cell counts in Figure 23.2. In all, 24 of the 72 subjects remained drug free. That's the fraction 24/72, or 1/3, of the subjects. If $H_0$ is true, there are no differences among the treatments,

```
Rows are levels of        Treatment
Columns are levels of     Drugfree
No Selector

                 No      Yes      total

Desi             10      14         24
                 41.7    58.3      100
                 16       8         24

Lithium          18       6         24
                 75      25        100
                 16       8         24

Placebo          20       4         24
                 83.3    16.7      100
                 16       8         24

total            48      24         72
                 66.7    33.3      100
                 48      24         72

table contents:
Count
Percent of Row Total
Expected Values

Chi-square =      10.50    with    2    df
p = 0.0052
```

FIGURE 23.2    *Data Desk* output for the two-way table in the cocaine study.

so we expect 1/3 of the subjects in each group to be drug-free. The expected count of successes is therefore 1/3 of 24, or 8, in each group. Here are the observed and expected counts of successes side-by-side:

|  | Observed | | Expected | |
|---|---|---|---|---|
|  | No | Yes | No | Yes |
| Desipramine | 10 | 14 | 16 | 8 |
| Lithium | 18 | 6 | 16 | 8 |
| Placebo | 20 | 4 | 16 | 8 |

In fact, desipramine has more successes (14) and fewer failures (10) than expected. The placebo has fewer successes (4) and more failures (20). That's another way of saying what the sample proportions in Example 23.1 say more directly: desipramine does much better than the placebo, with lithium in between. The chi-square statistic works with the counts rather than with proportions.

## THE CHI-SQUARE DISTRIBUTIONS

The chi-square statistic is quite different in form from the $z$ and $t$ statistics. It is not surprising that its sampling distribution is also quite different in form from the symmetric $z$ and $t$ distributions. The $P$-value for a chi-square test comes from comparing the value of the chi-square statistic with a *chi-square distribution.*

---

### THE CHI-SQUARE DISTRIBUTIONS

The **chi-square distributions** are a family of distributions that take only positive values and are skewed to the right. A specific chi-square distribution is specified by one parameter, called the **degrees of freedom.**

The chi-square test for a two-way table with $r$ rows and $c$ columns uses $P$-values from the chi-square distribution with $(r - 1)(c - 1)$ degrees of freedom. The $P$-value is the area to the right of $\chi^2$ under the chi-square density curve.

---

Figure 23.3 shows the density curves for three members of the chi-square family of distributions. As the degrees of freedom increase, the density curves become less skewed and larger values become more probable. The $3 \times 2$ table in Example 23.2 has $r = 3$ and $c = 2$, resulting in $2 \times 1 = 2$ degrees of freedom. Of course, software usually finds $P$-values for us so that we do not need to work directly with the chi-square distributions. Table C (inside the back cover of this book) gives tail probabilities for chi-square distributions. You can use this table to bracket the $P$-value when you do not get exact values from software.

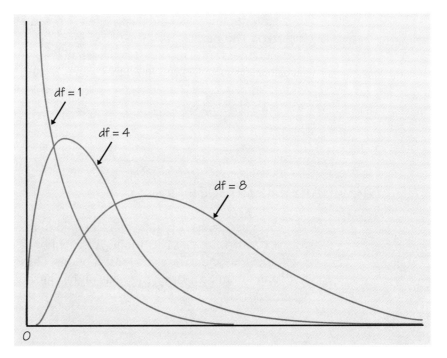

FIGURE 23.3    Density curves for the chi-square distributions with 1, 4, and 8 degrees of freedom. Chi-square distributions take only positive values.

## USING THE CHI-SQUARE TEST

To test equality of several proportions, we made a two-way table of the counts. The groups being compared (desipramine, lithium, and placebo in our example) form the rows of the table. The columns are just success or failure. The chi-square test asks whether the row variable and the column variable are related to each other.

We can make a two-way table of counts whenever we have data on two categorical variables. The categories for one variable label the rows of the table, and the categories for the other variable label the columns. The chi-square test asks whether the two variables in any such table are related to each other. Here is a 4 × 4 example.

### EXAMPLE 23.3

A study of the relationship between men's marital status and the level of their jobs used data on all 8235 male managers and professionals employed by a large manufacturing firm. Each man's job has a grade set by the company that reflects the value of that particular job to the company. The authors of the study grouped the many job grades into quarters. Grade 1 contains jobs in the lowest quarter of job grades, and grade 4 contains those in the highest quarter. Here are the data:[86]

|  |  | \multicolumn{4}{c}{Job grade} |  |  |  |
|  |  | 1 | 2 | 3 | 4 |
|  | Divorced | 15 | 70 | 34 | 7 |
| Marital | Married | 874 | 3927 | 2396 | 533 |
| status | Single | 58 | 222 | 50 | 7 |
|  | Widowed | 8 | 20 | 10 | 4 |

Do these data show a statistically significant relationship between marital status and job grade?

The chi-square test compares these observed counts with the 16 cell counts that would be expected under the null hypothesis

$H_0$ : there is no relationship between marital status and job grade

The *Data Desk* output in Figure 23.4 tells us that

$$\chi^2 = 67.40 \quad P \leq 0.0001$$

There is very strong evidence of some relationship between marital status and job grade. ◀

The chi-square test is an overall test. A small *P*-value tells us that there is *some* relationship between the row variable and the column variable. It tells us nothing about the nature of the relationship. You should always accompany the chi-square test by a description of what the data show. To describe the relationship between two categorical variables, you can:

- Calculate and compare appropriate percents.
- Look at the cells that contribute the most to the chi-square statistic. *Data Desk* reports the **standardized residuals**, which are the square roots (with sign) of the terms that we add to get chi-square:

$$\frac{\text{observed count} - \text{expected count}}{\sqrt{\text{expected count}}}$$

| | 1 | 2 | 3 | 4 | total |
|---|---|---|---|---|---|
| **Rows are levels of** | **Marital Status** | | | | |
| **Columns are levels of** | **Job Class** | | | | |
| No Selector | | | | | |
| **Divorced** | 15 | 70 | 34 | 7 | 126 |
| | 11.9 | 55.6 | 27.0 | 5.56 | 100 |
| | 0.101497 | 0.638353 | -0.663983 | -0.492708 | 0 |
| **Married** | 874 | 3927 | 2396 | 533 | 7730 |
| | 11.3 | 50.8 | 31.0 | 6.90 | 100 |
| | -0.749350 | -0.825134 | 1.21409 | 0.694272 | 0 |
| **Single** | 58 | 222 | 50 | 7 | 337 |
| | 17.2 | 65.9 | 14.8 | 2.08 | 100 |
| | 3.02625 | 3.68448 | -5.14124 | -3.27439 | 0 |
| **Widowed** | 8 | 20 | 10 | 4 | 42 |
| | 19.0 | 47.6 | 23.8 | 9.52 | 100 |
| | 1.41793 | -0.348339 | -0.757501 | 0.709750 | 0 |

table contents:
Count
Percent of Row Total
Standardized Residuals

Chi-square =    67.40    with    9    df
$p \leq 0.0001$

FIGURE 23.4    *Data Desk* output for the 4 × 4 table in Example 23.3.

## EXAMPLE 23.4

We think that marital status may help explain job grade. We therefore calculate the percent of men in each marital status whose jobs have each grade. These percents appear in the output displayed in Figure 23.4. Each row of percents gives the distribution of job grade among men with a specific marital status. Each row adds to 100% (up to roundoff error) because it accounts for all men in one marital status.    ◄

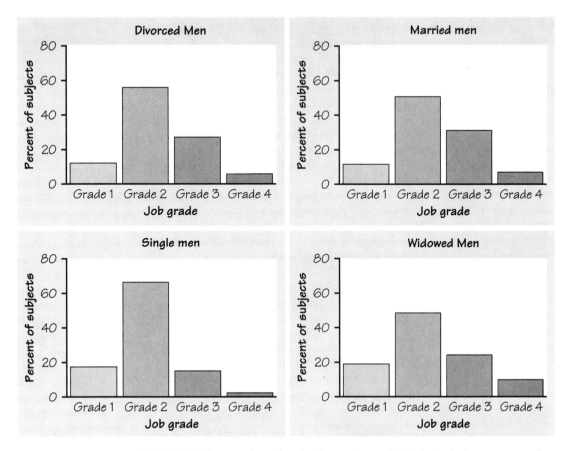

FIGURE 23.5    Bar charts for the data in Example 23.3. Each chart presents the percents of each job grade among men with one marital status.

The bar charts in Figure 23.5 show the distributions of job grade for the four marital states. We see at once that smaller percents of single men have jobs in the higher grades 3 and 4. Not only married men but men who were once married and are now divorced or widowed are more likely to hold higher-grade jobs.

Look now at the residuals, the last entry for each cell in the output in Figure 23.4. The sum of the squares of these residuals is the chi-square value. We see that all of the largest residuals are in the "Single" row. Single men have more grade 1 and 2 jobs and fewer grade 3 and 4 jobs than would be expected if job grade were not related to marital status. The squares of these four residuals contribute 59.9 of the total 67.4 chi-square.

Both the residuals and the comparison of percents point to single men as differing from the others. Of course, this association between marital status and job grade does not show that being single *causes* lower-grade jobs. The explanation might be as simple as the fact that single men tend to be younger and so have not yet advanced to higher grades.    ◄

## CELL COUNTS REQUIRED FOR THE CHI-SQUARE TEST

The chi-square test, like the $z$ procedures for comparing two proportions, is an approximate method that becomes more accurate as the counts in the cells of the table get larger. Fortunately, the approximation is accurate for quite modest counts. Here is a practical guideline based on the expected counts.[87]

---

CELL COUNTS REQUIRED FOR THE CHI-SQUARE TEST

You can safely use the chi-square test with $P$-values from the chi-square distribution when no more than 20% of the expected counts are less than 5 and all individual expected counts are 1 or greater. In particular, all four expected counts in a 2 × 2 table should be 5 or greater.

---

Example 23.3 easily passes this test. All the expected counts are greater than 1, and only 2 out of 16 (12.5%) are less than 5.

## THE CHI-SQUARE TEST AND THE Z TEST

We can use the chi-square test to compare any number of proportions. If we are comparing $r$ proportions and make the columns of the table "success" and "failure," the counts form an $r \times 2$ table. $P$-values come from the chi-square distribution with $r - 1$ degrees of freedom. If $r = 2$, we are comparing just two proportions. We have two ways to do this: the $z$ test from Lesson 22 and the chi-square test with 1 degree of freedom for a 2 × 2 table. *These two tests always agree.* In fact, the chi-square statistic $\chi^2$ is just the square of the $z$ statistic, and the $P$-value for $\chi^2$ is exactly the same as the two-sided $P$-value for $z$. We recommend using the $z$ test to compare two proportions, because it gives you the choice of a one-sided test and is related to a confidence interval for $p_1 - p_2$.

## EXERCISES

23.1    North Carolina State University studied student performance in a course required by its chemical engineering major. One question of interest is the relationship between time spent in extracurricular activities and whether a student earned a C or better in the course. Here are the data for the 119 students who answered a question about extracurricular activities:[88]

|  |  | Course grade A, B, C | D, F |
|---|---|---|---|
| Activity | Less than 2 | 11 | 9 |
| hours | 2 to 12 | 68 | 23 |
| per week | More than 12 | 3 | 5 |

(a) This is an $r \times c$ table. What are the numbers $r$ and $c$?

(b) Find the proportion of successful students (C or better) in each of the three extracurricular activity groups. What kind of relationship between extracurricular activities and succeeding in the course do these proportions seem to show?

(c) Make a bar chart to compare the three proportions of successes.

(d) The null hypothesis says that the proportions of successes are the same in all three groups if we look at the population of all students. Does the chi-square test give good evidence against this hypothesis?

(e) Which term contributes the most to $\chi^2$? What specific relation between extracurricular activities and academic success does this term point to?

(f) Does the North Carolina State study convince you that spending more or less time on extracurricular activities *causes* changes in academic success? Explain your answer.

23.2    How are the smoking habits of students related to their parents' smoking? Here are data from a survey of students in eight Arizona high schools:[89]

|  | Student smokes | Student does not smoke |
|---|---|---|
| Both parents smoke | 400 | 1380 |
| One parent smokes | 416 | 1823 |
| Neither parent smokes | 188 | 1168 |

(a) This is an $r \times c$ table. What are the numbers $r$ and $c$?

(b) Calculate the proportion of students who smoke in each of the three parent groups. Then describe in words the association between parent smoking and student smoking.

(c) Make a graph to display the association.

(d) Explain in words what the null hypothesis $H_0 : p_1 = p_2 = p_3$ says about student smoking. Carry out the chi-square test and report your conclusion.

(e) Which two terms contribute the most to $\chi^2$? What specific relation between parent smoking and student smoking do these terms point to?

(f) Does this study convince you that parent smoking *causes* student smoking? Explain your answer.

23.3    Exercise 22.11 (page 307) compared HMO members who filed complaints with an SRS of members who did not complain. The study actually broke the complainers into two subgroups: those who filed complaints about medical treatment and those who filed nonmedical complaints. Here are the data on the total number in each group and the number who voluntarily left the HMO:

|       | No complaint | Medical complaint | Nonmedical complaint |
|-------|--------------|-------------------|----------------------|
| Total | 743          | 199               | 440                  |
| Left  | 22           | 26                | 28                   |

(a) Find the percent of each group who left. What do the data show?

(b) Make a two-way table of complaint status by left or not.

(c) In plain language, what null and alternative hypotheses does the chi-square statistic test? What are its degrees of freedom? Carry out the test and report your conclusions.

(d) Check the expected counts to verify that we can safely use the chi-square test here.

23.4    Gastric freezing was once a recommended treatment for ulcers in the upper intestine. Use of gastric freezing stopped after experiments showed it had no effect. One randomized comparative experiment found that 28 of the 82 gastric freezing patients improved, while 30 of the 78 patients in the placebo group improved.[90] We can test the hypothesis of "no difference" between the two groups in two ways: using the two-sample $z$ statistic or using the chi-square statistic.

(a) State the null hypothesis with a two-sided alternative and carry out the $z$ test. What is the $P$-value?

(b) Present the data in a 2 × 2 table. Use the chi-square test to test the hypothesis from (a). Verify that the $\chi^2$ statistic is the square of the $z$ statistic. Verify that the chi-square $P$-value agrees with the $z$ result.

(c) What do you conclude about the effectiveness of gastric freezing as a treatment for ulcers?

23.5    A study of the career plans of young women and men sent questionnaires to all 722 members of the senior class in the College of Business Administration at the University of Illinois. One question asked which major within the business program the student had chosen. Here are the data from the students who responded:[91]

|  | Female | Male |
|---|---|---|
| Accounting | 68 | 56 |
| Administration | 91 | 40 |
| Economics | 5 | 6 |
| Finance | 61 | 59 |

(a) Test the null hypothesis that there is no relation between the sex of students and their choice of major. Give a $P$-value and state your conclusion.

(b) Describe the differences between the distributions of majors for women and men with percents, with a graph, and in words.

(c) Which two cells have the largest components of the chi-square statistic? How do the observed and expected counts differ in these cells? (This should strengthen your conclusions in (b).)

(d) Two of the observed cell counts are small. Do these data satisfy our guidelines for safe use of the chi-square test?

(e) What percent of the students did not respond to the questionnaire? The nonresponse weakens conclusions drawn from these data.

23.6    It seems that the attitude of cancer patients can influence the progress of their disease. We can't experiment with humans, but here is a rat experiment on this theme. Inject 60 rats with tumor cells and then divide them at random into two groups of 30. All the rats receive electric shocks, but rats in Group 1 can end the shock by pressing a lever. (Rats learn this sort of thing quickly.) The rats in Group 2 cannot control the shocks, which presumably makes them feel helpless and unhappy. We suspect that the rats in Group 1 will develop fewer tumors. The results: 11 of the Group 1 rats and 22 of the Group 2 rats developed tumors.[92]

(a) State the null and alternative hypotheses for this investigation. Explain why the $z$ test rather than the chi-square test for a $2 \times 2$ table is the proper test.

(b) Carry out the test and report your conclusion.

23.7 Do unregulated providers of child care in their homes follow different health and safety practices in different cities? A study looked at people who regularly provided care for someone else's children in poor areas of three cities. The numbers who required medical releases from parents to allow medical care in an emergency were 42 of 73 providers in Newark, N.J., 29 of 101 in Camden, N.J., and 48 of 107 in south Chicago, Ill.[93]

(a) Are there significant differences among the proportions of child-care providers who require medical releases in the three cities?

(b) How should the data be produced in order for your test to be valid? (In fact, the samples came in part from asking parents who were subjects in another study who provided their child care. The author of the study wisely did not use a statistical test. He wrote: "Application of conventional statistical procedures appropriate for random samples may produce biased and misleading results.")

23.8 Sample surveys on sensitive issues can give different results depending on how the question is asked. A University of Wisconsin study divided 2400 respondents into 3 groups at random. All were asked if they had ever used cocaine. One group of 800 was interviewed by phone; 21% said they had used cocaine. Another 800 people were asked the question in a one-on-one personal interview; 25% said "Yes." The remaining 800 were allowed to make an anonymous written response; 28% said "Yes."[94] Are there statistically significant differences among these proportions? (State the hypotheses, give the test statistic and its $P$-value, and state your conclusions.)

23.9 The success of a sample survey can depend on the season of the year. The Italian National Statistical Institute kept records of nonresponse to one of its national telephone surveys. All calls were made between 7 p.m. and 10 p.m. Here is a table of the percents of responses and of three types of nonresponse at different seasons. The percents in each row add to 100% (up to roundoff error).[95]

| Season | Calls made | Successful interviews | Nonresponse | | |
|---|---|---|---|---|---|
| | | | No answer | Busy signal | Refusal |
| Jan. 1 to Apr. 13 | 1558 | 68.5% | 21.4% | 5.8% | 4.3% |
| Apr. 21 to June 20 | 1589 | 52.4% | 35.8% | 6.4% | 5.4% |
| July 1 to Aug. 31 | 2075 | 43.4% | 41.5% | 8.6% | 6.5% |
| Sept. 1 to Dec. 15 | 2638 | 60.0% | 30.0% | 5.3% | 4.7% |

(a) What null and alternative hypotheses does chi-square test? The sample sizes are so large that the test is sure to be highly significant, so don't do it.

(b) Consider just the proportion of successful interviews. Describe how this proportion varies with the seasons, and assess the statistical significance of the changes. What do you think explains the changes? (Look at the full table for ideas.)

23.10 Continue the analysis of the data in the previous exercise by considering just the proportion of people called who refused to participate. We might think that the refusal rate changes less with the season than, for example, the rate of "no answer." State the hypothesis that the refusal rate does not change with the season. Check that you can safely use the chi-square test. Carry out the test. What do you conclude?

23.11 Return to the data on business students' choice of major, Exercise 23.5. We will use these data to explore expected cell counts.

(a) What does the null hypothesis say about the proportions of women in the four majors?

(b) There are 386 students in the sample. What proportion of these are women? If the null hypothesis is true, we expect this proportion to apply within each major as well as overall. There are 124 accounting students. What is the expected number of women among these 124 students if $H_0$ is true? What is the expected number of women among the 131 administration majors? Check your results against the expected counts given by software for these cells.

(c) Here is the general case. The overall proportion of observations falling in any column of a two-way table is

$$\frac{\text{column total}}{\text{table total}}$$

If this proportion applies in every row as well as overall, the expected count in any cell is

$$\text{expected count} = \frac{\text{row total} \times \text{column total}}{\text{table total}}$$

This is the rule for finding expected counts. Verify that this rule agrees with the results of your software in Exercise 23.5.

23.12 Aspirin is currently the only drug known to help prevent future strokes in people who have suffered a previous stroke. Dutch medical researchers compared the drug dipyridamole with aspirin and with a placebo. They also tested a combination of aspirin and dipyridamole. Here are the results after two years:[96]

| Treatment | Number of subjects | Number of strokes | Number of deaths |
|---|---|---|---|
| Placebo | 1649 | 250 | 202 |
| Aspirin | 1649 | 206 | 182 |
| Dipyridamole | 1654 | 211 | 188 |
| Both | 1650 | 157 | 185 |

(a) The researchers used a randomized comparative experimental design. Outline this design.

(b) Concentrate on the number of deaths. Is there significant evidence that the four treatments differ in the proportion of patients who die? In what ways do they differ?

(c) Critics said that it was not proper to give some subjects a placebo when aspirin was considered effective. Suppose that the experiment had not included the placebo group. Is there a significant difference among the three remaining groups?

(d) Suppose finally that the experiment had simply compared the combination treatment with aspirin. Is there good evidence that the combination treatment is superior to aspirin?

23.13   Answer the questions in the previous exercise when stroke, rather than death, is the outcome of interest.

23.14   Verify the results presented in Figure 23.2 for the data of Example 23.1. Starting from the 6 cell counts:

(a) Use the rule in Exercise 23.11 to find the 6 expected cell counts.

(b) Find the 6 residuals.

(c) Square the residuals and add them to find the value of the chi-square statistic.

## LESSON 24

# INFERENCE FOR REGRESSION

When a scatterplot shows a linear relationship between a quantitative explanatory variable $x$ and a quantitative response variable $y$, we can use correlation (Lesson 8) to describe the strength of the relationship. We can also use least-squares regression (Lesson 9) to predict $y$ for a given value of $x$. Now we want to do tests and confidence intervals in this setting.

EXAMPLE 24.1

Figure 24.1 is a scatterplot of data that played a central role in the discovery that the universe is expanding. The data appear in Table 24.1. They are the distances from earth of 24 spiral galaxies and the velocities at which these galaxies are moving away from us, reported by the astronomer Edwin Hubble in 1929.[97] (Distances in these data are measured in units of one million parsecs. A parsec is a common unit of distance in astronomy. It is equal to about 3.26 light-years, or about 19 million million miles.)

There is a positive linear relationship, $r = 0.790$, so that more distant galaxies are moving away more rapidly. The line on the scatterplot is the least-squares regression line for predicting the velocity of a galaxy from its distance. The

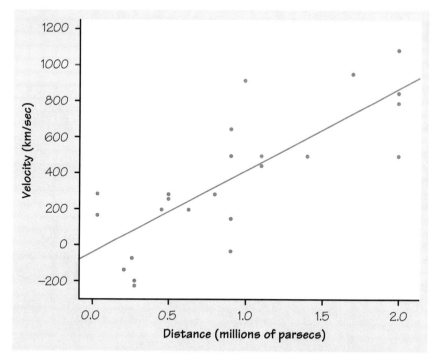

FIGURE 24.1    Scatterplot of Hubble's data on the velocity at which galaxies are moving away from us versus their distance from us.

TABLE 24.1    Hubble's data on the distance and velocity of galaxies

| Distance ($10^6$ parsecs) | Velocity (km/sec) | Distance ($10^6$ parsecs) | Velocity (km/sec) | Distance ($10^6$ parsecs) | Velocity (km/sec) |
|---|---|---|---|---|---|
| 0.032 | 170 | 0.5 | 270 | 1.1 | 450 |
| 0.034 | 290 | 0.63 | 200 | 1.1 | 500 |
| 0.214 | −130 | 0.8 | 300 | 1.4 | 500 |
| 0.263 | −70 | 0.9 | −30 | 1.7 | 960 |
| 0.275 | −185 | 0.9 | 650 | 2.0 | 500 |
| 0.275 | −220 | 0.9 | 150 | 2.0 | 850 |
| 0.45 | 200 | 0.9 | 500 | 2.0 | 850 |
| 0.50 | 290 | 1.0 | 920 | 2.0 | 1090 |

equation of this line is

$$\hat{y} = -40.7836 + 454.158x$$

The slope of this line tells us that on the average these galaxies move away 454 kilometers per second faster for every additional million parsecs of distance. For astronomers, this is a key number for determining how fast the universe is expanding and hence how old it is.

The regression of velocity on distance explains about $r^2 = 62\%$ of the observed variation in the velocities of these galaxies. The scatterplot shows moderate scatter about the fitted line but no very large residuals or potentially very influential observations.    ◄

Before you attempt inference in the regression setting, always examine your data as we did in Example 24.1:

1. Make a **scatterplot.** Plot the explanatory variable $x$ horizontally and the response variable $y$ vertically. Fitting a line only makes sense if the overall pattern of the scatterplot is roughly linear (straight-line).
2. Fit the **least-squares regression line** for the data. This line lies as close as possible to the points (in the sense of least squares) in the vertical ($y$) direction. We use it to predict $y$ for a given $x$.
3. Look for **outliers** and **influential observations.** Outliers are points that lie far from the overall pattern represented by the least-squares line. Influential observations are points that move the fitted line, usually points that are far out in the $x$ direction and isolated from other points. Inference is not safe if there are influential points, because the results will depend strongly on these few points.
4. Find the **correlation** $r$ and its square $r^2$. The squared correlation $r^2$ describes how well the regression line fits the data. It is the proportion of the observed variation in $y$ that is accounted for by the straight-line relationship of $y$ with $x$.

## THE REGRESSION INFERENCE MODEL

The slope $b = 454.158$ and intercept $a = -40.7836$ of the least-squares line are *statistics*. They are computed from the sample data and would no doubt be different if Hubble had observed a different sample of galaxies.

The *parameters* of interest are the slope $\beta$ and intercept $\alpha$ we would get if we regressed velocity on distance for the population of all galaxies in the universe. Of course, we don't expect all galaxies to lie exactly on a straight line, only that there is an "on the average" linear relationship. Here is a more detailed statement of the setting for regression inference.

---

ASSUMPTIONS FOR REGRESSION INFERENCE

We have $n$ observations on an explanatory variable $x$ and a response variable $y$. Our goal is to study or predict the behavior of $y$ for given values of $x$.

- For any fixed value of $x$, the response $y$ varies according to a normal distribution. Repeated responses $y$ are independent of each other.
- The mean response $\mu_y$ has a straight-line relationship with $x$:

$$\mu_y = \alpha + \beta x$$

The slope $\beta$ and intercept $\alpha$ are unknown parameters.
- The standard deviation of $y$ (call it $\sigma$) is the same for all values of $x$. The value of $\sigma$ is unknown.

---

The **true regression line** $\mu_y = \alpha + \beta x$ says that the *mean* response $\mu_y$ moves along a straight line as the explanatory variable $x$ changes. We can't observe the true regression line. The values of $y$ that we do observe vary about their means according to a normal distribution. If we hold $x$ fixed and take many observations on $y$, the normal pattern will eventually appear in a histogram. In practice, we observe $y$ for many different values of $x$, and so we see an overall linear pattern with points scattered about it. The standard deviation $\sigma$ determines whether the points fall close to the true regression line (small $\sigma$) or are widely scattered (large $\sigma$).

Figure 24.2 shows the regression model in picture form. The line in the figure is the true regression line. The mean of the response $y$ moves along this line as the explanatory variable $x$ takes different values. The normal curves show how $y$ varies when $x$ is held fixed at a single value. All of the curves have the same $\sigma$, so the variability of $y$ is the same for all values of $x$. You should check the assumptions for inference when you do inference about regression. We will see later how to do that.

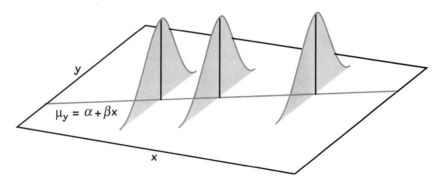

FIGURE 24.2    The regression model. The line is the true regression line, which shows how the mean response $\mu_y$ changes as the explanatory variable $x$ changes. For any fixed value of $x$, the observed response $y$ varies about this line according to a normal distribution.

## ESTIMATING PARAMETERS

The first step in inference is to estimate the unknown parameters $\alpha$, $\beta$, and $\sigma$. When the regression model describes our data and we calculate the least-squares line $\hat{y} = a + bx$:

- The slope $b$ of the least-squares line is an unbiased estimator of the true slope $\beta$.
- The intercept $a$ of the least-squares line is an unbiased estimator of the true intercept $\alpha$.

The remaining parameter of the model is the standard deviation $\sigma$, which describes the variability of the response $y$ about the true regression line. The least-squares line estimates the true regression line. So the **residuals** estimate how much $y$ varies about the true line. Recall that the residuals are the vertical deviations of the data points from the least-squares line:

$$\text{residual} = \text{observed } y - \text{predicted } y$$
$$= y - \hat{y}$$

There are $n$ residuals, one for each data point. Because $\sigma$ is the standard deviation of responses about the true regression line, we estimate it by a sample standard deviation of the residuals. We call this sample standard deviation a *standard error* to emphasize that it is estimated from data. The residuals from a least-squares line always have mean zero. That simplifies their standard error.

---

STANDARD ERROR ABOUT THE LEAST-SQUARES LINE

The **standard error about the line** is

$$s = \sqrt{\frac{1}{n-2}\sum \text{residual}^2}$$

$$= \sqrt{\frac{1}{n-2}\sum (y - \hat{y})^2}$$

Use $s$ to estimate the unknown $\sigma$ in the regression model.

---

Because we use the standard error about the line so often in regression inference, we just call it $s$. Software that fits the least-squares line will also calculate $s$ for you. Notice that $s^2$ is an average of the squared deviations of the data points from the line, so it qualifies as a variance. We average the squared deviations by dividing their sum by the **degrees of freedom** $n-2$. We first met the idea of degrees of freedom in the case of the ordinary sample standard deviation of $n$ observations (Lesson 4), which has $n-1$ degrees of freedom. The mean response now contains two parameters, $\alpha$ and $\beta$, rather than a single mean $\mu$. The proper degrees of freedom is now $n-2$ rather than $n-1$.

We will study several kinds of inference in the regression setting. The standard error $s$ about the line is the key measure of the variability of the responses in regression. It is part of the standard error of all the statistics we will use for inference.

## EXAMPLE 24.2

Figure 24.3 displays *Data Desk* regression output for Hubble's data. The estimates of the model parameters $\alpha$, $\beta$, and $\sigma$ are

$$a = -40.7836 \quad b = 454.158 \quad s = 232.9$$

and $s$ has 22 degrees of freedom.     ◄

```
┌─────────────────────────────────────────────────────────────── □ ▣ ┐
│ Dependent variable is:    Velocity                                  │
│ No Selector                                                         │
│ R squared = 62.4%    R squared (adjusted) = 60.6%                   │
│ s = 232.9  with  24 - 2 = 22  degrees of freedom                    │
│                                                                     │
│ Source        Sum of Squares     df    Mean Square    F-ratio       │
│ Regression    1976648             1       1976648      36.4          │
│ Residual      1193442            22       54247.4                    │
│                                                                     │
│ Variable      Coefficient    s.e. of Coeff    t-ratio      prob     │
│ Constant      -40.7836        83.44            -0.489       0.6298   │
│ Distance      454.158         75.24             6.04      ≤ 0.0001   │
│                                                                     │
└─────────────────────────────────────────────────────────────────────┘
```

FIGURE 24.3    *Data Desk* regression output for Hubble's data.

## INFERENCE ABOUT THE REGRESSION SLOPE

The slope $\beta$ of the true regression line is usually the most important parameter in a regression problem. The slope is the rate of change of the mean response as the explanatory variable increases. We often want to estimate $\beta$. The slope $b$ of the least-squares line is an unbiased estimator of $\beta$. A confidence interval is more useful because it shows how accurate the estimate $b$ is likely to be. The confidence interval for $\beta$ has the familiar form

$$\text{estimate} \pm t^* SE_{\text{estimate}}$$

Here are the details.

---

CONFIDENCE INTERVAL FOR REGRESSION SLOPE

A level $C$ confidence interval for the slope $\beta$ of the true regression line is

$$b \pm t^* SE_b$$

As usual, $t^*$ and $-t^*$ capture between them the middle area $C$ under the $t$ density curve with $n - 2$ degrees of freedom.

Although we give the recipe for the standard error of $b$ a bit later, you should rarely have to calculate it by hand. Regression software gives the standard error $SE_b$ along with $b$ itself.

---

**EXAMPLE 24.3**

The output in Figure 24.3 tells us that

$$b = 454.158 \qquad SE_b = 75.24$$

A 95% confidence interval for the true slope $\beta$ uses the value $t^* = 2.074$ from the $t$ distribution with 22 degrees of freedom. The confidence interval is

$$b \pm t^* SE_b = 454.158 \pm (2.074)(75.24)$$
$$= 454.16 \pm 156.05$$

We are 95% confident that

$$298.1 < \beta < 610.2$$

That is, the average velocity with which galaxies are receding increases by between 298.1 and 610.2 kilometers per second for each additional million parsecs of distance. (More recent measurements have greatly improved on Hubble's pioneer work, giving quite different estimates of $\beta$.)    ◀

---

We can also test hypotheses about the value of the slope $\beta$. The most common hypothesis is

$$H_0 : \beta = 0$$

A regression line with slope 0 is horizontal. That is, the mean of $y$ does not change at all when $x$ changes. So $H_0$ says that there is *no true linear relationship* between $x$ and $y$. Put another way, $H_0$ says that *straight-line dependence on $x$ is of no value for predicting $y$*. Put yet another way, $H_0$ says that there is *no correlation* between $x$ and $y$ in the population from which we drew our data. You can use the test for zero slope to test the hypothesis of zero correlation between any two quantitative variables. That's a useful trick.

The test statistic is just the standardized version of the least-squares slope $b$. It is another $t$ statistic. Here are the details.

---

**TESTS FOR REGRESSION SLOPE**

To test the hypothesis $H_0 : \beta = 0$, compute the $t$ statistic

$$t = \frac{b}{SE_b}$$

Find $P$-values for this statistic from the $t$ distribution with $n - 2$ degrees of freedom.

---

Regression output from statistics software usually gives $t$ and its *two-sided P*-value. For a one-sided test, divide the $P$-value in the output by 2.

**EXAMPLE 24.4**

Earlier work had suggested that galaxies farther from us are moving away faster. Hubble produced the first careful data to confirm this. We can use his data to test

$$H_0 : \beta = 0$$
$$H_a : \beta > 0$$

The one-sided alternative says that velocity increases with distance. From the output in Figure 24.3, we see that

$$t = 6.04 \qquad P \leq 0.0001$$

The one-sided $P$-value is half the software value. There is extremely strong evidence that velocity does increase with distance. ◀

## INFERENCE ABOUT PREDICTION

One of the most common reasons to fit a line to data is to predict the response to a particular value of the explanatory variable. The method is simple: just substitute the value of $x$ into the equation of the line.

**EXAMPLE 24.5**

> Given a galaxy at a distance of 1.5 million parsecs, we estimate its velocity to be
>
> $$\hat{y} = -40.7836 + 454.158 \times 1.5$$
> $$= -40.7836 + 681.237 = 640.45 \text{ km/sec}$$
>
> ◄

To give a confidence interval that describes how accurate this prediction is, you must answer these questions: Do you want to predict the *mean* velocity for *all* galaxies at this distance? Or do you want to predict the velocity of *one specific* galaxy located this distance from us? Both of these predictions may be interesting, but they are two different problems. The actual prediction is the same, $\hat{y} = 640.45$ kilometers per second. But the margin of error is different for the two kinds of prediction. Different galaxies at the same distance are not all receding at exactly the same speed. So we need a larger margin of error to pin down one galaxy than to estimate the mean velocity for all galaxies this distance away.

Write the given value of the explanatory variable $x$ as $x^*$. In Example 24.5, $x^* = 1.5$. The distinction between predicting a single outcome and predicting the mean of all outcomes when $x = x^*$ determines the correct margin of error. To emphasize the distinction, we use different terms for the two intervals.

- To estimate the *mean* response, we use an ordinary confidence interval for the parameter

$$\mu_y = \alpha + \beta x^*$$

The regression model says that $\mu_y$ is the mean of responses $y$ when $x$ has the value $x^*$. It is a fixed number whose value we don't know.

- To estimate an *individual* response $y$, we use a **prediction interval.** A prediction interval estimates a single random response $y$ rather than a parameter such as $\mu_y$. The response $y$ is not a fixed number. If we took several observations with $x = x^*$, we would get different responses.

Fortunately, the meaning of a prediction interval is very much like the meaning of a confidence interval. A 95% prediction interval, like a 95% confidence interval, is right 95% of the time in repeated use. "Repeated use" now means that we take an observation on $y$ for each of the $n$ values of $x$ in the original data, and then take one more observation $y$ at the point $x = x^*$. Form the prediction interval from the $n$ observations, then see if it covers the one more $y$. It will in 95% of all repetitions.

Both intervals have the usual form

$$\hat{y} \pm t^*\text{SE}$$

but the prediction interval is wider than the confidence interval. Here are the details.

---

CONFIDENCE AND PREDICTION INTERVALS
FOR REGRESSION RESPONSE

A level $C$ **confidence interval for the mean response** $\mu_y$ when $x$ takes the value $x^*$ is

$$\hat{y} \pm t^*\text{SE}_{\hat{\mu}}$$

Here $\text{SE}_{\hat{\mu}}$ is the standard error for predicting the mean response when $x = x^*$.

A level $C$ **prediction interval for a single observation** on $y$ when $x$ takes the value $x^*$ is

$$\hat{y} \pm t^*\text{SE}_{\hat{y}}$$

The standard error for predicting one response $\text{SE}_{\hat{y}}$ is larger than the standard error for predicting the mean response.

In both recipes, $t^*$ captures the middle area $C$ under the $t$ density curve with $n - 2$ degrees of freedom.

---

**EXAMPLE 24.6**

Software tells us that a 95% confidence interval, based on Hubble's data, for the mean velocity of all galaxies located 1.5 million parsecs away is

$$505.7 < \mu_y < 775.2$$

In contrast, the 95% prediction interval for one specific galaxy at that distance is

$$139.0 < y < 1141.9$$

As advertised, the prediction interval is much wider.    ◀

## OPTIONAL: SOME FORMULAS

The calculations required for regression inference are a bit unpleasant. In practice, we always use software to automate them. Here for reference are the actual formulas for the various standard errors used in regression inference. All are multiples of the standard error $s$ about the fitted line.

The standard error of the least-squares slope $b$ is

$$SE_b = \frac{s}{\sqrt{\sum(x - \bar{x})^2}}$$

This standard error gets larger when the observations are more widely scattered about the least-squares line (larger $s$). It gets smaller when the $x$-values observed are more spread out (larger denominator).

The standard error $SE_{\hat{\mu}}$ for predicting the mean response is

$$SE_{\hat{\mu}} = s\sqrt{\frac{1}{n} + \frac{(x^* - \bar{x})^2}{\sum(x - \bar{x})^2}}$$

This standard error shows the same behavior as $SE_b$. In addition, it gets larger when the $x^*$ for which we want to predict the mean response is far from the mean $\bar{x}$ of the $x$-values in our data.

The standard error $SE_{\hat{y}}$ for predicting a single response is[98]

$$SE_{\hat{y}} = s\sqrt{1 + \frac{1}{n} + \frac{(x^* - \bar{x})^2}{\sum(x - \bar{x})^2}}$$

This differs from the previous expression only in the extra "1" under the square root. That extra term makes the interval wider.

## CHECKING THE REGRESSION INFERENCE ASSUMPTIONS

You can fit a least-squares line to any set of *x-y* data. Regression inference, however, makes sense only in more restricted settings. Start by verifying that the scatterplot shows a **roughly linear relationship** and **no strongly influential observations.** The key to checking the other assumptions required for inference is to work with the **residuals,** the vertical deviations of the observations from the fitted line.

To check **normality,** make a normal probability plot of the residuals. For Hubble's data, the plot in Figure 24.4 shows that the residuals have no important deviations from normality. To check **constant variability** (a single $\sigma$ no matter what *x* is), plot the residuals against the explanatory variable *x*. The plot should show a random scatter about the line at residual 0, like the plot for the Hubble data in Figure 24.5. If it fans out as *x* increases, for example, the variability of *y* is increasing with *x* and standard regression inference isn't valid. Checking **independence** of the observations is difficult to do from the data—it is usually a matter of making sure the observations were not allowed to influence each other when you made them.

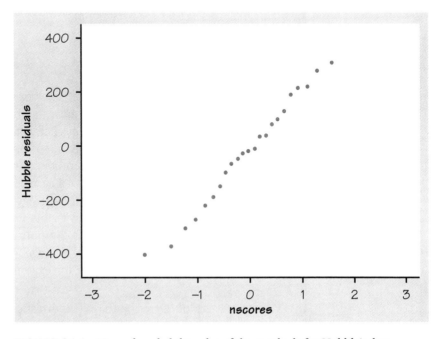

FIGURE 24.4   Normal probability plot of the residuals for Hubble's data.

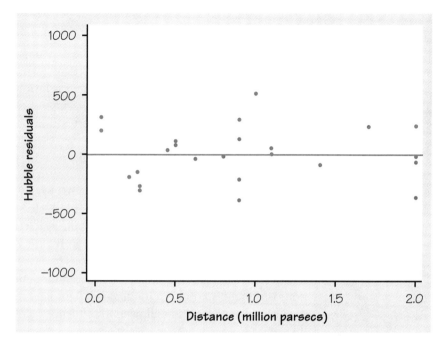

FIGURE 24.5    Plot of the Hubble residuals against the explanatory variable.

## EXERCISES

24.1    The Sanchez household is about to install solar panels to reduce the cost of heating their house. In order to know how much the solar panels help, they record their consumption of natural gas before the panels are installed. Gas consumption is higher in cold weather, so the relationship between outside temperature and gas consumption is important. The table on the next page displays data for 16 months.[99] The response variable $y$ is the average amount of natural gas consumed each day during the month, in hundreds of cubic feet. The explanatory variable $x$ is the average number of heating degree-days each day during the month. (One heating degree-day is accumulated for each degree a day's average temperature falls below $65°$ F. An average temperature of $20°$ F, for example, corresponds to 45 degree-days.)

(a) Make a scatterplot. Describe the form, direction, and strength of the relationship. Find the least-squares regression line of gas consumption on degree-days and add the line to your plot. What percent of the observed variation in gas used is explained by the regression?

| Month | Degree-days | Gas used (100 cu ft) | Month | Degree-days | Gas used (100 cu ft) |
|-------|-------------|----------------------|-------|-------------|----------------------|
| Nov.  | 24 | 6.3  | July  | 0  | 1.2  |
| Dec.  | 51 | 10.9 | Aug.  | 1  | 1.2  |
| Jan.  | 43 | 8.9  | Sept. | 6  | 2.1  |
| Feb.  | 33 | 7.5  | Oct.  | 12 | 3.1  |
| Mar.  | 26 | 5.3  | Nov.  | 30 | 6.4  |
| Apr.  | 13 | 4.0  | Dec.  | 32 | 7.2  |
| May   | 4  | 1.7  | Jan.  | 52 | 11.0 |
| June  | 0  | 1.2  | Feb.  | 30 | 6.9  |

(b) Make a normal probability plot of the residuals and also a plot of the residuals against the explanatory variable. Verify that there is no barrier to regression inference using these data.

(c) By how much does gas consumption increase on the average for every additional degree-day? (Give a 95% confidence interval.)

(d) The Sanchez household installs solar panels. The following March, there are 20 degree-days per day. What would their gas usage have been without the solar panels? (Give a 95% confidence interval.)

24.2    Investors ask about the relationship between returns on investments in the United States and on investments overseas. The table on the next page presents data on the total returns on U.S. and overseas common stocks over a 26-year period. (The total return is change in price plus any dividends paid, converted into U.S. dollars. Both returns are averages over many individual stocks.)[100]

(a) Make a scatterplot suitable for predicting overseas returns from U.S. returns. Find the least-squares regression line and add it to your plot. What percent of the variation in overseas returns can be explained by the regression on U.S. returns?

(b) Make a normal probability plot of the residuals and also plot the residuals against the U.S. returns. There is some suggestion that spread increases as $x$ increases, but the effect is not strong enough to bar regression inference.

(c) Is there good evidence that there is a linear relationship between U.S. and foreign returns? State hypotheses and give a test statistic and its $P$-value.

(d) You think U.S. stocks will return 15% next year. Give a 90% prediction interval for the return on overseas stocks next year if this is true. Is your prediction interval practically useful?

| Year | Overseas % return | U.S. % return | Year | Overseas % return | U.S. % return |
|------|-------------------|---------------|------|-------------------|---------------|
| 1971 | 29.6 | 14.6 | 1984 | 7.4 | 6.1 |
| 1972 | 36.3 | 18.9 | 1985 | 56.2 | 31.6 |
| 1973 | −14.9 | −14.8 | 1986 | 69.4 | 18.6 |
| 1974 | −23.2 | −26.4 | 1987 | 24.6 | 5.1 |
| 1975 | 35.4 | 37.2 | 1988 | 28.5 | 16.8 |
| 1976 | 2.5 | 23.6 | 1989 | 10.6 | 31.5 |
| 1977 | 18.1 | −7.4 | 1990 | −23.0 | −3.1 |
| 1978 | 32.6 | 6.4 | 1991 | 12.8 | 30.4 |
| 1979 | 4.8 | 18.2 | 1992 | −12.1 | 7.6 |
| 1980 | 22.6 | 32.3 | 1993 | 32.9 | 10.1 |
| 1981 | −2.3 | −5.0 | 1994 | 6.2 | 1.3 |
| 1982 | −1.9 | 21.5 | 1995 | 11.2 | 37.6 |
| 1983 | 23.7 | 22.4 | 1996 | 6.4 | 22.9 |

24.3   How well does a car's city gas mileage (as measured by the Environmental Protection Agency) predict its highway gas mileage? Table 24.2 records the city and highway mileage for a sample of 68 1997 car models.[101]

(a) Make a scatterplot of the data. Describe the form, direction, and strength of the relationship. Find the least-squares regression line and add it to your plot. The plot contains one unusual point. What car is this?

(b) Remove this point and again find the least-squares line. Was the un-usual point highly influential? Use all 68 cars in the rest of your work.

(c) Examine the residuals to check the assumptions of the regression model. Are the assumptions reasonably well met?

(d) Explain the meaning of the slope $\beta$ of the true regression line in this setting. Give a 95% confidence interval for $\beta$.

(e) The Volkswagen Passat, a midsize model not in the sample, gets 18 miles per gallon in the city. Predict the Passat's highway mileage. Then give an interval which covers the true highway mileage with 95% confidence.

24.4   The 1997 car models in Table 24.2 are classified by the EPA as compact, midsize, or large on the basis of their passenger and luggage volume. In the previous exercise we ignored this classification. Regress highway mileage $y$ on city mileage $x$ separately for each class of car. Plot all three regression lines on a scatterplot of the data. Are any of the lines substantially different

**TABLE 24.2**  City and highway gas mileage for 1997 car models

| Model | Size | City MPG | Highway MPG |
|---|---|---|---|
| Acura 2.5TL | Compact | 20 | 25 |
| Acura 3.5TL | Compact | 19 | 24 |
| Audi A4 quattro | Compact | 20 | 29 |
| BMW 318ti | Compact | 23 | 31 |
| BMW 528i | Compact | 19 | 26 |
| Buick Skylark | Compact | 22 | 32 |
| Chevrolet Cavalier | Compact | 24 | 31 |
| Chrysler Sebring | Compact | 21 | 30 |
| Dodge Neon | Compact | 25 | 34 |
| Ford Contour | Compact | 23 | 32 |
| Ford Escort | Compact | 26 | 34 |
| Honda Accord | Compact | 23 | 30 |
| Hyundai Accent | Compact | 27 | 36 |
| Hyundai Elantra | Compact | 23 | 31 |
| Jaguar XJ6 | Compact | 17 | 23 |
| Lexus ES300 | Compact | 19 | 26 |
| Mazda Millennia | Compact | 20 | 28 |
| Mazda Protegé | Compact | 25 | 33 |
| Mercedes-Benz C230 | Compact | 23 | 30 |
| Mercedes-Benz C280 | Compact | 20 | 27 |
| Mitsubishi Galant | Compact | 22 | 28 |
| Mitsubishi Mirage | Compact | 29 | 36 |
| Nissan Altima | Compact | 21 | 29 |
| Saturn SL | Compact | 24 | 34 |
| Subaru Legacy | Compact | 24 | 31 |
| Volkswagen Golf | Compact | 22 | 29 |
| Volkswagen Jetta GLX | Compact | 18 | 24 |
| Volvo 960 | Compact | 18 | 26 |
| Acura 3.5RL | Midsize | 19 | 25 |
| Audi A8 quattro | Midsize | 17 | 25 |
| Buick Century | Midsize | 20 | 29 |
| Cadillac Catera | Midsize | 18 | 25 |
| Cadillac Eldorado | Midsize | 17 | 26 |

(*continued*)

**TABLE 24.2**   City and highway gas mileage for 1997 car models *(continued)*

| Model | Size | City MPG | Highway MPG |
|---|---|---|---|
| Chevrolet Lumina | Midsize | 20 | 29 |
| Chrysler Cirrus | Midsize | 20 | 30 |
| Dodge Stratus | Midsize | 22 | 32 |
| Ford Taurus | Midsize | 19 | 28 |
| Ford Thunderbird | Midsize | 18 | 26 |
| Hyundai Sonata | Midsize | 20 | 27 |
| Infiniti I30 | Midsize | 21 | 28 |
| Infiniti Q45 | Midsize | 18 | 24 |
| Lexus GS300 | Midsize | 18 | 24 |
| Lexus LS400 | Midsize | 19 | 25 |
| Lincoln-Mercury Mark VIII | Midsize | 18 | 26 |
| Mazda 626 | Midsize | 23 | 31 |
| Mercedes-Benz E320 | Midsize | 20 | 27 |
| Mercedes-Benz E420 | Midsize | 18 | 25 |
| Mitsubishi Diamante | Midsize | 18 | 26 |
| Nissan Maxima | Midsize | 21 | 28 |
| Oldsmobile Aurora | Midsize | 17 | 26 |
| Rolls-Royce Silver Spur | Midsize | 11 | 16 |
| Saab 900 | Midsize | 18 | 26 |
| Toyota Camry | Midsize | 23 | 30 |
| Volvo 850 | Midsize | 19 | 26 |
| BMW 740iL | Large | 17 | 24 |
| Buick LeSabre | Large | 19 | 30 |
| Buick Park Avenue | Large | 19 | 28 |
| Buick Roadmaster | Large | 17 | 25 |
| Cadillac DeVille | Large | 17 | 26 |
| Chrysler Concorde | Large | 19 | 27 |
| Chrysler New Yorker | Large | 17 | 26 |
| Ford Crown Victoria | Large | 17 | 25 |
| Lincoln-Mercury Continental | Large | 17 | 25 |
| Mercedes-Benz S320 | Large | 17 | 24 |
| Mercedes-Benz S420 | Large | 16 | 22 |
| Mercedes-Benz S500 | Large | 15 | 21 |
| Saab 9000 | Large | 17 | 26 |
| Toyota Avalon | Large | 21 | 31 |

SOURCE: Environmental Protection Agency, *Model Year 1997 Fuel Economy Guide.*

from the others? Would you be willing to use a single regression line to predict highway mileage from city mileage for cars in all three classes?

24.5   Table 20.2 (page 261) contains data on the level of particulate air pollution in a city and in a rural area upwind from the city. We would like to predict the average pollution level in the city for days when the rural level is 50 grams. Check the assumptions for regression inference, then give a 90% interval for the desired prediction.

24.6   Manatees are large, gentle sea creatures that live along the Florida coast. Many manatees are killed or injured by powerboats. Here are data on powerboat registrations (in thousands) and the number of manatees killed by boats in Florida in the years 1977 to 1990:

| Year | Powerboat registrations (1000) | Manatees killed | Year | Powerboat registrations (1000) | Manatees killed |
|------|------|------|------|------|------|
| 1977 | 447 | 13 | 1984 | 559 | 34 |
| 1978 | 460 | 21 | 1985 | 585 | 33 |
| 1979 | 481 | 24 | 1986 | 614 | 33 |
| 1980 | 498 | 16 | 1987 | 645 | 39 |
| 1981 | 513 | 24 | 1988 | 675 | 43 |
| 1982 | 512 | 20 | 1989 | 711 | 50 |
| 1983 | 526 | 15 | 1990 | 719 | 47 |

(a) Make a scatterplot, including the least-squares regression line.

(b) Is there strong evidence that the count of powerboats is related to the number of manatees killed? How much does each additional thousand powerboats increase the number of manatees killed? (Give a 95% confidence interval.)

(c) If Florida were to freeze powerboat registrations at 716,000, what would be the average number of manatees killed each year? (Give a 95% confidence interval.)

(d) Here are four more years of manatee data:

    1991  716  53   1993  716  35
    1992  716  38   1994  735  49

Did the actual average manatee deaths for the three years that had 716,000 boats fall in your interval from (c)? Now add these four observations to your scatterplot in (a) and recalculate the regression line with the new data. Did the line change in a practically meaningful way? What would now be your 95% interval for the prediction of (c)?

24.7    Table 24.3 contains data on the size of perch caught in a lake in Finland.[102]

(a) We want to know how well we can predict the width of a perch from its length. Make a scatterplot of width against length. There is a strong linear pattern, as expected. The heaviest perch had six newly eaten fish in its stomach and is an outlier in weight. Find this fish on your scatterplot and circle the point. Is this fish an outlier in your plot of width against length?

**TABLE 24.3**    Measurements of 56 perch

| Length (centimeters) | Width (centimeters) | Weight (grams) | Length (centimeters) | Width (centimeters) | Weight (grams) |
|---|---|---|---|---|---|
| 8.4 | 2.02 | 16.0 | 13.7 | 3.29 | 13.6 |
| 15.0 | 3.58 | 15.2 | 16.2 | 4.33 | 15.3 |
| 17.4 | 4.32 | 15.9 | 18.0 | 4.90 | 17.3 |
| 18.7 | 5.01 | 16.1 | 19.0 | 5.30 | 15.1 |
| 19.6 | 4.84 | 14.6 | 20.0 | 4.84 | 13.2 |
| 21.0 | 5.31 | 15.8 | 21.0 | 5.52 | 14.7 |
| 21.0 | 5.31 | 16.3 | 21.3 | 5.96 | 15.5 |
| 22.0 | 5.72 | 14.5 | 22.0 | 5.28 | 15.0 |
| 22.0 | 5.72 | 15.0 | 22.0 | 5.50 | 15.0 |
| 22.0 | 5.17 | 17.0 | 22.5 | 5.49 | 15.1 |
| 22.5 | 6.37 | 15.1 | 22.7 | 5.58 | 15.0 |
| 23.0 | 4.90 | 14.8 | 23.5 | 5.90 | 14.9 |
| 24.0 | 6.86 | 14.6 | 24.0 | 6.00 | 15.0 |
| 24.6 | 6.32 | 15.9 | 25.0 | 6.08 | 13.9 |
| 25.6 | 6.22 | 15.7 | 26.5 | 6.78 | 14.8 |
| 27.3 | 7.92 | 17.9 | 27.5 | 6.82 | 15.0 |
| 27.5 | 6.71 | 15.0 | 27.5 | 6.93 | 15.8 |
| 28.0 | 7.45 | 14.3 | 28.7 | 7.23 | 15.4 |
| 30.0 | 7.23 | 15.1 | 32.8 | 9.68 | 17.7 |
| 34.5 | 9.69 | 17.5 | 35.0 | 10.78 | 20.9 |
| 36.5 | 10.18 | 17.6 | 36.0 | 9.97 | 17.6 |
| 37.0 | 10.18 | 15.9 | 37.0 | 9.95 | 16.2 |
| 39.0 | 10.49 | 18.1 | 39.0 | 10.49 | 14.5 |
| 39.0 | 11.74 | 17.8 | 40.0 | 11.28 | 16.8 |
| 40.0 | 11.04 | 17.0 | 40.0 | 11.68 | 17.6 |
| 40.0 | 10.48 | 15.6 | 42.0 | 12.05 | 15.4 |
| 43.0 | 11.35 | 16.1 | 43.0 | 11.82 | 16.3 |
| 43.5 | 11.92 | 17.7 | 44.0 | 11.79 | 16.3 |

**(b)** Find the least-squares regression line to predict width from length and add the line to your plot.

**(c)** The length of a typical perch is about $x^* = 25$ centimeters. Predict the mean width of such fish and give a 95% confidence interval.

**(d)** Examine the residuals. Is there any reason to mistrust inference?

24.8    We can also use the data in Table 24.3 to study the prediction of the weight of a perch from its length.

**(a)** Make a scatterplot of weight versus length, with length as the explanatory variable. Describe the pattern of the data and any clear outliers.

**(b)** It is more reasonable to expect the one-third power of the weight to have a straight-line relationship with length than to expect weight itself to have a straight-line relationship with length. Explain why this is true. (Hint: What happens to weight if length, width, and height all double?)

**(c)** Use your software to create a new variable that is the one-third power of weight. Make a scatterplot of this new response variable against length. Describe the pattern and any clear outliers.

**(d)** Is the straight-line pattern in (c) stronger or weaker than that in (a)? Compare the plots and also the values of $r^2$.

**(e)** Find the least-squares regression line to predict the new weight variable from length. Predict the mean of the new variable for perch 25 centimeters long, and give a 95% confidence interval.

**(f)** Examine the residuals from your regressions. Does it appear that any of the regression assumptions are not met?

24.9    To demonstrate the effect of nematodes (microscopic worms) on plant growth, a botanist prepares 16 identical planting pots, then introduces different numbers of nematodes into the pots. A tomato seedling is transplanted into each pot. Here are data on the increase in height of the seedlings (in centimeters) during the 16 days after planting:[103]

| Nematodes | Seedling growth | | | |
|---|---|---|---|---|
| 0 | 10.8 | 9.1 | 13.5 | 9.2 |
| 1,000 | 11.1 | 11.1 | 8.2 | 11.3 |
| 5,000 | 5.4 | 4.6 | 7.4 | 5.0 |
| 10,000 | 5.8 | 5.3 | 3.2 | 7.5 |

**(a)** An unthinking student regresses growth on nematode count and uses the regression for inference. He says that "with 90% confidence, each additional 1000 nematodes reduces mean seedling growth by between $L$ and $U$." What are the numbers $L$ and $U$?

(b) Make a scatterplot of the data and show the regression line on the plot. Did the student's work make sense? Explain your answer.

24.10 Table 3.2 (page 24) contains data on education in the states. We expect that how much states spend on education helps explain how much they pay their teachers.

(a) Make a scatterplot and do the regression. What percent of the variation among the states in teachers' salaries is explained by regression on state education spending?

(b) Examine the residuals. Are there any states with very large residuals or that may be strongly influential? If so, redo the regression without these states and report whether they were very influential.

(c) The data analysis in (a) and (b) requires no assumptions. It does not make sense to do inference for these data. For example, it makes no sense to give a 95% confidence interval for the slope in the population of all states. Why not?

24.11 *Archaeopteryx* is an extinct beast having feathers like a bird but teeth and a long bony tail like a reptile. Here are the lengths in centimeters of the femur (a leg bone) and the humerus (a bone in the upper arm) for the five fossil specimens that preserve both bones:[104]

| Femur | 38 | 56 | 59 | 64 | 74 |
|---|---|---|---|---|---|
| Humerus | 41 | 63 | 70 | 72 | 84 |

The strong linear relationship between the lengths of the two bones helped persuade scientists that all five specimens belong to the same species. We will use this small data set to verify that some of the formulas we have given do lie behind software results.

(a) Enter the data into your software and carry out the regression of humerus length $y$ on femur length $x$. We will start with the equation of the fitted line.

(b) Calculate the predicted humerus lengths and residuals for the five data points. Check that the sum of the residuals is 0 (up to roundoff error). Use the residuals to estimate the standard deviation $\sigma$ in the regression model. Verify that your result agrees with $s$ given by software.

(c) Calculate the mean $\bar{x}$ of the five femur lengths and the sum of the five squared deviations $(x - \bar{x})^2$. Then find the standard error of the slope $b$ and the $t$ statistic for testing $H_0 : \beta = 0$. Verify that your results agree with those given by the software.

# PART VI REVIEW

Part VI presents several topics in statistical inference. All apply the basic reasoning of sampling distributions, confidence intervals, and statistical tests. Many are based on $z$ or $t$ statistics obtained by standardizing an estimate of the parameter in question. The topics are otherwise distinct, however. You can study them in any order or study some but not others. Here are the most important skills you should acquire from your study of these topics.

## A. RECOGNITION

1. Recognize from the design of a study whether one-sample, matched pairs, or two-sample procedures are needed.

2. Recognize when a study concerns the relationship between two variables. Distinguish quantitative explanatory and response variables from relationships between two categorical variables.

3. Recognize what parameter or parameters an inference problem concerns. In particular, distinguish among settings that require inference about a mean, comparing two means, a proportion, comparing two proportions, or comparing more than two proportions.

4. Recognize which type of inference you need in a particular regression setting: inference about the slope, predicting a mean response, or predicting an individual response.

B.  TWO-SAMPLE $t$ PROCEDURES

1.  Recognize when the two-sample $t$ procedures are appropriate in practice.

2.  Use a $t$ procedure to give a confidence interval for the difference between two means.

3.  Use a $t$ procedure to test the hypothesis that two populations have equal means against either a one-sided or a two-sided alternative.

4.  Know that procedures for comparing the standard deviations of two normal populations are available, but that these procedures are risky because they are not at all robust against nonnormal distributions.

C.  INFERENCE ABOUT ONE PROPORTION

1.  Use a $z$ procedure to give a confidence interval for a population proportion $p$.

2.  Use a $z$ procedure to carry out a test of significance for the hypothesis $H_0 : p = p_0$ about a population proportion $p$ against either a one-sided or a two-sided alternative.

3.  Check that you can safely use these $z$ procedures in a particular setting.

D.  COMPARING TWO PROPORTIONS

1.  Use the two-sample $z$ procedure to give a confidence interval for the difference $p_1 - p_2$ between proportions in two populations based on independent samples from the populations.

2.  Use a $z$ procedure to test the hypothesis $H_0 : p_1 = p_2$ that proportions in two distinct populations are equal.

3.  Check that you can safely use these $z$ procedures in a particular setting.

E.  TWO-WAY TABLES AND CHI-SQUARE TESTS

1.  Arrange data on successes and failures in several groups into a two-way table of counts of successes and failures in all groups.

2.  Use percents to describe the relationship between two categorical variables starting from the counts in a two-way table.

3.  Explain what null hypothesis the chi-square statistic tests in a specific two-way table.

4.  Carry out the chi-square test. Verify from the expected counts that use of the test is safe.

5. If the test is significant, use percents and comparison of expected and observed counts via cell residuals to see which deviations from the null hypothesis are most important.

## F. REGRESSION INFERENCE

1. Inspect the data to recognize situations in which inference isn't safe: a nonlinear relationship, influential observations, strongly skewed residuals in a small sample, or nonconstant variation of the data points about the regression line.

2. Explain in any specific regression setting the meaning of the slope $\beta$ of the true regression line. Calculate a confidence interval for $\beta$. Test the hypothesis that $\beta = 0$ and explain the meaning of this hypothesis.

3. Explain the distinction between a confidence interval for the mean response and a prediction interval for an individual response. Calculate either a confidence interval or a prediction interval.

## REVIEW EXERCISES

**Review 1**  Nitrites are often added to meat products as preservatives. In a study of the effect of nitrites on bacteria, researchers measured the rate of uptake of an amino acid for 60 cultures of bacteria: 30 growing in a medium to which nitrites had been added and another 30 growing in a standard medium as a control group. Here are the data from this study:

| Control | | | Nitrite | | |
|---|---|---|---|---|---|
| 6,450 | 8,709 | 9,361 | 8,303 | 8,252 | 6,594 |
| 9,011 | 9,036 | 8,195 | 8,534 | 10,227 | 6,642 |
| 7,821 | 9,996 | 8,202 | 7,688 | 6,811 | 8,766 |
| 6,579 | 10,333 | 7,859 | 8,568 | 7,708 | 9,893 |
| 8,066 | 7,408 | 7,885 | 8,100 | 6,281 | 7,689 |
| 6,679 | 8,621 | 7,688 | 8,040 | 9,489 | 7,360 |
| 9,032 | 7,128 | 5,593 | 5,589 | 9,460 | 8,874 |
| 7,061 | 8,128 | 7,150 | 6,529 | 6,201 | 7,605 |
| 8,368 | 8,516 | 8,100 | 8,106 | 4,972 | 7,259 |
| 7,238 | 8,830 | 9,145 | 7,901 | 8,226 | 8,552 |

Examine the data and briefly describe their distribution. Carry out a test of the research hypothesis that nitrites decrease amino acid uptake, and report your results.

**Review 2**    Around the year 1900, the English statistician Karl Pearson tossed a coin 24,000 times. He obtained 12,012 heads.

(a) Test the null hypothesis that Pearson's coin had probability 0.5 of coming up heads versus the two-sided alternative. Give the $P$-value. Do you reject $H_0$ at the 1% significance level?

(b) Find a 99% confidence interval for the probability of heads for Pearson's coin. This is the range of probabilities that cannot be rejected at the 1% significance level.

**Review 3**    An experiment on the side effects of pain relievers assigned arthritis patients to one of several over-the-counter pain medications. Of the 440 patients who took one brand of pain reliever, 23 suffered some "adverse symptom." Does the experiment provide strong evidence that fewer than 10% of patients who take this medication have adverse symptoms?

**Review 4**    High levels of cholesterol in the blood are not healthy in either humans or dogs. Because a diet rich in saturated fats raises the cholesterol level, it is plausible that dogs owned as pets have higher cholesterol levels than dogs owned by a veterinary research clinic. "Normal" levels of cholesterol based on the clinic's dogs would then be misleading. A clinic compared healthy dogs it owned with healthy pets brought to the clinic to be neutered. The summary statistics for blood cholesterol levels (milligrams per deciliter of blood) appear below:[105]

| Group | $n$ | $\bar{y}$ | $s$ |
|---|---|---|---|
| Pets | 26 | 193 | 68 |
| Clinic | 23 | 174 | 44 |

(a) Is there strong evidence that pets have higher mean cholesterol level than clinic dogs? State the $H_0$ and $H_a$ and carry out an appropriate test. Give the $P$-value and state your conclusion.

(b) Give a 95% confidence interval for the difference in mean cholesterol levels between pets and clinic dogs.

(c) Give a 95% confidence interval for the mean cholesterol level in pets.

(d) What assumptions must be satisfied to justify the procedures you used in (a), (b), and (c)? Assuming that the cholesterol measurements have no outliers and are not strongly skewed, what is the chief threat to the validity of the results of this study?

**Review 5** Do various occupational groups differ in their diets? A British study compared the food and drink intake of 98 drivers and 83 conductors of London double-decker buses. The conductors' jobs require more physical activity. The article reporting the study gives the data as "Mean daily consumption (± s. e.)." Some of the study results appear below:[106]

|  | Drivers | Conductors |
| --- | --- | --- |
| Total calories | 2821 ± 44 | 2844 ± 48 |
| Alcohol (grams) | 0.24 ± 0.06 | 0.39 ± 0.11 |

(a) What does "s. e." stand for? Give $\bar{y}$ and $s$ for each of the four sets of measurements.

(b) Is there significant evidence at the 5% level that conductors and drivers consume different numbers of calories per day?

(c) How significant is the observed difference in mean alcohol consumption?

(d) Give a 90% confidence interval for the mean daily alcohol consumption of London double-decker bus conductors.

(e) Give an 80% confidence interval for the difference in mean daily alcohol consumption between drivers and conductors.

**Review 6** Physical fitness is related to personality characteristics. In one study of this relationship, middle-aged college faculty who had volunteered for a fitness program were divided into low-fitness and high-fitness groups based on a physical examination. The subjects then took the Cattell Sixteen Personality Factor Questionnaire. Here are the data for the "ego strength" personality factor:[107]

| Group | Fitness | $n$ | $\bar{y}$ | $s$ |
| --- | --- | --- | --- | --- |
| 1 | Low | 14 | 4.64 | 0.69 |
| 2 | High | 14 | 6.43 | 0.43 |

(a) Is the difference in mean ego strength significant at the 5% level? At the 1% level? Be sure to state $H_0$ and $H_a$.

(b) You should be hesitant to generalize these results to the population of all middle-aged men. Explain why.

**Review 7**   One way of checking the effect of undercoverage, nonresponse, and other sources of error in a sample survey is to compare the sample with known facts about the population. About 10.5% of American adults are black. The proportion $\hat{p}$ of blacks in an SRS of adults should therefore be close to 10.5%. It is unlikely to be exactly 10.5% because of sampling variability. If a national sample of size 1500 contains only 9.2% blacks, should we suspect that the sampling procedure is somehow underrepresenting blacks?

**Review 8**   North Carolina State University looked at the factors that affect the success of students in a required chemical engineering course. Students must get a C or better in the course in order to continue as chemical engineering majors. There were 65 students from urban or suburban backgrounds, and 52 of these students succeeded. Another 55 students were from rural or small-town backgrounds; 30 of these students succeeded in the course.[108]

(a) Is there good evidence that the proportion of students who succeed is different for urban/suburban versus rural/small-town backgrounds? (State hypotheses, give a test statistic and its $P$-value, and state your conclusion.)

(b) Give a 90% confidence interval for the size of the difference.

**Review 9**   Different kinds of companies compensate their key employees in different ways. Established companies may pay higher salaries, while new companies may offer stock options that will be valuable if the company succeeds. Do high-tech companies tend to offer stock options more often than other companies? One study looked at a random sample of 200 companies. Of these, 91 were listed in the *Directory of Public High Technology Corporations* and 109 were not listed. Treat these two groups as SRSs of high-tech and non-high-tech companies. Seventy-three of the high-tech companies and 75 of the non-high-tech companies offered incentive stock options to key employees.[109]

(a) Is there evidence that a higher proportion of high-tech companies offer stock options?

(b) Give a 95% confidence interval for the difference in the proportions of the two types of companies that offer stock options.

Review 10   The North Carolina State University study (see Exercise 8) also looked at possible differences in the proportions of female and male students who succeeded in the course. They found that 23 of the 34 women and 60 of the 89 men succeeded. Is there evidence of a difference between the proportions of women and men who succeed?

Review 11   A bank compares two proposals to increase the amount that its credit card customers charge on their cards. (The bank earns a percentage of the amount charged, paid by the stores that accept the card.) Proposal A offers to eliminate the annual fee for customers who charge $2400 or more during the year. Proposal B offers a small percent of the total amount charged as a cash rebate at the end of the year. The bank offers each proposal to an SRS of 150 of its credit card customers. At the end of the year, the total amount charged by each customer is recorded. Here are the summary statistics:

| Group | $n$ | $\bar{y}$ | $s$ |
|-------|-----|-----------|-----|
| A | 150 | $1987 | $392 |
| B | 150 | $2056 | $413 |

(a) Do the data show a significant difference between the mean amounts charged by customers offered the two plans? Give the null and alternative hypotheses, the test statistic, and its $P$-value. State your practical conclusions.

(b) The distributions of amounts charged are skewed to the right, but outliers are prevented by the limits that the bank imposes on credit balances. Do you think that skewness threatens the validity of the test that you used in (a)? Explain your answer.

(c) Is the bank's study an experiment? Why? How does this affect the conclusions the bank can draw from the study?

Review 12   A large study of child care used samples from the data tapes of the Current Population Survey over a period of several years. The result is close to an SRS of child-care workers. The Current Population Survey has three classes of child-care workers: private household, non-household, and preschool teacher. Here are data on the number of blacks among women workers in these three classes:[110]

|              | Total | Black |
|--------------|-------|-------|
| Household    | 2455  | 172   |
| Nonhousehold | 1191  | 167   |
| Teachers     | 659   | 86    |

(a) We want to compare the proportions of black workers in the three classes of child-care workers. What percent of each class is black?

(b) Are the proportions in (a) significantly different? Carry out a test, give the $P$-value, and report your conclusions about what these data show.

Review 13   A major study of alternative welfare programs randomly assigned women on welfare to one of two programs, called "WIN" and "Options." WIN was the existing program. The new Options program gave more incentives to work. An important question was how much more (on the average) women in Options earned than those in WIN. Here is descriptive output for earnings in dollars over a three-year period:[111]

|      | N    | MEAN | STDEV | SE MEAN |
|------|------|------|-------|---------|
| OPT  | 1362 | 7638 | 289   | 7.8309  |
| WIN  | 1395 | 6595 | 247   | 6.6132  |

(a) Give a 99% confidence interval for the amount by which the mean earnings of Options participants exceeded the mean earnings of WIN subjects.

(b) The distribution of incomes is strongly skewed to the right but includes no extreme outliers because all the subjects were on welfare. What fact about these data allows us to use $t$ procedures despite the strong skewness?

Review 14   The Physicians' Health Study examined the effects of taking an aspirin every other day. Earlier studies suggested that aspirin might reduce the risk of heart attacks. The subjects were 22,071 healthy male physicians at least 40 years old. The study assigned 11,037 of the subjects at random to take aspirin. The others took a placebo pill. The study was double-blind. Here are the counts for some of the outcomes of interest to the researchers:

|                      | Aspirin group | Placebo group |
|----------------------|:-------------:|:-------------:|
| Fatal heart attacks  | 10            | 26            |
| Nonfatal heart attacks | 129         | 213           |
| Strokes              | 119           | 98            |

For which outcomes is the difference between the aspirin and placebo groups significant? (Do a separate test for each outcome. Use two-sided alternatives. Write a brief summary of your conclusions.)

**Review 15**   Some people think that chemists are more likely than other parents to have female children. (Perhaps chemists are exposed to something in their laboratories that affects the sex of their children.) The Washington State Department of Health lists the parents' occupations on birth certificates. Between 1980 and 1990, 555 children were born to fathers who were chemists. Of these births, 273 were girls. During this period, 48.8% of all births in Washington State were girls.[112] Is there evidence that the proportion of girls born to chemists is higher than the state proportion?

**Review 16**   The first law requiring child restraints in motor vehicles went into effect in 1978, in Tennessee. "Before enactment of child restraint laws, most children in the United States were unrestrained when traveling in motor vehicles." The writer of this statement cites a 1974 survey of drivers in Maryland and Virginia that found that "about 93% ($n = 8,933$) of children under the age of 10 were traveling unrestrained."[113] Assuming that the children observed are an SRS from the areas involved, give a 99% confidence interval for the proportion of unrestrained children in 1974.

**Review 17**   Sickle-cell trait is a hereditary condition that is common among blacks and can cause medical problems. Some biologists suggest that sickle-cell trait protects against malaria. That would explain why it is found in people who originally came from Africa, where malaria is common. A study in Africa tested 543 children for the sickle-cell trait and also for malaria. In all, 136 of the children had the sickle-cell trait, and 36 of these had heavy malaria infections. The other 407 children lacked the sickle-cell trait, and 152 of them had heavy malaria infections.[114]

(a) Give a 95% confidence interval for the proportion of all children in the population studied who have the sickle-cell trait.

(b) Is there good evidence that the proportion of heavy malaria infections is lower among children with the sickle-cell trait?

**Review 18**   A study comparing American and Australian corporations examined a sample of 133 American and 63 Australian corporations. There are the usual practical difficulties involving nonresponse and the question of what population the samples represent. Ignore these issues and treat the samples as SRSs from the United States and Australia. The average percent of revenues from "highly regulated businesses" was 27% for the Australian companies and 41% for the American companies.[115]

(a) The data are given as percents. Explain carefully why comparing the average percent of revenues from highly regulated businesses for U.S. and Australian corporations is *not* a comparison of two population proportions.

(b) What test would you use to make the comparison? (Don't try to carry out a test.)

**Review 19**   To study the export activity of manufacturing firms on Taiwan, researchers mailed questionnaires to an SRS of firms in each of five industries that export many of their products. The response rate was only 12.5%, because private companies don't like to fill out long questionnaires from academic researchers. Here are data on the planned sample sizes and the actual number of responses received from each industry:[116]

|                          | Sample size | Responses |
|--------------------------|-------------|-----------|
| Metal products           | 185         | 17        |
| Machinery                | 301         | 35        |
| Electrical equipment     | 552         | 75        |
| Transportation equipment | 100         | 15        |
| Precision instruments    | 90          | 12        |

If the response rates differ greatly, comparisons among the industries may be difficult. Is there good evidence of unequal response rates among the five industries?

**Review 20**   Shopping at secondhand stores is becoming more popular and has even attracted the attention of business schools. A study of customers' attitudes toward secondhand stores interviewed samples of shoppers at two secondhand stores of the same chain in two cities. The breakdown of the respondents by gender is as follows:[117]

|        | City 1 | City 2 |
|--------|--------|--------|
| Men    | 38     | 68     |
| Women  | 203    | 150    |
| Total  | 241    | 218    |

Is there a significant difference between the proportions of women customers in the two cities?

(a) State the null hypothesis, find the sample proportions of women in both cities, do a two-sided $z$ test, and give a $P$-value.

(b) Calculate the chi-square statistic $\chi^2$ and show that it is the square of the $z$ statistic. Show that the $P$-value agrees with your result from (a).

(c) Give a 95% confidence interval for the difference between the proportions of women customers in the two cities.

**Review 21** The study of shoppers in secondhand stores cited in the previous exercise also compared the income distributions of shoppers in the two stores. Here is the two-way table of counts:

| Income                | City 1 | City 2 |
|-----------------------|--------|--------|
| Under $10,000         | 70     | 62     |
| $10,000 to $19,999    | 52     | 63     |
| $20,000 to $24,999    | 69     | 50     |
| $25,000 to $34,999    | 22     | 19     |
| $35,000 or more       | 28     | 24     |

Is there good evidence that customers at the two stores have different income distributions?

**Review 22** A study of genetic influences on diabetes compared normal mice with similar mice genetically altered to remove the gene called $aP2$. Mice of both types were allowed to become obese by eating a high-fat diet. The researchers then measured the levels of insulin and glucose in their blood plasma. Here are some excerpts from their findings.[118] The normal mice are called "wild-type" and the altered mice are called "$aP2^{-/-}$."

Each value is the mean $\pm$ SEM of measurements on at least 10 mice. Mean values of each plasma component are compared between $aP2^{-/-}$ mice and wild-type controls by Student's $t$ test ($*P < 0.05$ and $**P < 0.005$).

| Parameter | Wild type | $aP2^{-/-}$ |
|---|---|---|
| Insulin (ng/ml) | $5.9 \pm 0.9$ | $0.75 \pm 0.2$** |
| Glucose (mg/dl) | $230 \pm 25$ | $150 \pm 17$* |

Despite much greater circulating amounts of insulin, the wild-type mice had higher blood glucose than the $aP2^{-/-}$ animals. These results indicate that the absence of $aP2$ interferes with the development of dietary obesity-induced insulin resistance.

Other biologists are supposed to understand the statistics reported so tersely.

(a) What does "SEM" mean? What is the expression for SEM based on $n$, $\bar{y}$, and $s$ from a sample?

(b) Which of the $t$ tests we have studied did the researchers apply?

(c) Explain to a biologist who knows no statistics what $P < 0.05$ and $P < 0.005$ mean. Which is stronger evidence of a difference between the two types of mice?

(d) The report says only that the sample sizes were "at least 10." Suppose that the results are based on exactly 10 mice of each type. Use the values in the table to find $\bar{y}$ and $s$ for the two insulin concentrations and carry out a test to assess the significance of the difference in means. What $P$-value do you obtain?

(e) Do the same thing for the glucose concentrations.

Review 23　The Third International Mathematics and Science Study gave a mathematics test to random samples of eighth-grade students in 41 countries in 1995. Here are the average test scores for three nations: Sweden, 519; Thailand, 522; United States, 500.[119] A report from the study lists Sweden among "nations with average scores significantly higher than the U.S." and Thailand among "nations with average scores not significantly different from the U.S." Explain carefully how this can be true, even though Thailand's average was higher than Sweden's. (Hint: Think about using the two-sample $t$ test to compare the mean scores in two countries.)

　Table VI.1 presents data on 78 seventh-grade students in a rural midwestern school.[120] The variables are

$$IQ = \text{student's score on a standard IQ test}$$
$$GPA = \text{student's grade point average}$$
$$Gender = \text{student's gender (F or M)}$$

| TABLE VI.1 | | IQ scores and grade point average for seventh graders | | | |
|---|---|---|---|---|---|
| GPA | IQ | Gender | GPA | IQ | Gender |
| 7.940 | 111 | M | 9.999 | 119 | F |
| 8.292 | 107 | M | 10.760 | 123 | M |
| 4.643 | 100 | M | 9.763 | 124 | M |
| 7.470 | 107 | M | 9.410 | 126 | M |
| 8.882 | 114 | F | 9.167 | 116 | M |
| 7.585 | 115 | M | 9.348 | 127 | M |
| 7.650 | 111 | M | 8.167 | 119 | M |
| 2.412 | 97 | M | 3.647 | 97 | M |
| 6.000 | 100 | F | 3.408 | 86 | F |
| 8.833 | 112 | M | 3.936 | 102 | M |
| 7.470 | 104 | F | 7.167 | 110 | M |
| 5.528 | 89 | F | 7.647 | 120 | M |
| 7.167 | 104 | M | 0.530 | 103 | M |
| 7.571 | 102 | F | 6.173 | 115 | M |
| 4.700 | 91 | F | 7.295 | 93 | M |
| 8.167 | 114 | F | 7.295 | 72 | F |
| 7.822 | 114 | F | 8.938 | 111 | F |
| 7.598 | 103 | F | 7.882 | 103 | F |
| 4.000 | 106 | M | 8.353 | 123 | M |
| 6.231 | 105 | F | 5.062 | 79 | M |
| 7.643 | 113 | M | 8.175 | 119 | M |
| 1.760 | 109 | M | 8.235 | 110 | M |
| 6.419 | 108 | F | 7.588 | 110 | M |
| 9.648 | 113 | M | 7.647 | 107 | M |
| 10.700 | 130 | F | 5.237 | 74 | F |
| 6.938 | 106 | M | 7.825 | 105 | M |
| 10.580 | 128 | M | 7.333 | 112 | F |
| 9.429 | 128 | M | 9.167 | 105 | M |
| 8.000 | 118 | M | 7.996 | 110 | M |
| 9.585 | 113 | M | 8.714 | 107 | F |
| 9.571 | 120 | F | 7.833 | 103 | F |
| 8.998 | 132 | F | 4.885 | 77 | M |
| 8.333 | 111 | F | 7.998 | 98 | F |
| 8.175 | 124 | M | 3.820 | 90 | M |
| 8.000 | 127 | M | 5.936 | 96 | F |
| 9.333 | 128 | F | 9.000 | 112 | F |
| 9.500 | 136 | M | 9.500 | 112 | F |
| 9.167 | 106 | M | 6.057 | 114 | M |
| 10.140 | 118 | F | 6.057 | 93 | F |

It isn't clear that these students are a random sample from any well-defined population. Research in education and other fields often uses "available subjects," such as students in a single school. It is usual to apply statistical inference to such data, but the generality of the conclusions can be challenged. Exercises 24 to 29 are based on these data.

**Review 24**   You want to estimate the mean GPA of all seventh-grade students for a study of grade inflation. Make a histogram and a normal probability plot of the sample GPAs to verify that we can use our inference methods. Then give a 99% confidence interval for the population mean GPA.

**Review 25**   It is common to find that boys have slightly higher scores on standardized tests, but that girls do better in school. Is this true of the seventh graders in this school?

(a)  Make a side-by-side graphical comparison of female/male IQ scores and female/male GPAs. What do the data show? Are there any suspected outliers by the $1.5 \times IQR$ criterion in any of the four distributions?

(b)  Make normal quantile plots of all four distributions. Is there any reason not to apply our inference methods?

(c)  Carry out two tests to examine whether (on the average) boys do better on IQ tests and whether girls do better in school. What do you conclude?

**Review 26**   Do the IQ test scores of students predict their GPA? Carry out a complete regression analysis to answer this question. (Your work should include a plot of the data, examination of the assumptions for regression inference, and a test for the significance of IQ as a predictor of GPA.)

**Review 27**   Continue the work of the previous exercise by examining *how well* IQ predicts GPA.

(a)  What percent of the observed variation in the GPAs of these students is explained by regressing on IQ?

(b)  By how much does GPA increase (on the average) for each additional point of IQ? (Give a 95% confidence interval.)

(c)  Judy scores 100 on the IQ test. Give a 95% interval for predicting Judy's GPA. Do you think the interval is narrow enough to be of practical use?

**(d)** Now give a 95% interval for predicting the mean GPA of all students who score 100 on the IQ test. Explain in simple language why this interval is not as wide as your interval in (c).

**Review 28**   Does IQ predict GPA markedly better for boys or for girls than for the other gender? Do separate regressions for the two genders. Compare $r^2$ and $s$ to answer the question. Then explain what the numbers $r^2$ and $s$ tell you.

**Review 29**   Examination of the data shows one boy with a very low GPA of 0.53. We wonder how much this unusual observation has affected our work.

**(a)** Repeat the comparison of female/male mean GPAs in part (c) of Exercise 25, removing this one boy. How much does the P-value change? Does your practical conclusion change?

**(b)** Repeat the regression in Exercise 26, removing this one boy. How much does the least-squares line change (make a plot with both lines to compare them)? How much does $r^2$ change? How much does the P-value of your test change? Did this individual have any practical effect on your conclusions?

# NOTES
# AND DATA
# SOURCES

1. The Shakespeare data appear in C. B. Williams, *Style and Vocabulary: Numerical Studies,* Griffin, London, 1970.

2. Based on data summaries in G. L. Cromwell et al., "A comparison of the nutritive value of *opaque-2, floury-2* and normal corn for the chick," *Poultry Science,* 57 (1968), pp. 840–847.

3. The gas mileage data are from the Environmental Protection Agency's *Model Year 1997 Fuel Economy Guide,* U.S. Department of Energy, October 1996. This annual publication is available online at http://www.epa.com. In the table, the data are for the basic engine/transmission combination for each model, and models that are essentially identical (such as the Ford Taurus and Mercury Sable) appear only once.

4. Data from John K. Ford, "Diversification: how many stocks will suffice?" *American Association of Individual Investors Journal,* January 1990, pp. 14–16.

5. Data on frosts from C. E. Brooks and N. Carruthers, *Handbook of Statistical Methods in Meteorology,* H. M. Stationery Office, 1953.

6. Data from T. Bjerkedal, "Acquisition of resistance in guinea pigs infected with different doses of virulent tubercle bacilli," *American Journal of Hygiene,* 72 (1960), pp. 130–148.

7. Data from *The Baseball Encyclopedia,* 3rd ed., Macmillan, New York, 1976. Maris's home run data are from the same source.

8. Cavendish's data and the background information about his work appear in S. M. Stigler, "Do robust estimators work with real data?" *Annals of Statistics,* 5 (1977), pp. 1055–1078.

9. Data released by Major League Baseball, reported in *USA Today,* November 15, 1996.

10. Data provided by Darlene Gordon, Purdue University, from a study of students' self-concept that collected data on IQ test scores, grade point average, gender, and other information.

11. Data from Stephen Jay Gould, "Entropic homogeneity isn't why no one hits .400 anymore," *Discover,* August 1986, pp. 60–66.

12. See Note 3.

13. *Consumer Reports,* June 1986, pp. 366–367. A more recent study of hot dogs appears in *Consumer Reports,* July 1993, pp. 415–419. The newer data cover few brands of poultry hot dogs and take calorie counts mainly from the package labels, resulting in suspiciously round numbers.

14. Data provided by Robert Dale, Purdue University.

15. Based on T. N. Lam, "Estimating fuel consumption from engine size," *Journal of Transportation Engineering,* 111 (1985), pp. 339–357. The data for 10 to 50 km/h are measured; those for 60 and higher are calculated from a model given in the paper and are therefore smoothed.

16. The data are from W. L. Colville and D. P. McGill, "Effect of rate and method of planting on several plant characters and yield of irrigated corn," *Agronomy Journal,* 54 (1962), pp. 235–238.

17. A careful study of this phenomenon is W. S. Cleveland, P. Diaconis, and R. McGill, "Variables on scatterplots look more highly correlated when the scales are increased," *Science,* 216 (1982), pp. 1138–1141.

18. The data are from M. A. Houck et al., "Allometric scaling in the earliest fossil bird, *Archaeopteryx lithographica,*" *Science,* 247 (1990), pp. 195–198. The authors conclude from a variety of evidence that all specimens represent the same species.

19. These data were originally collected by L. M. Linde of UCLA but were first published by M. R. Mickey, O. J. Dunn, and V. Clark, "Note on the use of stepwise regression in detecting outliers," *Computers and Biomedical Research,* 1 (1967), pp. 105–111. The data have been used by several authors. I found them in N. R. Draper and J. A. John, "Influential observations and outliers in regression," *Technometrics,* 23 (1981), pp. 21–26.

20. From W. M. Lewis and M. C. Grant, "Acid precipitation in the western United States," *Science,* 207 (1980), pp. 176–177.

21. The U.S. returns are for the Standard & Poor's 500 stock index. The overseas returns are for the Morgan Stanley Europe, Australia, Far East (EAFE) index.

22. Frank J. Anscombe, "Graphs in statistical analysis," *The American Statistician,* 27 (1973), pp. 17–21.

23. From a survey by the Wheat Industry Council reported in *USA Today,* October 20, 1983.

24. Data released by the National Basketball Association, reported in *USA Today,* November 15, 1996.

25. The data appear in the *New York Times,* December 28, 1994. The *Times* credits Dr. M. H. Criqui of the University of California, San Diego, School of Medicine.

26. Modified from data provided by Dana Quade, University of North Carolina.

27. For more detail on the material of this section, along with references, see P. E. Converse and M. W. Traugott, "Assessing the accuracy of polls and surveys," *Science,* 234 (1986), pp. 1094–1098.

28. The estimates of the census undercount come from Howard Hogan, "The 1990 post-enumeration survey: operations and results," *Journal of the American Statistical Association,* 88 (1993), pp. 1047–1060. The information about nonresponse appears in Eugene P. Eriksen and Teresa K. DeFonso, "Beyond the net undercount: how to measure census error," *Chance,* 6, no. 4 (1993), pp. 38–43 and 14.

29. The Levi jeans and disposable diaper examples are taken from Cynthia Crossen, "Margin of error: studies galore support products and positions, but are they reliable?" *Wall Street Journal,* November 14, 1991.

30. See Note 10.

31. Giuliana Coccia, "An overview of non-response in Italian telephone surveys," *Proceedings of the 49th Session of the International Statistical Institute, 1993,* Book 3, pp. 271–272.

32. The first question was asked by Ross Perot, and the second by a Time/CNN poll, both in March 1993. The example comes from W. Mitofsky, "Mr. Perot, you're no pollster," *New York Times,* March 27, 1993.

33. L. L. Miao, "Gastric freezing: an example of the evaluation of medical therapy by randomized clinical trials," in J. P. Bunker, B. A. Barnes, and F. Mosteller (eds.), *Costs, Risks and Benefits of Surgery,* Oxford University Press, New York, 1977, pp. 198–211.

34. L. E. Moses and F. Mosteller, "Safety of anesthetics," in J. Tanur et al. (eds.), *Statistics: A Guide to the Unknown,* 3rd ed., Wadsworth, Belmont, Calif., 1989, pp. 15–24.

35. Based on Christopher Anderson, "Measuring what works in health care," *Science,* 263 (1994), pp. 1080–1082.

36. Details appear in Steering Committee of the Physicians' Health Study Research Group, "Final report on the aspirin component of the ongoing Physicians' Health Study," *New England Journal of Medicine,* 321 (1989), pp. 129–135.

37. Such an experiment is described in J. P. Newhouse, "A health insurance experiment," in J. M. Tanur et al. (eds.), *Statistics: A Guide to the Unknown,* 3rd edition, Wadsworth, Belmont, Calif., 1989, pp. 31–40.

38. See Note 10.

39. Information from Warren McIsaac and Vivek Goel, "Is access to physician services in Ontario equitable?" Institute for Clinical Evaluative Sciences in Ontario, October 18, 1993.

40. Data as of March 1995, retrieved January 1997 from the Census Bureau web site at http://www.census.gov.

41. Strictly speaking, the recipe we give for the standard deviation of $\bar{y}$ assumes that an SRS of size $n$ is drawn from an *infinite* population. If the population has finite size $N$, the standard deviation in the recipe is multiplied by $\sqrt{1 - n/N}$. This "finite population correction" approaches 1 as $N$ increases. When the population is at least 10 times as large as the sample, the correction is at least as large as 0.948, so that you can ignore the correction for most practical purposes.

42. Information from Francisco L. Rivera-Batiz, "Quantitative literacy and the likelihood of employment among young adults," *Journal of Human Resources*, 27 (1992), pp. 313–328.

43. Based on information in D. L. Shankland et al., "The effect of 5-thio-D-glucose on insect development and its absorption by insects," *Journal of Insect Physiology*, 14 (1968), pp. 63–72.

44. Data from D. L. Shankland, "Involvement of spinal cord and peripheral nerves in DDT-poisoning syndrome in albino rats," *Toxicology and Applied Pharmacology*, 6 (1964), pp. 197–213.

45. Data from Joan M. Susic, "Dietary Phosphorus Intakes, Urinary and Peritoneal Phosphate Excretion and Clearance in Continuous Ambulatory Peritoneal Dialysis Patients," M.S. thesis, Purdue University, 1985.

46. Based on I. Cuellar, L. C. Harris, and R. Jasso, "An acculturation scale for Mexican American normal and clinical populations," *Hispanic Journal of Behavioral Sciences*, 2 (1980), pp. 199–217.

47. Data provided by Drina Iglesia, Purdue University. The study results are reported in D. D. S. Iglesia, E. J. Cragoe, Jr., and J. W. Vanable, "Electric field strength and epithelization in the newt (*Notophthalmus viridescens*)," *Journal of Experimental Zoology*, 274 (1996), pp. 56–62.

48. See Note 8.

49. See Note 10.

50. Data from Orit E. Hetzroni, "The effects of active versus passive computer-assisted instruction on the acquisition, retention, and generalization of Blissymbols while using elements for teaching compounds," Ph.D. thesis, Purdue University, 1995.

51. Gökhan S. Hotamisligil et al., "Uncoupling of obesity from insulin resistance through a targeted mutation in *aP*2, the adipocyte fatty acid binding protein," *Science*, 274 (1996), pp. 1377–1379.

**52.** William T. Robinson, "Sources of market pioneer advantages: the case of industrial goods industries," *Journal of Marketing Research,* 25 (February 1988), pp. 87–94.

**53.** Michael B. Maziz et al., "Perceived age and attractiveness of models in cigarette advertisements," *Journal of Marketing,* 56 (January 1992), pp. 22–37.

**54.** Data provided by Timothy Sturm.

**55.** These recommendations are based on extensive computer work. See, for example, Harry O. Posten, "The robustness of the one-sample *t*-test over the Pearson system," *Journal of Statistical Computation and Simulation,* 9 (1979), pp. 133–149, and E. S. Pearson and N. W. Please, "Relation between the shape of population distribution and the robustness of four simple test statistics," *Biometrika,* 62 (1975), pp. 223–241.

**56.** Data from R. A. Berner and G. P. Landis, "Gas bubbles in fossil amber as possible indicators of the major gas composition of ancient air," *Science,* 239 (1988), pp. 1406–1409.

**57.** See Note 47.

**58.** Based loosely on D. R. Black et al., "Minimal interventions for weight control: a cost-effective alternative," *Addictive Behaviors,* 9 (1984), pp. 279–285.

**59.** Data provided by Matthew Moore.

**60.** Data from the appendix of D. A. Kurtz (ed.), *Trace Residue Analysis,* American Chemical Society Symposium Series, No. 284, 1985.

**61.** Data provided by Chris Olsen, who found the information in scuba-diving magazines.

**62.** Data from S. C. Hand and E. Gnaiger, "Anaerobic dormancy quantified in *Artemia* embryos," *Science,* 239 (1988), pp. 1425–1427.

**63.** Data from Lianng Yuh, "A biopharmaceutical example for undergraduate students," unpublished manuscript.

**64.** Based on Charles W. L. Hill and Phillip Phan, "CEO tenure as a determinant of CEO pay," *The Academy of Management Journal,* 34 (1991), pp. 707–717.

**65.** See Note 13.

**66.** The approximation is often called "Welch's modified two-sample *t*." You can find detailed information about it, and also about the conservative *t* procedures mentioned later in this lesson, in Paul Leaverton and John J. Birch, "Small sample power curves for the two sample location problem," *Technometrics,* 11 (1969), pp. 299–307; in Henry Scheffé, "Practical solutions of the Behrens-Fisher problem," *Journal of the American Statistical Association,* 65 (1970), pp. 1501–1508; and in D. J. Best and J. C. W. Rayner, "Welch's approximate solution for the Behrens-Fisher problem," *Technometrics,* 29 (1987), pp. 205–210.

    The Welch approximation often gives a degrees of freedom that is not a whole number. Large-scale statistical software (e.g., SAS and S-PLUS) uses algorithms for *t* with non-whole-number degrees of freedom. Smaller systems (e.g., *Data Desk* and *Minitab*) round to the next smaller degrees of freedom, a very small change. I suggest that you not use statistical software that does not offer the Welch version of the two-sample *t* procedures—that's a sign that the software is not up-to-date.

67. See the extensive simulation studies in Harry O. Posten, "The robustness of the two-sample *t*-test over the Pearson system," *Journal of Statistical Computation and Simulation,* 6 (1978), pp. 295–311, and in Harry O. Posten, H. Yeh, and Donald B. Owen, "Robustness of the two-sample *t*-test under violations of the homogeneity assumption," *Communications in Statistics,* 11 (1982), pp. 109–126.

68. The problem of comparing spreads is difficult even with advanced methods. Common distribution-free procedures do not offer a satisfactory alternative, because they are sensitive to unequal shapes when comparing two distributions. A good introduction to the available methods is W. J. Conover, M. E. Johnson, and M. M. Johnson, "A comparative study of tests for homogeneity of variances, with applications to outer continental shelf bidding data," *Technometrics,* 23 (1981), pp. 351–361. Modern resampling procedures often work well. See Dennis D. Boos and Colin Brownie, "Bootstrap methods for testing homogeneity of variances," *Technometrics,* 31 (1989), pp. 69–82.

69. From H. G. Gough, *The Chapin Social Insight Test,* Consulting Psychologists Press, Palo Alto, Calif., 1968.

70. Based on M. C. Wilson et al., "Impact of cereal leaf beetle larvae on yields of oats," *Journal of Economic Entomology,* 62 (1969), pp. 699–702.

71. From Costas Papoulias and Panayiotis Theodossiou, "Analysis and modeling of recent business failures in Greece," *Managerial and Decision Economics,* 13 (1992), pp. 163–169.

72. See Note 2.

73. This example is loosely based on the source cited in Note 44.

74. See Note 50.

75. From a news article in *Science,* 224 (1983), pp. 1029–1031.

76. From M. A. Zlatin and R. A. Koenigsknecht, "Development of the voicing contrast: a comparison of voice onset time in stop perception and production," *Journal of Speech and Hearing Research,* 19 (1976), pp. 93–111.

77. Data from Joseph H. Catania et al., "Prevalence of AIDS-related risk factors and condom use in the United States," *Science,* 258 (1992), pp. 1101–1106.

78. The study is reported in William Celis III, "Study suggests Head Start helps beyond school," *New York Times,* April 20, 1993.

79. Strictly speaking, the recipe we give for the standard deviation of $\hat{p}$ assumes that an SRS of size $n$ is drawn from an *infinite* population. If the population has finite size $N$, the standard deviation in the recipe is multiplied by $\sqrt{1 - n/N}$. This "finite population correction" approaches 1 as $N$ increases. When $n/N \leq 0.1$, it is $\geq 0.948$. All methods in this lesson apply in practice only when the population is at least 10 times as large as the sample.

**80.** Sara J. Solnick and David Hemenway, "Complaints and disenrollment at a health maintenance organization," *The Journal of Consumer Affairs,* 26 (1992), pp. 90–103.

**81.** Arne L. Kalleberg and Kevin T. Leicht, "Gender and organizational performance: determinants of small business survival and success," *The Academy of Management Journal,* 34 (1991), pp. 136–161.

**82.** The Detroit Area Study is described in Gerhard Lenski, *The Religious Factor,* Doubleday, New York, 1961.

**83.** See Note 31.

**84.** See Note 42.

**85.** D. M. Barnes, "Breaking the cycle of addiction," *Science,* 241 (1988), pp. 1029–1030.

**86.** Sanders Korenman and David Neumark, "Does marriage really make men more productive?" *Journal of Human Resources,* 26 (1991), pp. 282–307.

**87.** There are many computer studies of the accuracy of chi-square critical values for $\chi^2$. For a brief discussion and some references, see Section 3.2.5 of David S. Moore, "Tests of chi-squared type," in Ralph B. D'Agostino and Michael A. Stephens (eds.), *Goodness-of-Fit Techniques,* Marcel Dekker, New York, 1986, pp. 63–95. If the expected cell counts are roughly equal, the chi-square approximation is adequate when the average expected counts are as small as 1 or 2. The guideline given in the text protects against unequal expected counts. For a survey of inference for smaller samples, see Alan Agresti, "A survey of exact inference for contingency tables," *Statistical Science,* 7 (1992), pp. 131–177.

**88.** Richard M. Felder et al., "Who gets it and who doesn't: a study of student performance in an introductory chemical engineering course," *1992 ASEE Annual Conference Proceedings,* American Society for Engineering Education, Washington, D.C., 1992, pp. 1516–1519.

**89.** S. V. Zagona (ed.), *Studies and Issues in Smoking Behavior,* University of Arizona Press, Tucson, 1967, pp. 157–180.

**90.** Lillian Lin Miao, "Gastric freezing: an example of the evaluation of medical therapy by randomized clinical trials," in John P. Bunker, Benjamin A. Barnes, and Frederick Mosteller (eds.), *Costs, Risks, and Benefits of Surgery,* Oxford University Press, New York, 1977, pp. 198–211.

**91.** Francine D. Blau and Marianne A. Ferber, "Career plans and expectations of young women and men," *Journal of Human Resources,* 26 (1991), pp. 581–607.

**92.** Adapted from M. A. Visintainer, J. R. Volpicelli, and M. E. P. Seligman, "Tumor rejection in rats after inescapable or escapable shock," *Science,* 216 (1982), pp. 437–439.

**93.** James R. Walker, "New evidence on the supply of child care," *Journal of Human Resources,* 27 (1991), pp. 40–69.

**94.** Modified from Felicity Barringer, "Measuring sexuality through polls can be shaky," *New York Times,* April 25, 1993.

95. See Note 31.

96. Martin Enserink, "Fraud and ethics charges hit stroke drug trial," *Science,* 274 (1996), pp. 2002–2005.

97. E. P. Hubble, "A relation between distance and radial velocity among extra-galactic nebulae," *Proceedings of the National Academy of Sciences,* 15 (1929), pp. 168–173.

98. Strictly speaking, this quantity is the estimated standard deviation of $\hat{y} - y$, where $y$ is the additional observation taken at $x = x^*$.

99. Data provided by Robert Dale, Purdue University.

100. See Note 21.

101. See Note 3.

102. The data in Table 24.3 are part of a larger data set in the *Journal of Statistics Education* archive, accessible via Internet. The original source is Pekka Brofeldt, "Bidrag till kaennedom on fiskbestondet i vaara sjoear. Laengel-maevesi," in T. H. Jaervi, *Finlands Fiskeriet,* Band 4, *Meddelanden utgivna av fiskerifoereningen i Finland,* Helsinki, 1917. The data were contributed to the archive (with information in English) by Juha Puranen of the University of Helsinki.

103. Data provided by Matthew Moore.

104. See Note 18.

105. From V. D. Bass, W. E. Hoffmann, and J. L. Dorner, "Normal canine lipid profiles and effects of experimentally induced pancreatitis and hepatic necrosis on lipids," *American Journal of Veterinary Research,* 37 (1976), pp. 1355–1357.

106. From J. W. Marr and J. A. Heady, "Within- and between-person variation in dietary surveys: number of days needed to classify individuals," *Human Nutrition: Applied Nutrition,* 40A (1986), pp. 347–364.

107. From A. H. Ismail and R. J. Young, "The effect of chronic exercise on the personality of middle-aged men," *Journal of Human Ergology,* 2 (1973), pp. 47–57.

108. See Note 88.

109. Based on Greg Clinch, "Employee compensation and firms' research and development activity," *Journal of Accounting Research,* 29 (1991), pp. 59–78.

110. David M. Blau, "The child care labor market," *Journal of Human Resources,* 27 (1992), pp. 9–39.

111. Based on D. Friedlander, *Supplemental Report on the Baltimore Options Program,* Manpower Demonstration Research Corporation, 1987.

112. Eric Ossiander, letter to the editor, *Science,* 257 (1992), p. 1461.

113. William N. Evans and John D. Graham, "An estimate of the lifesaving benefit of child restraint use legislation," *Journal of Health Economics,* 9 (1990), pp. 121–132.

114. A. C. Allison and D. F. Clyde, "Malaria in African children with deficient erythrocyte dehydrogenase," *British Medical Journal,* 1 (1961), pp. 1346–1349.

115. From Noel Capon et al., "A comparative analysis of the strategy and structure of United States and Australian corporations," *Journal of International Business Studies,* 18 (1987), pp. 51–74.

116. Erdener Kaynak and Wellington Kang-yen Kuan, "Environment, strategy, structure, and performance in the context of export activity: an empirical study of Taiwanese manufacturing firms," *Journal of Business Research,* 27 (1993), pp. 33–49.

117. William D. Darley, "Store-choice behavior for pre-owned merchandise," *Journal of Business Research,* 27 (1993), pp. 17–31.

118. See Note 51.

119. U.S. Department of Education, National Center for Education Statistics, *Pursuing Excellence: A Study of U.S. Eighth-Grade Mathematics and Science Teaching, Learning, Curriculum, and Achievement in International Context,* NCES 97–198, Government Printing Office, Washington, D.C., 1996.

120. See Note 10.

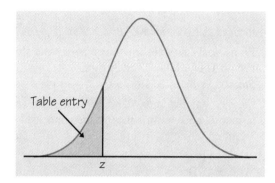

Table entry for $z$ is the area under the standard normal curve to the left of $z$.

| **TABLE A** | **Standard normal probabilities** | | | | | | | | | |
|---|---|---|---|---|---|---|---|---|---|---|
| $z$ | .00 | .01 | .02 | .03 | .04 | .05 | .06 | .07 | .08 | .09 |
| −3.4 | .0003 | .0003 | .0003 | .0003 | .0003 | .0003 | .0003 | .0003 | .0003 | .0002 |
| −3.3 | .0005 | .0005 | .0005 | .0004 | .0004 | .0004 | .0004 | .0004 | .0004 | .0003 |
| −3.2 | .0007 | .0007 | .0006 | .0006 | .0006 | .0006 | .0006 | .0005 | .0005 | .0005 |
| −3.1 | .0010 | .0009 | .0009 | .0009 | .0008 | .0008 | .0008 | .0008 | .0007 | .0007 |
| −3.0 | .0013 | .0013 | .0013 | .0012 | .0012 | .0011 | .0011 | .0011 | .0010 | .0010 |
| −2.9 | .0019 | .0018 | .0018 | .0017 | .0016 | .0016 | .0015 | .0015 | .0014 | .0014 |
| −2.8 | .0026 | .0025 | .0024 | .0023 | .0023 | .0022 | .0021 | .0021 | .0020 | .0019 |
| −2.7 | .0035 | .0034 | .0033 | .0032 | .0031 | .0030 | .0029 | .0028 | .0027 | .0026 |
| −2.6 | .0047 | .0045 | .0044 | .0043 | .0041 | .0040 | .0039 | .0038 | .0037 | .0036 |
| −2.5 | .0062 | .0060 | .0059 | .0057 | .0055 | .0054 | .0052 | .0051 | .0049 | .0048 |
| −2.4 | .0082 | .0080 | .0078 | .0075 | .0073 | .0071 | .0069 | .0068 | .0066 | .0064 |
| −2.3 | .0107 | .0104 | .0102 | .0099 | .0096 | .0094 | .0091 | .0089 | .0087 | .0084 |
| −2.2 | .0139 | .0136 | .0132 | .0129 | .0125 | .0122 | .0119 | .0116 | .0113 | .0110 |
| −2.1 | .0179 | .0174 | .0170 | .0166 | .0162 | .0158 | .0154 | .0150 | .0146 | .0143 |
| −2.0 | .0228 | .0222 | .0217 | .0212 | .0207 | .0202 | .0197 | .0192 | .0188 | .0183 |
| −1.9 | .0287 | .0281 | .0274 | .0268 | .0262 | .0256 | .0250 | .0244 | .0239 | .0233 |
| −1.8 | .0359 | .0351 | .0344 | .0336 | .0329 | .0322 | .0314 | .0307 | .0301 | .0294 |
| −1.7 | .0446 | .0436 | .0427 | .0418 | .0409 | .0401 | .0392 | .0384 | .0375 | .0367 |
| −1.6 | .0548 | .0537 | .0526 | .0516 | .0505 | .0495 | .0485 | .0475 | .0465 | .0455 |
| −1.5 | .0668 | .0655 | .0643 | .0630 | .0618 | .0606 | .0594 | .0582 | .0571 | .0559 |
| −1.4 | .0808 | .0793 | .0778 | .0764 | .0749 | .0735 | .0721 | .0708 | .0694 | .0681 |
| −1.3 | .0968 | .0951 | .0934 | .0918 | .0901 | .0885 | .0869 | .0853 | .0838 | .0823 |
| −1.2 | .1151 | .1131 | .1112 | .1093 | .1075 | .1056 | .1038 | .1020 | .1003 | .0985 |
| −1.1 | .1357 | .1335 | .1314 | .1292 | .1271 | .1251 | .1230 | .1210 | .1190 | .1170 |
| −1.0 | .1587 | .1562 | .1539 | .1515 | .1492 | .1469 | .1446 | .1423 | .1401 | .1379 |
| −0.9 | .1841 | .1814 | .1788 | .1762 | .1736 | .1711 | .1685 | .1660 | .1635 | .1611 |
| −0.8 | .2119 | .2090 | .2061 | .2033 | .2005 | .1977 | .1949 | .1922 | .1894 | .1867 |
| −0.7 | .2420 | .2389 | .2358 | .2327 | .2296 | .2266 | .2236 | .2206 | .2177 | .2148 |
| −0.6 | .2743 | .2709 | .2676 | .2643 | .2611 | .2578 | .2546 | .2514 | .2483 | .2451 |
| −0.5 | .3085 | .3050 | .3015 | .2981 | .2946 | .2912 | .2877 | .2843 | .2810 | .2776 |
| −0.4 | .3446 | .3409 | .3372 | .3336 | .3300 | .3264 | .3228 | .3192 | .3156 | .3121 |
| −0.3 | .3821 | .3783 | .3745 | .3707 | .3669 | .3632 | .3594 | .3557 | .3520 | .3483 |
| −0.2 | .4207 | .4168 | .4129 | .4090 | .4052 | .4013 | .3974 | .3936 | .3897 | .3859 |
| −0.1 | .4602 | .4562 | .4522 | .4483 | .4443 | .4404 | .4364 | .4325 | .4286 | .4247 |
| −0.0 | .5000 | .4960 | .4920 | .4880 | .4840 | .4801 | .4761 | .4721 | .4681 | .4641 |

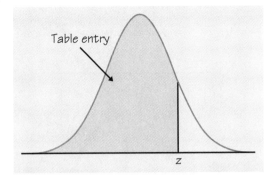

Table entry

Table entry for $z$ is the area under the standard normal curve to the left of $z$.

### TABLE A    Standard normal probabilities (*continued*)

| z | .00 | .01 | .02 | .03 | .04 | .05 | .06 | .07 | .08 | .09 |
|---|-----|-----|-----|-----|-----|-----|-----|-----|-----|-----|
| 0.0 | .5000 | .5040 | .5080 | .5120 | .5160 | .5199 | .5239 | .5279 | .5319 | .5359 |
| 0.1 | .5398 | .5438 | .5478 | .5517 | .5557 | .5596 | .5636 | .5675 | .5714 | .5753 |
| 0.2 | .5793 | .5832 | .5871 | .5910 | .5948 | .5987 | .6026 | .6064 | .6103 | .6141 |
| 0.3 | .6179 | .6217 | .6255 | .6293 | .6331 | .6368 | .6406 | .6443 | .6480 | .6517 |
| 0.4 | .6554 | .6591 | .6628 | .6664 | .6700 | .6736 | .6772 | .6808 | .6844 | .6879 |
| 0.5 | .6915 | .6950 | .6985 | .7019 | .7054 | .7088 | .7123 | .7157 | .7190 | .7224 |
| 0.6 | .7257 | .7291 | .7324 | .7357 | .7389 | .7422 | .7454 | .7486 | .7517 | .7549 |
| 0.7 | .7580 | .7611 | .7642 | .7673 | .7704 | .7734 | .7764 | .7794 | .7823 | .7852 |
| 0.8 | .7881 | .7910 | .7939 | .7967 | .7995 | .8023 | .8051 | .8078 | .8106 | .8133 |
| 0.9 | .8159 | .8186 | .8212 | .8238 | .8264 | .8289 | .8315 | .8340 | .8365 | .8389 |
| 1.0 | .8413 | .8438 | .8461 | .8485 | .8508 | .8531 | .8554 | .8577 | .8599 | .8621 |
| 1.1 | .8643 | .8665 | .8686 | .8708 | .8729 | .8749 | .8770 | .8790 | .8810 | .8830 |
| 1.2 | .8849 | .8869 | .8888 | .8907 | .8925 | .8944 | .8962 | .8980 | .8997 | .9015 |
| 1.3 | .9032 | .9049 | .9066 | .9082 | .9099 | .9115 | .9131 | .9147 | .9162 | .9177 |
| 1.4 | .9192 | .9207 | .9222 | .9236 | .9251 | .9265 | .9279 | .9292 | .9306 | .9319 |
| 1.5 | .9332 | .9345 | .9357 | .9370 | .9382 | .9394 | .9406 | .9418 | .9429 | .9441 |
| 1.6 | .9452 | .9463 | .9474 | .9484 | .9495 | .9505 | .9515 | .9525 | .9535 | .9545 |
| 1.7 | .9554 | .9564 | .9573 | .9582 | .9591 | .9599 | .9608 | .9616 | .9625 | .9633 |
| 1.8 | .9641 | .9649 | .9656 | .9664 | .9671 | .9678 | .9686 | .9693 | .9699 | .9706 |
| 1.9 | .9713 | .9719 | .9726 | .9732 | .9738 | .9744 | .9750 | .9756 | .9761 | .9767 |
| 2.0 | .9772 | .9778 | .9783 | .9788 | .9793 | .9798 | .9803 | .9808 | .9812 | .9817 |
| 2.1 | .9821 | .9826 | .9830 | .9834 | .9838 | .9842 | .9846 | .9850 | .9854 | .9857 |
| 2.2 | .9861 | .9864 | .9868 | .9871 | .9875 | .9878 | .9881 | .9884 | .9887 | .9890 |
| 2.3 | .9893 | .9896 | .9898 | .9901 | .9904 | .9906 | .9909 | .9911 | .9913 | .9916 |
| 2.4 | .9918 | .9920 | .9922 | .9925 | .9927 | .9929 | .9931 | .9932 | .9934 | .9936 |
| 2.5 | .9938 | .9940 | .9941 | .9943 | .9945 | .9946 | .9948 | .9949 | .9951 | .9952 |
| 2.6 | .9953 | .9955 | .9956 | .9957 | .9959 | .9960 | .9961 | .9962 | .9963 | .9964 |
| 2.7 | .9965 | .9966 | .9967 | .9968 | .9969 | .9970 | .9971 | .9972 | .9973 | .9974 |
| 2.8 | .9974 | .9975 | .9976 | .9977 | .9977 | .9978 | .9979 | .9979 | .9980 | .9981 |
| 2.9 | .9981 | .9982 | .9982 | .9983 | .9984 | .9984 | .9985 | .9985 | .9986 | .9986 |
| 3.0 | .9987 | .9987 | .9987 | .9988 | .9988 | .9989 | .9989 | .9989 | .9990 | .9990 |
| 3.1 | .9990 | .9991 | .9991 | .9991 | .9992 | .9992 | .9992 | .9992 | .9993 | .9993 |
| 3.2 | .9993 | .9993 | .9994 | .9994 | .9994 | .9994 | .9994 | .9995 | .9995 | .9995 |
| 3.3 | .9995 | .9995 | .9995 | .9996 | .9996 | .9996 | .9996 | .9996 | .9996 | .9997 |
| 3.4 | .9997 | .9997 | .9997 | .9997 | .9997 | .9997 | .9997 | .9997 | .9997 | .9998 |

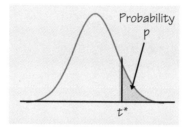

Table entry for $p$ and $C$ is the value $t^*$ with probability $p$ lying to its right and probability $C$ lying between $-t^*$ and $t^*$.

## TABLE B  The $t$ distributions

| df | \multicolumn{11}{c}{Upper tail probability $p$} |
| --- | --- | --- | --- | --- | --- | --- | --- | --- | --- | --- | --- | --- |
| | .25 | .20 | .15 | .10 | .05 | .025 | .02 | .01 | .005 | .0025 | .001 | .0005 |
| 1 | 1.000 | 1.376 | 1.963 | 3.078 | 6.314 | 12.71 | 15.89 | 31.82 | 63.66 | 127.3 | 318.3 | 636.6 |
| 2 | 0.816 | 1.061 | 1.386 | 1.886 | 2.920 | 4.303 | 4.849 | 6.965 | 9.925 | 14.09 | 22.33 | 31.60 |
| 3 | 0.765 | 0.978 | 1.250 | 1.638 | 2.353 | 3.182 | 3.482 | 4.541 | 5.841 | 7.453 | 10.21 | 12.92 |
| 4 | 0.741 | 0.941 | 1.190 | 1.533 | 2.132 | 2.776 | 2.999 | 3.747 | 4.604 | 5.598 | 7.173 | 8.610 |
| 5 | 0.727 | 0.920 | 1.156 | 1.476 | 2.015 | 2.571 | 2.757 | 3.365 | 4.032 | 4.773 | 5.893 | 6.869 |
| 6 | 0.718 | 0.906 | 1.134 | 1.440 | 1.943 | 2.447 | 2.612 | 3.143 | 3.707 | 4.317 | 5.208 | 5.959 |
| 7 | 0.711 | 0.896 | 1.119 | 1.415 | 1.895 | 2.365 | 2.517 | 2.998 | 3.499 | 4.029 | 4.785 | 5.408 |
| 8 | 0.706 | 0.889 | 1.108 | 1.397 | 1.860 | 2.306 | 2.449 | 2.896 | 3.355 | 3.833 | 4.501 | 5.041 |
| 9 | 0.703 | 0.883 | 1.100 | 1.383 | 1.833 | 2.262 | 2.398 | 2.821 | 3.250 | 3.690 | 4.297 | 4.781 |
| 10 | 0.700 | 0.879 | 1.093 | 1.372 | 1.812 | 2.228 | 2.359 | 2.764 | 3.169 | 3.581 | 4.144 | 4.587 |
| 11 | 0.697 | 0.876 | 1.088 | 1.363 | 1.796 | 2.201 | 2.328 | 2.718 | 3.106 | 3.497 | 4.025 | 4.437 |
| 12 | 0.695 | 0.873 | 1.083 | 1.356 | 1.782 | 2.179 | 2.303 | 2.681 | 3.055 | 3.428 | 3.930 | 4.318 |
| 13 | 0.694 | 0.870 | 1.079 | 1.350 | 1.771 | 2.160 | 2.282 | 2.650 | 3.012 | 3.372 | 3.852 | 4.221 |
| 14 | 0.692 | 0.868 | 1.076 | 1.345 | 1.761 | 2.145 | 2.264 | 2.624 | 2.977 | 3.326 | 3.787 | 4.140 |
| 15 | 0.691 | 0.866 | 1.074 | 1.341 | 1.753 | 2.131 | 2.249 | 2.602 | 2.947 | 3.286 | 3.733 | 4.073 |
| 16 | 0.690 | 0.865 | 1.071 | 1.337 | 1.746 | 2.120 | 2.235 | 2.583 | 2.921 | 3.252 | 3.686 | 4.015 |
| 17 | 0.689 | 0.863 | 1.069 | 1.333 | 1.740 | 2.110 | 2.224 | 2.567 | 2.898 | 3.222 | 3.646 | 3.965 |
| 18 | 0.688 | 0.862 | 1.067 | 1.330 | 1.734 | 2.101 | 2.214 | 2.552 | 2.878 | 3.197 | 3.611 | 3.922 |
| 19 | 0.688 | 0.861 | 1.066 | 1.328 | 1.729 | 2.093 | 2.205 | 2.539 | 2.861 | 3.174 | 3.579 | 3.883 |
| 20 | 0.687 | 0.860 | 1.064 | 1.325 | 1.725 | 2.086 | 2.197 | 2.528 | 2.845 | 3.153 | 3.552 | 3.850 |
| 21 | 0.686 | 0.859 | 1.063 | 1.323 | 1.721 | 2.080 | 2.189 | 2.518 | 2.831 | 3.135 | 3.527 | 3.819 |
| 22 | 0.686 | 0.858 | 1.061 | 1.321 | 1.717 | 2.074 | 2.183 | 2.508 | 2.819 | 3.119 | 3.505 | 3.792 |
| 23 | 0.685 | 0.858 | 1.060 | 1.319 | 1.714 | 2.069 | 2.177 | 2.500 | 2.807 | 3.104 | 3.485 | 3.768 |
| 24 | 0.685 | 0.857 | 1.059 | 1.318 | 1.711 | 2.064 | 2.172 | 2.492 | 2.797 | 3.091 | 3.467 | 3.745 |
| 25 | 0.684 | 0.856 | 1.058 | 1.316 | 1.708 | 2.060 | 2.167 | 2.485 | 2.787 | 3.078 | 3.450 | 3.725 |
| 26 | 0.684 | 0.856 | 1.058 | 1.315 | 1.706 | 2.056 | 2.162 | 2.479 | 2.779 | 3.067 | 3.435 | 3.707 |
| 27 | 0.684 | 0.855 | 1.057 | 1.314 | 1.703 | 2.052 | 2.158 | 2.473 | 2.771 | 3.057 | 3.421 | 3.690 |
| 28 | 0.683 | 0.855 | 1.056 | 1.313 | 1.701 | 2.048 | 2.154 | 2.467 | 2.763 | 3.047 | 3.408 | 3.674 |
| 29 | 0.683 | 0.854 | 1.055 | 1.311 | 1.699 | 2.045 | 2.150 | 2.462 | 2.756 | 3.038 | 3.396 | 3.659 |
| 30 | 0.683 | 0.854 | 1.055 | 1.310 | 1.697 | 2.042 | 2.147 | 2.457 | 2.750 | 3.030 | 3.385 | 3.646 |
| 40 | 0.681 | 0.851 | 1.050 | 1.303 | 1.684 | 2.021 | 2.123 | 2.423 | 2.704 | 2.971 | 3.307 | 3.551 |
| 50 | 0.679 | 0.849 | 1.047 | 1.299 | 1.676 | 2.009 | 2.109 | 2.403 | 2.678 | 2.937 | 3.261 | 3.496 |
| 60 | 0.679 | 0.848 | 1.045 | 1.296 | 1.671 | 2.000 | 2.099 | 2.390 | 2.660 | 2.915 | 3.232 | 3.460 |
| 80 | 0.678 | 0.846 | 1.043 | 1.292 | 1.664 | 1.990 | 2.088 | 2.374 | 2.639 | 2.887 | 3.195 | 3.416 |
| 100 | 0.677 | 0.845 | 1.042 | 1.290 | 1.660 | 1.984 | 2.081 | 2.364 | 2.626 | 2.871 | 3.174 | 3.390 |
| 1000 | 0.675 | 0.842 | 1.037 | 1.282 | 1.646 | 1.962 | 2.056 | 2.330 | 2.581 | 2.813 | 3.098 | 3.300 |
| $z^*$ | 0.674 | 0.841 | 1.036 | 1.282 | 1.645 | 1.960 | 2.054 | 2.326 | 2.576 | 2.807 | 3.091 | 3.291 |
| | 50% | 60% | 70% | 80% | 90% | 95% | 96% | 98% | 99% | 99.5% | 99.8% | 99.9% |
| | \multicolumn{11}{c}{Confidence level $C$} |

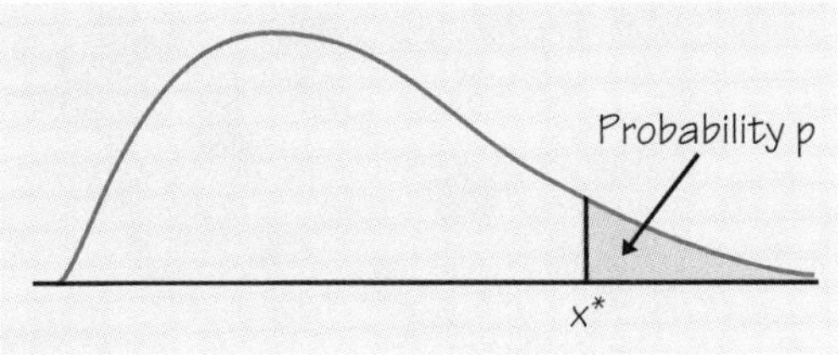


Table entry for $p$ is the value $x^*$ with probability $p$ lying to its right.

**TABLE C The chi-square distributions**

| | $p$ | | | | | | | | | | | |
|---|---|---|---|---|---|---|---|---|---|---|---|---|
| df | .25 | .20 | .15 | .10 | .05 | .025 | .02 | .01 | .005 | .0025 | .001 | .0005 |
| 1 | 1.32 | 1.64 | 2.07 | 2.71 | 3.84 | 5.02 | 5.41 | 6.63 | 7.88 | 9.14 | 10.83 | 12.12 |
| 2 | 2.77 | 3.22 | 3.79 | 4.61 | 5.99 | 7.38 | 7.82 | 9.21 | 10.60 | 11.98 | 13.82 | 15.20 |
| 3 | 4.11 | 4.64 | 5.32 | 6.25 | 7.81 | 9.35 | 9.84 | 11.34 | 12.84 | 14.32 | 16.27 | 17.73 |
| 4 | 5.39 | 5.99 | 6.74 | 7.78 | 9.49 | 11.14 | 11.67 | 13.28 | 14.86 | 16.42 | 18.47 | 20.00 |
| 5 | 6.63 | 7.29 | 8.12 | 9.24 | 11.07 | 12.83 | 13.39 | 15.09 | 16.75 | 18.39 | 20.51 | 22.11 |
| 6 | 7.84 | 8.56 | 9.45 | 10.64 | 12.59 | 14.45 | 15.03 | 16.81 | 18.55 | 20.25 | 22.46 | 24.10 |
| 7 | 9.04 | 9.80 | 10.75 | 12.02 | 14.07 | 16.01 | 16.62 | 18.48 | 20.28 | 22.04 | 24.32 | 26.02 |
| 8 | 10.22 | 11.03 | 12.03 | 13.36 | 15.51 | 17.53 | 18.17 | 20.09 | 21.95 | 23.77 | 26.12 | 27.87 |
| 9 | 11.39 | 12.24 | 13.29 | 14.68 | 16.92 | 19.02 | 19.68 | 21.67 | 23.59 | 25.46 | 27.88 | 29.67 |
| 10 | 12.55 | 13.44 | 14.53 | 15.99 | 18.31 | 20.48 | 21.16 | 23.21 | 25.19 | 27.11 | 29.59 | 31.42 |
| 11 | 13.70 | 14.63 | 15.77 | 17.28 | 19.68 | 21.92 | 22.62 | 24.72 | 26.76 | 28.73 | 31.26 | 33.14 |
| 12 | 14.85 | 15.81 | 16.99 | 18.55 | 21.03 | 23.34 | 24.05 | 26.22 | 28.30 | 30.32 | 32.91 | 34.82 |
| 13 | 15.98 | 16.98 | 18.20 | 19.81 | 22.36 | 24.74 | 25.47 | 27.69 | 29.82 | 31.88 | 34.53 | 36.48 |
| 14 | 17.12 | 18.15 | 19.41 | 21.06 | 23.68 | 26.12 | 26.87 | 29.14 | 31.32 | 33.43 | 36.12 | 38.11 |
| 15 | 18.25 | 19.31 | 20.60 | 22.31 | 25.00 | 27.49 | 28.26 | 30.58 | 32.80 | 34.95 | 37.70 | 39.72 |
| 16 | 19.37 | 20.47 | 21.79 | 23.54 | 26.30 | 28.85 | 29.63 | 32.00 | 34.27 | 36.46 | 39.25 | 41.31 |
| 17 | 20.49 | 21.61 | 22.98 | 24.77 | 27.59 | 30.19 | 31.00 | 33.41 | 35.72 | 37.95 | 40.79 | 42.88 |
| 18 | 21.60 | 22.76 | 24.16 | 25.99 | 28.87 | 31.53 | 32.35 | 34.81 | 37.16 | 39.42 | 42.31 | 44.43 |
| 19 | 22.72 | 23.90 | 25.33 | 27.20 | 30.14 | 32.85 | 33.69 | 36.19 | 38.58 | 40.88 | 43.82 | 45.97 |
| 20 | 23.83 | 25.04 | 26.50 | 28.41 | 31.41 | 34.17 | 35.02 | 37.57 | 40.00 | 42.34 | 45.31 | 47.50 |
| 21 | 24.93 | 26.17 | 27.66 | 29.62 | 32.67 | 35.48 | 36.34 | 38.93 | 41.40 | 43.78 | 46.80 | 49.01 |
| 22 | 26.04 | 27.30 | 28.82 | 30.81 | 33.92 | 36.78 | 37.66 | 40.29 | 42.80 | 45.20 | 48.27 | 50.51 |
| 23 | 27.14 | 28.43 | 29.98 | 32.01 | 35.17 | 38.08 | 38.97 | 41.64 | 44.18 | 46.62 | 49.73 | 52.00 |
| 24 | 28.24 | 29.55 | 31.13 | 33.20 | 36.42 | 39.36 | 40.27 | 42.98 | 45.56 | 48.03 | 51.18 | 53.48 |
| 25 | 29.34 | 30.68 | 32.28 | 34.38 | 37.65 | 40.65 | 41.57 | 44.31 | 46.93 | 49.44 | 52.62 | 54.95 |
| 26 | 30.43 | 31.79 | 33.43 | 35.56 | 38.89 | 41.92 | 42.86 | 45.64 | 48.29 | 50.83 | 54.05 | 56.41 |
| 27 | 31.53 | 32.91 | 34.57 | 36.74 | 40.11 | 43.19 | 44.14 | 46.96 | 49.64 | 52.22 | 55.48 | 57.86 |
| 28 | 32.62 | 34.03 | 35.71 | 37.92 | 41.34 | 44.46 | 45.42 | 48.28 | 50.99 | 53.59 | 56.89 | 59.30 |
| 29 | 33.71 | 35.14 | 36.85 | 39.09 | 42.56 | 45.72 | 46.69 | 49.59 | 52.34 | 54.97 | 58.30 | 60.73 |
| 30 | 34.80 | 36.25 | 37.99 | 40.26 | 43.77 | 46.98 | 47.96 | 50.89 | 53.67 | 56.33 | 59.70 | 62.16 |
| 40 | 45.62 | 47.27 | 49.24 | 51.81 | 55.76 | 59.34 | 60.44 | 63.69 | 66.77 | 69.70 | 73.40 | 76.09 |
| 50 | 56.33 | 58.16 | 60.35 | 63.17 | 67.50 | 71.42 | 72.61 | 76.15 | 79.49 | 82.66 | 86.66 | 89.56 |
| 60 | 66.98 | 68.97 | 71.34 | 74.40 | 79.08 | 83.30 | 84.58 | 88.38 | 91.95 | 95.34 | 99.61 | 102.7 |
| 80 | 88.13 | 90.41 | 93.11 | 96.58 | 101.9 | 106.6 | 108.1 | 112.3 | 116.3 | 120.1 | 124.8 | 128.3 |
| 100 | 109.1 | 111.7 | 114.7 | 118.5 | 124.3 | 129.6 | 131.1 | 135.8 | 140.2 | 144.3 | 149.4 | 153.2 |

# SOLUTIONS TO SELECTED EXERCISES

## LESSON 2

2.1 **(a)** The complete data set describes employees of a company.

  **(b)** The categorical variables are Gender, Race, and Job type.

  **(c)** The quantitative variables are Age and Salary. The units of measurement are years and dollars respectively.

  **(d)** Some purposes to gather these data could include:

    **1.** Assessing the needs of its various groups based on age.

    **2.** Determining which employees are close to retirement age.

    **3.** Determining gender and racial composition of its employees.

    **4.** Keeping track of salary outlays.

    **5.** Determining the nature and composition of its employees by the type of work performed.

2.3 **(a)** Gender: Categorical

  **(b)** Age: Quantitative

  **(c)** Race: Categorical

  **(d)** Smoker: Categorical

  **(e)** Systolic blood pressure: Quantitative

  **(f)** Level of calcium in blood: Quantitative

## LESSON 3

3.1 **(a)** Plot the fields of study on the horizontal axis and the percentage of women earning doctorates in each field on the vertical axis. The bar graph should be scaled to show visually demonstrable differences in the various fields of study. It should be clear from the bar graph the two fields of study with the highest percentage of females are education and psychology. The middle are life sciences and physical sciences. The fields with the lowest percentages are computer sciences and engineering.

(b) No, it would not be correct to use a pie chart to display these data. The data provide us with the percentage of females earning doctorates in each field. Pie charts require the sum of all percentages to equal 100%.

3.3 There is no clear mode. Rather, the data appear to be multimodal. This can, however, be due to the relatively small sample size.

There appears to be one potential outlier; a score of 200. The center which produces about half the observations above and half below is the score 137. The spread, ignoring the outlier, is about 77 i.e., $(178 - 101)$.

3.5 (a) The shapes of the two distributions are clearly different. The distribution of the weights of chicks fed the normal corn has an approximate center of 355. The distribution of the weights of chicks fed the new corn has an approximate center of 395. These differences in center and shape suggest that the new corn has increased the typical weight and affected the distribution of the chicks which fed on it compared to the chicks which were fed the normal corn.

(b) The distribution of chick weights fed the normal corn is roughly symmetric. The distribution of chick weights fed the new corn is left skewed. Both histograms show a unimodal distribution.

3.7 (a) Histogram: Shows a single peaked roughly symmetric distribution. There is a gap between 22.5 and 16 mpg with a potential outlier at 16 mpg. The approximate center is about 26 mpg.

Dotplot: Shows a main body of data between 23 and 32 miles per gallon with one potential outlier at 16 mpg. The center of the data, excluding the outlier, is approximately 26 mpg. The data are evenly spread out with the exception of the outlier.

(b) Cars subject to the "gas guzzler" tax are probably cars with fuel efficiency which is less than the typical miles per gallon. In our data set these cars would most likely include those cars which get less than 26 highway miles per gallon. Cars which get less than this typical value are: Acura 3.5RL, Audi Quattro, Cadillac Catera, Infiniti Q45, Lexus GS300, Lexus LS400, Mercedes-Benz E420, and the Rolls Royce Silver Spur.

3.9 (a) The histogram shows a clearly right skewed distribution with a possible second mode. There are no clear gaps or outliers.

**(b)** There are 15 observations in which there was no frost and there are 65 total observations. Therefore, about 23% of the 65 years never fell below freezing.

3.11 The distribution of survival times is right skewed, single peaked and unimodal. There are two gaps, one between 280–300 and another between 440–480. The distribution shows the expected right skew.

3.13 The distribution of the percent of students taking the SAT is bimodal. There are no obvious outliers or gaps. The second mode could be an indication of a lurking variable.

3.15 The distribution of dollars spent on education per pupil is roughly normal and unimodal. There are five potential outliers separated by a gap beyond 7500. The outliers may give the impression that the data have a slight right skew, but we should consider these points as deviations from the overall pattern.

## LESSON 4

4.1 Due to the nature of Forbes magazine, we can suspect that the distribution of the readers' wealth contains several high outliers. These high outliers would pull up the mean considerably, but have little or no effect on the median. Therefore the mean is most likely $2.2 million and the median is most likely $800,000.

4.3 Mean = $62,500, Median = $25,000, Midrange = $140,000, Mode = $25,000. Seven of the eight people earn less than the mean.

4.5 **(a)** 5, 5, 5, 5.

**(b)** 0, 0, 10, 10.

**(c)** Any four numbers which are the same will produce a standard deviation of zero. Although a formal proof for (b) is beyond the scope of this text, values which are spaced farthest apart are likely to produce the largest variability. Trial and error verifies that two values should be at each extreme.

**(d)** Yes, (c) would change. Any two numbers between 0 and 10 can be chosen for the center two numbers in order to produce the largest range, but the min and the max must remain the same.

4.7 The feature which makes these two spans approximately equal is the presence of an outlier. The outlier causes the mean to be pulled up and the standard deviation to increase.

4.9  (a) mean = 5.4
     (b) standard deviation = .642
     (c) The results agree with the exception of some rounding error.

4.11  (a) A histogram shows that the distribution of the percentages of
          winning elections is roughly symmetric, unimodal with no
          outliers.
      (b) The median percent of the vote won by a successful candidate
          is 50.7%.
      (c) The upper quartile is 58.8. The years in which successful
          candidates captured at least that percent of the vote are: 1984,
          1972 and 1964.

4.13  According to the 1.5 IQR criterion there is one outlier—an
      unusually low value of 16 miles per gallon.

4.15  Although the distribution is symmetric there are several outliers,
      which suggests that a five-number summary is preferred:

```
Summary of  Percent 65 and over
Percentile  25
Median       12.65
Min    4.2
Max    18.3
Lower ith %tile    11.4
Upper ith %tile    13.7
```

According to the 1.5 IQR criterion, there are 2 outliers—an
unusually high value of 18.3% and an unusually low value of
4.2%.

4.17  The data are strongly right skewed, unimodal with one high
      outlier. Due to the extreme skew and outlier, a five-number
      summary is preferred:

```
Summary of  Salaries
Median       1250290
Min          109000
Max          9237500
Q1           205000
Q2           2300000
```

4.19  (a)

```
Summary of  HWYMPG
Mean         26.5385
Median       26
StdDev       3.03619
IntQRange    3
```

**(b)**                     Liters = 3.785   (Gallons)

                  Summary of  Liters
                  Mean        100.448
                  Median      98.41
                  StdDev      11.492
                  IntQRange   11.355

## LESSON 5

5.3 **(a)** The mean and median both fall at the same value which is .5. The quartiles Q1 and Q3 fall at .25 and .75 respectively.

5.5 **(a)** 2.5% of the men are taller than 74 inches.
   **(b)** 95% of the heights of men will fall between 65 and 74 inches.
   **(c)** 16% of the men are shorter than 66.5 inches.

5.7 **(a)** 95% of human pregnancies will last between 234 and 298 days.
   **(b)** The shortest 2.5% of all pregnancies is 234 days.

5.9 Ty Cobb falls into the 1910s. Therefore, his standardized score can be expressed: $z = 4.15$
   Ted Williams falls into the 1940s. Therefore, his standardized score can be expressed: $z = 4.26$
   George Brett falls into the 1970s. Therefore, his standardized score can be expressed: $z = 4.07$
   When compared to each other and relative to the peers of their eras, Ted Williams had the highest average, Ty Cobb had the second highest average, and George Brett had the third highest average.

5.11 **(a)** The middle 95% lie between $-21\%$ and 45%.
   **(b)** $z = -.73$; $RF\,(Z < -.73) = .2327$ or 23.27%.
   **(c)** $z = .79$; $RF\,(Z > .79) = (1 - .7852) = .2148$ or 21.48%.

5.15 The normal probability plot of Cavendish's measurements of the density of the Earth shows a roughly linear trend between Earth density and their n-scores. Although there are a few points which deviate from this linear trend, the overall distribution can be described as being roughly normal.

5.17 A normal distribution will tend to have a roughly straight line trend between the individual values of a variable and their respective n-scores. The normal probability plot of our uniform distribution shows an S-shaped pattern which is typical of this type of distribution.

## PART I REVIEW

R1 (a) A histogram shows a roughly normal distribution with three clear outliers.

(b) The three unusual data points appear as outliers which are more extreme than the main body of the data.

After removing the outliers, the numerical summaries are:

```
Summary of Drive Times
42 total cases of which 2 are missing
Count              40
Mean            8.408
Median           8.42
StdDev       0.540248
IntQRange        0.75
```

(c) According to the numerical summaries the mean and median are very close. This feature indicates that the data are symmetric. While this is a necessary condition for normality, it is not sufficient. In order to assess normality we should examine a normal probability plot. If the data still appear normal, the appropriate numerical summaries are the mean (8.408) and standard deviation (.540) for measures of center and spread respectively.

R3 Both the stem plot and the histogram reveal a multimodal distribution with no clear pattern. There is one value which is unusually low in calories relative to the rest. This value is 107 which is most likely the *Eat Slim Veal Hot Dogs* brand.

R5

| Summary of | Home runs | | | | | |
|------------|-----------|--------|-----|-----|-------|-----|
| Group      | Count     | Median | Min | Max | Q1    | Q3  |
| DiMaggio   | 13        | 30     | 12  | 46  | 20.75 | 32  |
| Mantle     | 18        | 28.5   | 13  | 54  | 21    | 37  |

The back-to-back stemplot shows a roughly symmetric shape for both players. It also reveals potential outliers for both players. DiMaggio has two low outliers, 12 and 14, and one high outlier, 46. Mantle has two high outliers, 52 and 54. The five-number summary shows a slightly higher median value of 30 for DiMaggio than

Mantle's median of 28.5. There appears to be more variability in Mantle's distribution of home runs than for DiMaggio as evidenced by their respective IQR; 16 and 11.25. The comparison suggests that DiMaggio typically hit slightly higher home runs per year than Mantle. In addition, the data suggest that DiMaggio was slightly more consistent in his number of home runs per year.

R7 Proportion of population scoring below 1.7: $RF(ARSMA < 1.7) = RF(Z < -1.625) = .0576 \ or \ 5.76\%$.
Proportion of population scoring between 1.7 and 2.1: $RF(1.7 < ARSMA < 2.1) = .0776$ or 7.76%.

R9 In order to be among the top 30% of the population used to develop the test who are the most Anglo/English in cultural orientation a person would have to score at least 3.424. $RF(X > x^*) = RF(Z > .53) = .30 \ or \ 30\%; x^* = 3.424$.

R11 The histogram and boxplot show a clearly right skewed, unimodal distribution with several unusually high values (outliers). According to the 1.5 IQR criterion there are nine counties which can be considered outliers. The appropriate measure of center for this distribution would be the median which is 30,000 people. The appropriate measure of spread would be the IQR which is 41. These measures of center and spread are appropriate due to the skewed distribution and the presence of outliers.

R13 Sarah's standardized score: $z = \frac{135-110}{25} = 1$. Sarah scored 1 standard deviation above the mean. Using either the Z-table or the 68%-95%-99.7% rule, we can estimate that Sarah scored in the 84.13th percentile.

Mother's standardized score: $z = \frac{120-90}{25} = 1.2$. Sarah's mother scored 1.2 standard deviations above the mean. Using the Z-table, we can estimate that Sarah's mother scored in the 88.49th percentile.

It appears that relative to their age cohorts Sarah's mother obtained the higher score.

R15 No, even if the mean and standard deviation of two distributions are identical they do not necessarily share the same shape. The mean and standard deviation are non-resistant measures. The presence of outliers can affect the shape of a distribution as well as affecting numerical summaries. Although five-number summaries are resistant to outliers the shape of two distributions with identical five-number summaries can have quite different shapes. For example, a uniform and normal distribution can have identical numerical summaries but by definition are different in their shapes.

## LESSON 6

6.1 A side-by-side boxplot of the data shows that Babe Ruth typically as measured by medians hit the most number of home runs during his career followed by Mantle, DiMaggio, and Maris. DiMaggio hit most consistently as measured by the IQR followed by Ruth, Mantle and Maris. An unusually high value according to the 1.5 IQR criterion for DiMaggio is the year in which he hit 46 home runs.

Summaries

| Variable | Median | Min | Max | IntQRange | Q1 | Q3 |
|---|---|---|---|---|---|---|
| Ruth | 46 | 22 | 60 | 16.25 | 36.5 | 52.75 |
| Maris | 24.5 | 8 | 61 | 19 | 14 | 33 |
| DiMaggio | 30 | 12 | 46 | 11.25 | 20.75 | 32 |
| Mantle | 28.5 | 13 | 54 | 16 | 21 | 37 |

6.3 The four Census Bureau regions differ in regards to median in the following ways: The Midwest has the highest median followed by West, South and Northeast. There are no outliers according to the 1.5 IQR criterion in any of the four regions.

6.5 The data show systematic differences in the distribution of calories across the three types of hot dogs. Both meat and beef hot dogs show right skewed distributions, whereas poultry shows a left skew. These differences in shape can be seen by the respective median positions relative to their quartiles. Both meat and beef have close medians 152.5 and 153 respectively. Poultry has the lowest median calories of 129. The spread in the distribution of calories as indicated by the IQR are close for all three types of hot dogs—38.5 (beef), 41 (meat) and 42 (poultry). There are no extreme caloric values for any of the three types of hot dogs.

A histogram of the distributions of calories for the 17 different brands of meat hot dogs shows an almost uniform distribution (with the exception of the outlier). This feature is not revealed by a boxplot. In addition, there is an unusually low caloric value which is not seen in the box plot.

6.7 The distribution of pitchers' salaries is right skewed with no clear outliers. The distribution of position players' salaries is roughly symmetric and has two clear outliers. The median of pitchers' salaries is markedly lower than that of position players. In addition the spread as measured by IQR is greater for pitchers than for position players.

## LESSON 7

7.1 **(a)** Plot city mileage on the horizontal axis and highway mileage on the vertical axis. If plotted correctly, the bulk of data will appear in the upper right with one point clearly separate from the rest in the lower left.

**(b)** The direction of the relationship is positive.

**(c)** The form of the relationship is approximately linear. There is one clear outlier; the Rolls Royce Silver Spur which only gets 11 city-mpg and 16 highway-mpg.

**(d)** The strength of the relationship is strong. There is tight linear cluster of points which indicates that highway mileage can be predicted fairly well from city mileage.

7.3 The variable Percent taking should be plotted on the $X$-axis (horizontally), because it is the explanatory variable which we believe median math SAT scores to be dependent upon. The direction is negative, the form is curved and the strength is moderate to strong.

7.5 The two types of vehicles have different relationships. The relationship between city and highway mileage for sports and utility vehicles is curved, whereas the same relationship for mid-size cars is linear. One line drawn through all the points would be an inaccurate way to predict highway mileage from city mileage for all cars.

7.7 **(a)** The variable which should be placed on the $X$-axis is Plants per acre, because it is the independent variable from which we suspect yield is dependent.

**(b)** The overall pattern shown in the scatterplot is not linear, but somewhat parabolic. There is neither a clear positive nor negative direction in the overall pattern.

**(c)** The relationship between the average yield and the five different plants-to-acre ratios is clearly parabolic. The optimal plants-to-acre ratio occurs at 20 plants per acre which produces an average yield of 146.225 bushels. Farmers with similar conditions should use this planting rate to get the largest yield of corn.

7.9 **(a)** The pattern of Europe's population growth has been positive, strong and curved.

**(b)** The pattern of Europe's population growth in log scale is again positive and strong, but the form is now almost perfectly linear. Transforming the data by using the logarithms of population, changed the form of the relationship from curved to linear.

## LESSON 8

8.1 (a) The relationship is linear, positive and strong. There are no clear outliers.

(b) $r = 3.9766/4 = .99415$

| Femur | Humerus | $\frac{y_i - \bar{y}}{s_y}$ | $\frac{x_i - \bar{x}}{s_x}$ | $\left(\frac{y_i - \bar{y}}{s_y}\right)\left(\frac{x_i - \bar{x}}{s_x}\right)$ |
|---|---|---|---|---|
| 38 | 41 | −1.53048 | −1.57329 | 2.40789 |
| 56 | 63 | −0.16669 | −0.1888 | 0.03147 |
| 59 | 70 | 0.06061 | 0.25173 | 0.01526 |
| 64 | 72 | 0.43944 | 0.37759 | 0.16593 |
| 74 | 84 | 1.19711 | 1.13277 | 1.35605 |
| | | | | $\sum = 3.9766$ |

(c)     Pearson Product-Moment Correlation

| | Femur | Humerus |
|---|---|---|
| Femur | 1.000 | |
| Humerus | 0.994 | 1.000 |

8.3 (a) The correlation for women will be higher than for men. The data points for women are clustered tighter together along a line. Data points for men fan out and are not clustered along a line.

(b)     Pearson Product-Moment Correlation **Women**

| | Metabolic Rate | Body Mass |
|---|---|---|
| Metabolic Rate | 1.000 | |
| Body Mass | 0.926 | 1.000 |

Pearson Product-Moment Correlation **Men**

| | Metabolic Rate | Body Mass |
|---|---|---|
| Metabolic Rate | 1.000 | |
| Body Mass | 0.519 | 1.000 |

(c)     Summary of Body Mass
For categories in Sex

| Group | Mean |
|---|---|
| F | 1235.08 |
| M | 1626.83 |

The fact that men are heavier on average does not affect the correlation.

8.5 The correlations for both the original data and the transformed data are identical; $r = .253$. This is to be expected because both transformations were linear and the correlation coefficient is not affected by transformations of this nature.

8.7 The correlation would be 1.0, because the height of the husbands would be a perfect linear combination of the wives.

8.9 (a) Correlation is only applicable to quantitative variables; sex is a categorical variable.
   (b) A correlation can only take values from $-1.0$ to $1.0$.
   (c) A correlation has no units.

# LESSON 9

9.1 The explanatory variable is the height of a child at age six. The response variable is the height of a child at age sixteen. These variables are both quantitative.

9.3 (a) The graph shows a straight line which is positive.
   (b) Savings $= \$500 + \$100(20) = \$2,500$
   (c) Savings $= \$500 + \$200x$

9.5 (a) According to the regression line the pH levels at the beginning and the end of the study were:

$$\text{Week} = 1: \quad \text{pH} = 5.43 - .0053(1) = 5.425$$
$$\text{Week} = 2: \quad \text{pH} = 5.43 - .0053(150) = 4.635$$

   (b) The slope of the regression line is $-.0053$. The slope of the regression line states that for every passing week during the study the pH level decreased by .0053. This implies that over time the rain increases in acidity by a rate of .0053 pH per week. The association between pH and time is negative.

9.7 $\bar{y}_{gas} = 615.0$; $\bar{x}_{Temp} = 40.76$; $S_{gas} = 231.6$; $S_{Temp} = 11.45$; $r_{(gas,temp)} = -.983$. Slope estimate: $b = r\frac{S_y}{S_x} = -.983\left(\frac{231.6}{11.45}\right) = -19.88$. Intercept estimate: $a = \bar{y} - b\bar{x} = 615 - [-19.88(40.76)] = 615 + 810.31 = 1452.31$. Regression equation: $\hat{y} = a + bx = 1452.31 - 19.88x$. The results agree up to roundoff error.

9.9 (a) The slope can be expressed: $b = r\frac{S_y}{S_x} = .6\left(\frac{8}{30}\right) = .16$. The intercept can be expressed: $a = \bar{y} - b\bar{x} = 75 - .16(280) = 30.2$.

(b) To predict Julie's final exam score: Final exam scores = 30.2 + .16(300) = 78.2.

(c) Recall: R-Squared = $r^2$ = $.6^2$ = .36. This indicates that only 36% percent of the total observed variation in final exam scores has been explained by the least-squares regression of final exam scores on pre-exam totals. Without an actual plot of the data or the residuals it is hard to determine which direction Julie's predicted score is off. The low value of the proportion of variation in final exam scores explained the regression, however, suggests a poor fit of the data to the regression line.

9.11 (a) The correlation coefficient $r$ = $-.64$, given in example 9.3, is valid. The claim that $r^2$ falls to 11% after child 18 is omitted is also valid. In this data set, the presence of an influential observation makes the data appear more linear than the bulk of data suggest.

(b) The dotted line indicates the regression line after removing child 19 and the solid line indicates the regression line with child 19. The positions of these two lines are not very different from each other. Although child 19 is an outlier, it does not seem to have much influence on the overall regression. It appears that child 18 has substantially more influence. This is indicated by the greater change in the position of the regression after removing child 18 has than after removing child 19.

(c) Removing child 19 increases the $r^2$ from 49% to 57.16%. This increase is explained by the unusually high Gesell score child 19 obtained relative to his/her age at first speech. The unusually high Gesell score is an outlier which has the effect of decreasing the linear association between Score and Age and lowers the overall fit of the line. Removing child 19 improves the fit of the line and strengthens the linear relationship.

9.13 (a) All four correlation coefficients and regression equations are virtually identical, $r$ = .816 and $\hat{y}$ = 3 + .5x. Here, for example, are the results for data A:

```
Pearson Product-Moment Correlation
            Y(A)      X(A)
Y(A)    1.000
X(A)    0.816    1.000
```

```
Dependent variable is: Y(A)
Variable   Coefficient   s.e. of Coeff   t-ratio     prob
Constant     3.00009                      1.125      2.67   0.0257
X(A)         0.5000091                    0.1179     4.24   0.0022
```

(b) The scatterplot of data set A shows a roughly linear, positive and moderate relationship. The scatterplot of data set B shows a curved, positive and strong relationship. The scatterplot of data set C shows a linear, positive and strong relationship, but there is one influential outlier. The scatterplot of data set D shows a vertical set of data points with one influential outlier.

(c) Although the correlations and regression equations for each of the data sets are almost identical, only the regression line of data set A should be used to describe the dependence of $y$ on $x$. Data set A is the only set in which $x$ and $y$ are linearly related and/or not altered by outliers. In data set B, the relationship is clearly curved. In data set C, the presence of an extreme $y$-value (outlier) pulls the regression line up and away from the main body of data. In data set D, the presence of a point which is extreme in both $y$-axis and $x$-axis exerts extreme influence. This influential observation gives the numerical illusion of a linear relationship between $x$ and $y$, when in fact, none exists.

9.15 The reasoning is incorrect, because association does not imply causation. The correlation between the number of firefighters and the amount of damage is a measure of linear association. Therefore, regardless of the strength of the linear association, it is not plausible to imply causation.

## REVIEW II

R1 (a) By convention, the response variable is plotted on the vertical axis and the explanatory variable on the horizontal axis. Plot correct calories on the $x$-axis and guessed calories on the $y$-axis.

(b) The direction of the relationship between correct calories and guessed calories is positive. The form is roughly linear and there are two clear outliers. The outliers affect the relationship.

(c) The correlation coefficient; $r = .8245$ is reasonable because the bulk of the data is roughly linear. It should be noted, however, that the outliers affect the correlation coefficient.

(d) The fact that all the guesses are higher than the correct calories is not reflected in the correlation or the regression equation. Since the correlation coefficient is not affected by linear transformations, adding 100 calories to all guessed calories would have no effect on $r$. The slope of the regression line is unaffected by transformations which do not change the standard deviation of either the dependent or independent variables. Since adding or subtracting a constant does not alter the standard deviation, there will be no change in the slope of the regression equation. It should be noted, however, that the intercept would change.

(e) Omitting the outliers caused the correlation coefficient to become higher from .8245 to .9837. The rise in $r$ is due to the fact that $r$ is not resistant to outliers which often distort the linear relationship between two variables. In our data, the outliers caused the data points to appear more spread out, which reduced the strength of the linear relationship between guessed and correct calories. The position of the regression line with the outliers omitted is considerably lower and more centered to the bulk of the data than the original regression line. This suggests that the outliers pulled the line up towards them which in turn suggests that the outliers had a significant amount of influence.

R3 (a) The relationship between deaths from heart disease and consumption of alcohol is approximately linear in form, negative in direction and the strength of the relationship is strong. There are no countries in our sample which clearly deviate from the overall pattern of the relationship. The correlation coefficient: $r = -.8428$. The regression equation: Predicted Deaths/100K $= 260.563 - 22.97$ (Liters of Alcohol Consumed).

(b) The numerical value of the slope which describes the linear relationship between deaths from heart disease and alcohol consumption states: For every additional liter of alcohol consumed per person, there is a corresponding reduction in deaths from heart disease of about 23 people per 100,000 people.

(c) Although the relationship we observed is strong, it does not give good evidence of a cause and effect relationship. Correlation does not imply causation; there may exist other variables such as standards of living, diet, exercise etc . . . which are the real variables explaining heart disease.

R5 (a) The regression equation is: Predicted kills $= -41.4304 + 0.124862$ (Boats). In a year with 716,000 registered boats the equation is: Predicted kills $= -41.4304 + 0.124862\ (716)$. The predicted value is: 47.97 deaths.

(b) Unfortunately, there does not appear to be a trend indicating a decline in the number of manatee deaths. The measures may have had some marginal effects in 1992 and 1993, but the overall trend remains the same.

(c) The mean deaths for years which had 716,000 registered boats is 43.75. The predicted deaths for 716,000 registered boats is 47.97. The difference (residual) is about 4 manatee deaths. Within the context of our data this prediction is fairly accurate.

R7 (a) Hawaii is the outlier. The state has an unusually high median math SAT score given its median verbal SAT score.

(b) The regression equation is: Math $= 13.82 + 1.08\ (\text{Verbal})$.
 The predicted math score for a state which has a median verbal score of 455: Math $= 13.82 + 1.08\ (\text{Verbal}) = 13.82 + 1.08\ (455) = 504.68$.

(c) Hawaii has a median verbal of 404. The predicted median math score based on a median verbal of 404 is: Math $= 13.82 + 1.08\ (\text{Verbal}) = 13.82 + 1.08(404) = 450.14$. The actual median math score for Hawaii is 481.
 The residual for Hawaii $= y_{Hawaii} - \hat{y}_{Hawaii} = 481 - 450.14 = 30.86$.
 The outlier was not very influential as it did not change the position of the regression line significantly.

R9 There are substantial differences among the positions of the three lines. Although the direction is positive in all cases, it appears that the form of the relationship between city and highway mileage is different for the two vehicle types. Mid-size cars have an approximately linear relationship, whereas the relationship is slightly curved for sports and utility vehicles. The differences in form between the two types of vehicles are reflected in the slightly lower $r^2$ for sports and utility vehicles. This difference in $r^2$ is not substantial, but reflects that slightly more of the total variation in highway mileage has been explained by the linear relationship for mid-size cars than for sports and utility vehicles.

 The relationship between city and highway mileage for all cars is roughly linear with three potential outliers which correspond to some sports and utility vehicles. It is apparent from the scatterplot that if only one line is used the predictions will be systematically high for sports and utility vehicles and systematically low for mid-size cars.

R11 (a) The plot reveals no clear form in the relationship between returns and month and the direction is slightly positive. The association is weak and contains outliers. The behavior of average monthly returns appears somewhat erratic and does not follow a systematic pattern based on month, but increases slightly with time.

(b) The decreased variability is evident in the scatterplot by a tightening of the data points towards the end of the 228-month period.

# LESSON 10

10.1 The population are all employed adult women. The sample are the 48 surveys which were returned.

10.3 (a) All U.S. residents.
(b) All U.S. households.
(c) All voltage regulators produced by a particular manufacturer.

10.5 Hite's sampling method is very likely to produce strong bias because: (1) Distributed surveys only to women's groups—not an SRS of the population (all women). (2) Response was voluntary. (3) Only 4.5% of the sample responded—likely to suffer from undercoverage. This method is most likely to produce results much higher than that of the population, because voluntary respondents are more likely to have stronger feelings than non-respondents.

10.7 (a)
```
Summary of Percent 65 and over
Mean of all 50 states = 12.544
Sample Mean = 12.34
```

The mean of the sample and the mean of all 50 states are close. They would be closer if the sample size were larger.
*Individual samples will vary.

(b) The sample means are clustered around the mean for all 50 states. The sample means vary around the population mean in a non-systematic, unbiased fashion. They are neither consistently high nor low. There are roughly an equal amount of sample means above and below the population mean.

10.9 (a)
```
Summary of Percent Return
Count        228
Mean      3.06395
StdDev    11.4933
```

**(b)**

```
        Summary of Mean(n = 12)
Count            25
Mean          3.985
StdDev      3.61008
Min        -1.65917
Max          14.575
```

The person with the highest average return did 11.51% better than the mean of all 228 monthly returns. The person with the lowest average return did 4.723% worse than the mean of all 228 monthly returns.

**(c)**

```
        Summary of Mean(n = 48)
Count            25
Mean         3.6444
StdDev      1.78144
Min        0.917292
Max          7.3725
```

The shape for all three histograms are a roughly normal and unimodal. The largest and smallest sample means taken from a sample size of 48 are .917292 and 7.3725 respectively. There is less difference in the highest and lowest sample means from the mean of all 228 monthly returns than from the samples of size 12. The means of the various sample sizes are fairly close approximations of the mean of all 228 monthly returns. The variability as measured by the standard deviation is less for the sample means than for the population. In addition, standard deviation of the sample means decreases as the sample size increases.

10.11 **(a)** Households without telephones and those who choose not to list their phone numbers are omitted. These households include people who cannot afford telephones, people who choose not to have telephones and people who choose not to list their numbers.

**(b)** The random digit method will include those who choose not to list their numbers in the sampling frame.

10.13 (A) produced the 80% response favoring banning contributions and (B) produced the 40% response with the same opinion. The marked difference in results can be explained by the wording of the questions. The first question offers no alternative opinion and uses words such as "huge sums" and "special interest". The second question offers an alternative opinion and uses less emotional wording.

10.15 Although 186,000 people is a large sample size, they still constitute a voluntary response sample. Voluntary respondents usually have stronger feeling towards a particular subject than non-respondents. These stronger feelings are a source of bias. Therefore the opinions of 500 people who represent a simple random sample are still more representative of the population than the opinions of 186,000 people who voluntarily offer their opinions.

## LESSON 11

11.1 This study is an observation. Although the political scientist is asking for a response, no systematic treatment is imposed upon the group. The explanatory variable in this case is sex; the response variable is the type of party.

11.3 (a) The National Halothane Study is an observation, because the type of anesthetic was not intentionally imposed to the patients. Rather, the type of anesthetic used in each operation was simply recorded and a death rate was computed. Comparing different types of treatments does not necessarily qualify a study as an experiment, if no treatment was intentionally imposed.

(b) Some variables which could be confounded with the type of anesthetic a patient receives are severity of surgery or the weight and general health of a patient. These variables may be the true cause of the different death rates which have been mistaken as the effects of the different anesthetics.

11.5 (a) The experimental units are the package liners.

(b) The treatments are the four different temperatures at which liners are sealed.

(c) The response variable is the amount of force required to open the seals.

11.7 Randomly divide the sample into four equal groups of five; one for each temperature, i.e., Group 1 = 250°, Group 2 = 275°, Group 3 = 300°, Group 4 = 325°. Seal the four groups at the appropriate temperatures and compare the average amount of force needed to separate each group.

11.9    i. Outline of Experiment: Randomly assign 20 subjects (rats) to each of the two treatment groups (new product and standard diet). Feed Group 1 only the new product for 28 days; record weight daily. Feed Group 2 only the standard diet for 28 days; record weight daily. At the end of 28 days, compare the weights of the two groups over the cycle of the experiment.

**ii.** Labeling and Randomization: Label rats A–AN. Use software to randomly assign subjects into two groups.

11.11 The second design will produce better results, because it is an experiment as opposed to an observational study. The first design is susceptible to the effects of lurking variables as the true source of variability may be due to systematic differences between men who exercise on a regular basis versus men who do not.

In addition the variation within the group which exercises could be due to systematic differences in the type or style of exercise. The second design controls these effects by using a pool of 4000 men who are similar in the beginning of the study and separated randomly into two groups, leaving only the effects of a supervised exercise program to separate them.

11.13 **(a)** Fizz Laboratories should not simply administer the drug and record patient responses, because this method does not control for the effects of lurking variables such as a placebo effect, degree of illness, etc.

**(b)** Randomly assign 30 patients into two groups; Group 1 will receive the new pain-relief medication and Group 2 will receive a sugar pill (placebo). Administer the treatments and ask one hour later, "What degree of pain relief did you experience?" Compare the results of the two groups.

**(c)** Patients should not be told which treatment they receive, because this could alter their responses to reflect a placebo effect as to the effectiveness of the medication.

**(d)** This experiment should also be double blind, because the researchers who ask, "What degree of pain relief did you experience?" must rate the responses. The ratings could be biased by a knowledge of which treatment the patient received. In addition, by knowing the treatment group of the patient a researcher could inadvertently elicit a particular response.

11.15 **(a)** * Individual results will vary.

**(b)**

| Group | Mean |
|-------|------|
| A | 108.333 |
| B | 109.513 |

The two methods have similar mean IQs.
    * Individual results will vary.

**(c)** Randomization will usually produce similar means.
    * Individual results will vary.

## REVIEW III

R1 The taste test is an experiment. The people who are asked to taste the muffins are the subjects. The treatment is that they are asked to taste two different types of muffins.

R3 The call-in polls suffer from *voluntary response bias,* because viewers are asked to call-in their opinions. The call-in polls also suffer from *undercoverage* bias, which may also account for the results favoring Reagan over Carter. The undercoverage bias stems from the fact that the number a viewer must call to voice their opinion is a 900 number which requires payment. This method is likely to miss portions of the population which have less disposable income. Since there is a correlation between higher incomes and the Republican party, the viewers most likely to call in are not only wealthier, but comparatively more likely to be Republicans.

R5 (a) This question is slanted towards a response against gun control. The wording of the question attaches a negative connotation with gun control and positive connotation to gun ownership.

(b) This question is slanted towards a response of disarmament. The wording of the question does not allow for alternative positions other than the one stated.

(c) This question is slanted towards a response in favor of economic incentives for recycling. The wording of the question sets the conditions and elicits a particular response.

R7 (a) The response variable is survival times after surgery and the explanatory variable is the type of treatment the patient received.

(b) This study is not an experiment because there was no intentional treatment imposed upon the patients. Although there are two different treatments, the patients or the doctors chose the particular type of treatment.

(c) Since the reasons for the type of treatment a patient receives are not known and experimenters did not control for other variables, we cannot determine if the type of treatment alone accounted for the variability in survival times. Therefore, the effects of outside variables such as the severity of the cancer could be confounded as the effects of a particular treatment.

R9 (a) The population of interest is all students at a given school. This should include part-time students as they are also affected by a school's sexual harassment policy.

(b) The sample should include a *block design.* It is suspected that men and women will systematically differ in their opinions about a sexual harassment policy. In order to control for these systematic differences, we should separate the sample using sex as the blocking variable and randomize, if necessary, within each group.

(c) The practical difficulties in such an experiment could include: willingness of subjects to participate, finding a representative sample and truthfulness of responses from the subjects. Contacting the students raises a particular concern, because the method we choose could affect the response. If we call the students, they may feel that their identities are not anonymous which may lead to lack of cooperation or truthfulness. On the other hand, if we post an ad soliciting anonymous responses the experiment may suffer from voluntary response bias. One way to contact the subjects which would minimize the above effects could be through a double blind experiment where neither the students nor the people conducting the experiment know the true subject matter or the identities of the people involved. This could be achieved by shrouding the true questions among other questions regarding school policy and by randomly allocating only phone numbers to the researchers. Although this method is not perfect, it illustrates some of the compromises and shortcomings of experiments and surveys.

R11 (a) The results of the study could suffer from adverse selection, where particular groups of students would self select into the two options based on systematic differences in ability or other outside variables. Therefore, the two groups would not have proper randomization which would lead our results not to be a trustworthy indicator of the differences between the two teaching methods.

(b) Randomly assign 75 students to each of the two treatment groups; conduct the two classes; at the end of the semester administer and compare the results of the same final.

(c) Use software to randomly divide the 150 subjects into two equally sized groups.

## LESSON 12

12.1 Based on only 50 spins, an estimate of the probability of heads is $x/50$—where $x$ represents the number of heads.

12.3 (a) An estimate of the probability of heads on first toss is $\frac{1}{2}$. The expected value is $1 \times \frac{1}{2} = \frac{1}{2}$.

(b) An estimate of the probability that the first head lands on an odd number of tosses = $x/50$, where $x$ represents the count of odd numbers.

12.5 **(a)** An estimate of the probability that at least one tack lands point up is $x/50$, where $x$ represents the number of times at least one tack landed point up.

**(b)** An estimate of the probability that more than one tack lands point up is $x/50$, where $x$ represents the number of times at least two tacks landed point up.

12.7 Over the course of many hands of five-card poker, you will get three of a kind about 2% of the time.

12.9 **(a)** 51% of the shots resulted in hits. This is represented by the count of 1's in a Bernoulli trial.

**(b)** The longest succession of hits was 9. The longest succession of misses was 6.

  \* Individual responses will vary.

## LESSON 13

13.1 **(a)** 0

**(b)** 1

**(c)** .01

**(d)** .60

13.3 **(a)** $P(\text{Blue}) = 1 - [P(\text{Brown}) + P(\text{Red}) + P(\text{Yellow}) + P(\text{Green}) + P(\text{Orange})] = 1 - [.3 + .2 + .2 + .1 + .1] = .1$ or 10%

**(b)** $P(\text{Blue}) = 1 - [P(\text{Brown}) + P(\text{Red}) + P(\text{Yellow}) + P(\text{Green}) + P(\text{Orange})] = 1 - [.2 + .1 + .2 + .1 + .1] = .3$ or 30%

**(c)** Recall: Addition rule for disjoint events: $P(A \text{ or } B) = P(A) + P(B)$.

  Plain M&M : $P(\text{Red, Yellow or Orange}) = P(\text{Red}) + P(\text{Yellow}) + P(\text{Orange}) = .2 + .2 + .1 = .5$ or 50%.

  Peanut M&M: $P(\text{Red, Yellow or Orange}) = P(\text{Red}) + P(\text{Yellow}) + P(\text{Orange}) = .1 + .2 + .1 = .3$ or 30%.

13.5 **(a)** The probability that a woman states that her husband does less than his fair share must be 27%, because a probability model must sum to 1.0. Therefore, $1 - [P(\text{Does more}) + P(\text{Does fair})] = 1 - [.12 + .61] = .27$ or 27%.

**(b)** $P(\text{At least his fair share}) = P(\text{Does more than his fair share}) + P(\text{Does his fair share}) = .12 + .61 = .73$ or 73%.

13.7 **(a)** $P(\text{Red}) = .4737$

**(b)** $P(\text{Green}) = .0562$

**(c)** The probability of winning the first spin and losing the second spin is: $(.4327)(.5673) = .2455$ or 24.55%.

13.9 (a) $P(\text{Cardiovascular}) + P(\text{Cancer}) = .42 + .24 = .66$ or 66%.

(b) $P(\text{Cardiovascular and Cardiovascular}) = P(\text{Cardiovascular}) \cdot P(\text{Cardiovascular}) = (.42)(.42) = .1764$ or 17.64%

13.11 (a) $S = \{\text{Grows, Fails to Grow}\}$

(b) $S = \{0 \ldots \text{Death}\}$

(c) $S = \{F, D, C, B, A\}$ Assuming no partial letter grades.

(d) $S = \{\text{Basket, Miss})$

(e) $S = \{1, 2, 3, 4, 5, 6, 7\}$ Assuming only integers.

13.13 (a) Since the total area under the curve must equal 1.0 the height must be the reciprocal of the range. The range is 2, the height must be $\frac{1}{2}$ or .5.

(b) $P(y \leq 1) = .5(2 - 1) = .5$ or 50%

(c) $P(0.5 < y < 1.3) = .5(1.3 - .5) = .4$ or 40%

(d) $P(y \geq .8) = .5(2 - .8) = .6$ or 60%

## LESSON 14

14.1 (a) $P(\text{At least 65}) = .1834$

(b) $P(\text{Married} \mid \text{at least 65}) = .4253$

(c) i. 7,767 women are both married and in the 65 and over age group.

ii. $P(\text{Married and At least 65}) = \frac{7,767}{99,585} = .078$

(d) $P(\text{At least 65})P(\text{Married} \mid \text{At least 65}) = P(\text{Married and At least 65}) = (.1834)(.4253) = .078$

14.3 (a) $P(\text{Age 18 to 24} \mid \text{Married}) = .052$

(b) married; 18 to 24 years of age.

(c) .052 is the proportion of women who are 18–24 years of age among those women who are married.

14.5 (a) $P(\text{Male}) = .4736$

(b) $P(\text{Bachelors} \mid \text{Male}) = .687$

(c) $P(\text{Bachelors and Male}) = .3253$

14.7 $P(\text{Defective chip}) = .05; P(\text{Non-defective chip}) = 1 - .05 = .95;$ $P(\text{12 Non-defective chips}) = .95^{12} = .54$ or 54%

14.9 $P(\text{Failing bulb}) = .02; P(\text{Bulb staying lit}) = 1 - .02 = .98;$ $P(\text{String of lights staying lit}) = P(\text{20 Bulbs staying lit}) = .98^{20} = .668$ or 66.8%

14.11 (a) i. $P(\text{Under 65}) = .321 + .124 = .445$

ii. $P(\text{65 or Over}) = .365 + .190 = .555$

(b) i. $P(\text{Tests done}) = .321 + .365 = .686$

ii. $P(\text{Tests not done}) = .124 + .190 = .314$

(c) If the fact that the patients are 65 or over is independent of whether the test was done:

$P(\text{Test done}) = P(\text{Test done} \mid 65 \text{ or over})$

$P(\text{Test done}) = .686, \ P(\text{Test done} \mid 65 \text{ or over}) = .6577$

or

$P(\text{Test done and } 65 \text{ or over}) = P(\text{Test done})P(65 \text{ or over})$

$P(\text{Test done})P(65 \text{ or over}) = (.686)(.555) = .381.$

The actual $P(\text{Test done and } 65 \text{ or over}) = .365.$
Since the two probabilities in either method are not equal the two events are not independent. Since the probability of the tests being done is higher when assuming independence, the tests are done less frequently for older patients than if the testing were independent of age.

14.13 (a)  i. $P(\text{Firearm} \mid \text{Male}) = .646$
    ii. $P(\text{Firearm} \mid \text{Female}) = .393$

(b) The results in (a) indicate that men are more likely to use a firearm when committing suicide than women.

14.15 $P(\text{Success and No Infection}) = .86$

## LESSON 15

15.1 (a) 1%

(b) 1) Each probability is between 0 and 1. 2) $.48 + .38 + .08 + .05 + .01 = 1.0$

(c) $P(X \le 3) = P(X = 1) + P(X = 2) + P(X = 3) = .48 + .38 + .08 = .94$

(d) $P(X < 3) = P(X = 1) + P(X = 2) = .48 + .38 = .86$

(e) Probability that "a son of a lower-class father reaches one of the two highest classes": $P(X \ge 4) = P(X = 4) + P(X = 5) = .05 + .01 = .06$

15.3 $P(X < 3) = P(X = 1) + P(X = 2) = \frac{1}{6} + \frac{1}{6} = \frac{2}{6} = \frac{1}{3}$

15.5 (a) $(.5)(1 - 0)(1) + (.5)(2 - 1)(1) = 1.0$

(b) $P(Y < 1) = (.5)(1)(1) = .5$

(c) $P(Y < .5) = (.5)(.5)(.5) = .125$

15.7 7.2%; statistic.

15.9 48%; statistic, 52%; parameter.

15.11  The respective means are: 64.21, 64.28, 64.63, 64.42, 63.58, 63.72. The mean of each successive number should be plotted on the vertical axis and the numbers 1–6 which represent $n$ should be plotted on the horizontal axis.

15.13  On average the probability of a run of three or more heads or tails in ten tosses is .80. The probability of no runs of three or more heads or tails in ten tosses is .20. The mean outcome is $.80(-1) + .20(2) = -.40$. Since the mean outcome is negative, it is not advantageous to play.

## LESSON 16

16.1  **(a)** $\hat{p} = x/20$, where $x$ represents the number of heads.
**(b)** Yes, the center of the distribution is close to .50.

16.3  **(a)** Use software to collect 100 simple random samples from the survival time data.
**(b)**
```
Summary of Survival Times*
Count          72
Mean      141.847

Summary of Mean(Summary1)
Count          100
Mean      143.958
```

The center of the sampling distribution of $\bar{y}$ is within 2 days of $\mu$.
   * Individual responses will vary.

**(c)**
```
Summary of Survival Times*
StdDev       109.209

Summary of Mean(Summary1)
StdDev        27.011
```

The expected standard deviation of the 100 sample means is expected to be: $\sigma_{\bar{y}} = \frac{\sigma}{\sqrt{n}} = \frac{109.209}{\sqrt{12}} = 31.53$. The standard deviation of the 100 sample means came within 4.5 days of the expected standard deviation.
   * Individual responses will vary.

**(d)** The normal probability plot of the sample means shows a roughly normal distribution. The normal probability plot for the population shows a distinct right skew.

16.5 (a) $P(\text{Returns} < 0) = P(Z < .13) = .5517$ or 55.17%.
Approximately 55.17% of the stocks lost money.

(b) The mean and standard deviation of 5 stocks chosen at random from the N.Y.S.E would be $-.035$ and $.26/\sqrt{5}$ respectively.

(c) $P(\bar{y} < 0) = P(Z < .30) = .6179$ or 61.79%.
Approximately 61.79% of the returns in portfolios of 5 random stocks lost money.

The difference between (a) and (c) is that (a) describes the percent of all stocks which lost money, whereas (c) describes the percent of the average of five randomly chosen stocks which lost money.

16.7 Filter Weight $\sim N(123, .08)$

(a) $\bar{y}_{Weight} \sim N\left(123, \frac{.08}{\sqrt{3}}\right)$

(b) $P(\text{Reported Filter Weight} \geq 124) = P(Z \geq 21.17) = 0$

16.9 Egg Weight $\sim N(65, 5)$

$P(750\text{ g} < \text{Carton of Eggs} < 825\text{ g}) = P(62.5 < \bar{y} < 68.75) = P(-1.73 < Z < 2.6). = .9953 - .0418 = .9535$ or 95.35%

The probability that a mean weight from a random sample of size 12 falls between 750 and 825 is 95.35%.

16.11 Although the distribution of NOX is not stated, the distribution of $\bar{y}$ is assumed to be normal because of the "large" sample size $n = 125$. $\bar{y} \sim N\left(.9, \frac{.15}{\sqrt{125}}\right)$,   $P(Z > z*) = .01$,   $z* = 2.33$,   $L = .93126$.

## REVIEW IV

R1 (a) $S = \{M, F\}$

(b) $S = \{1, 2, 3\}$

R3 Model A is a legitimate probability model, because all probabilities are between 0–1 and the probabilities of all possible outcomes sum to 1.0.

Although Model B is probably an unfair coin, it is a legitimate probability model, because all probabilities are between 0–1 and the probabilities of all possible outcomes sum to 1.0.

Model C is not a legitimate probability model, because of the presence of negative values for two outcomes.

Model D is not a legitimate probability model, because the sum of the probabilities of all possible outcomes is 1.5 which is greater than 1.0.

**R5** **(a)** Using the complement rule, the probability of not forested land is $1 - P(\text{forest}) = 1 - .35 = .65$ or 65%.

**(b)** Using the addition rule for disjoint events the probability that the land is either forest or pasture $= P(\text{Forest}) + P(\text{Pasture}) = .03 + .35 = .38$ or 38%.

**(c)** Using both the rule for complements and the addition rule for disjoint events the probability that a randomly chosen acre in Canada is something other than forest or pasture $= 1 - P(\text{forest or pasture}) = 1 - .38 = .62$ or 62%.

**R7** Using the complement rule, the probability of moving on to one of the higher classes is $= 1 - P(\text{remains in the lower class}) = 1 - .46 = .54$ or 54%.

**R9** **(a)** The standard deviation of the mean result is 5.77.

**(b)** $n = \left(\dfrac{\sigma}{\text{desired standard deviation of sample means}}\right)^2 = \left(\dfrac{10}{5}\right)^2 = 4$

**(c)** Since there is less variation from sample to sample than from individual to individual, we can obtain a more precise prediction of the true population mean.

**R11** The bold numbers are statistics, because they are taken from a sample and are not the mean and standard deviation of the scores of all students of high academic ability.

**R13** **(a)** $P(\text{WAIS} \geq 105) = P(Z \geq .33) = 1 - .6293 = .3707$ or 37.07%

**(b)** The mean of the sample means will be centered at the true population mean $= 100$ and the standard deviation of the sample means $= 15/\sqrt{60} = 1.936$.

**(c)** $P(\text{Average WAIS} \geq 105) = P(Z \geq 2.58) = 1 - .9951 = .0049$ or .49%

**(d)** The answer to (a) would be affected dramatically, because the math used in determining the probability of the event in question is based upon a normal density curve as the mathematical model for unit to unit variation. The answers to (b) and (c) would not be affected to the same degree if at all, because the sampling distribution of sample means remains normal regardless of the population distribution—as long as the sample size is sufficiently large.

**R15** $\bar{y} \sim N(.75°, .5°/\sqrt{4})$

$P(74.5 \geq \bar{y} \text{ or } \bar{y} \leq 75.5) = P(Z \leq -2 \text{ or } Z \geq 2) = .0228 + (1 - .0228) = .0456$ or 4.56%

The same answer can be found using the 68-95-99.7 rule.

**R17** RMPT $\sim N(13.6, 3.1/\sqrt{22})$

$P(\bar{y} < \bar{y}^*) = P(Z < -1.65) = .05, L = 12.51$

R19  $P$(renegotiation and dollar falls) $= .40(.80) = .32$ or 32%

R21

| Event | Definition | Probability |
|-------|------------|-------------|
| A | B, B, G, G | .0625 |
| B | B, G, B, G | .0625 |
| C | G, G, B, B | .0625 |
| D | G, B, G, B | .0625 |
| E | G, B, B, G | .0625 |
| F | B, G, G, B | .0625 |
| G | B, G, G, G | .0625 |
| H | G, G, G, B | .0625 |
| I | G, B, G, G | .0625 |
| J | G, G, B, B | .0625 |
| K | B, B, B, G | .0625 |
| L | B, B, G, B | .0625 |
| M | B, G, B, B | .0625 |
| N | G, B, B, B | .0625 |
| O | B, B, B, B | .0625 |
| P | G, G, G, G | .0625 |
|   | Total:     | 1.0 |

(a)  $P$(oldest child is a girl) $= P(A) + P(B) + P(E) + P(G) + P(I) + P(K) + P(P) = .4375$ or 43.75%

(b)  $P$(at least 3 boys) $= P(K) + P(L) + P(M) + P(N) + P(O) = .3125$ or 31.25%

(c)  $P$(at least 3 same sex children) $= P(G) + P(H) + P(I) + P(J) + P(K) + P(L) + P(M) + P(N) + P(O) + P(P) = .625$ or 62.5%

## LESSON 17

17.1  For a, b and c use the ActivStats normal density tool. Set the mean to 0 and the standard deviation to 1. Maintaining symmetry, slide the flags until the desired level of confidence is achieved and select show $z$-score. The $z$-scores should agree with those in the book (up to roundoff error).

17.3  99% C.I. in example 17.3 $= (.8303, .8505)$
The interval created is: $.8505 - .8303 = .0202$
When $n = 12$, a 99% C.I. $= (.83534, .84546)$
The interval becomes $.84546 - .83534 = .0101$
$.0101/.0202 = .5$

Therefore the interval created when $n = 12$ is half the interval created when $n = 3$. This process occurs because the square root of 12 is double the square root of 3, i.e.,
$$2\sqrt{3} = \sqrt{12}$$

17.5  The student is incorrect; the student is using the confidence interval to make an assessment about the probability of individuals. A confidence interval is based upon the *sampling distribution* of a particular statistic.

17.7  **(a)** The mean of the sampling distribution for $\bar{y}$ is the population parameter $\mu$. The standard deviation of the sampling distribution of $\bar{y}$ is $\sigma/\sqrt{n} = .4/\sqrt{50} \approx .057$.

**(b)** Sketch a bell shaped curve which is unimodal and symmetric. Label the sketch as below:

$$\mu - 3\tfrac{.4}{\sqrt{50}} \quad \mu - 2\tfrac{.4}{\sqrt{50}} \quad \mu - \tfrac{.4}{\sqrt{50}} \quad \mu \quad \mu + \tfrac{.4}{\sqrt{50}} \quad \mu + 2\tfrac{.4}{\sqrt{50}} \quad \mu + 3\tfrac{.4}{\sqrt{50}}$$

**(c)** $m = z^* \tfrac{\sigma}{\sqrt{n}} = 2\left(\tfrac{.4}{\sqrt{50}}\right) = .113$

**(d)** 95%

**(e)** Individual results will vary.

17.9  **(a)** The displays show a unimodal and slightly right skewed distribution.

**(b)** With 90.00% Confidence, $223.97726 < \mu(\text{crank}) < 224.02661$

17.11  **(a)** 95% C.I. $= 275 \pm 1.96\tfrac{60}{\sqrt{1077}} = 275 \pm 3.58 = (271.42, 278.58)$

**(b)** 90% C.I. $= 275 \pm 1.65\tfrac{60}{\sqrt{1077}} = 275 \pm 3.02 = (271.99, 278.02)$

   99% C.I. $= 275 \pm 2.576\tfrac{60}{\sqrt{1077}} = 275 \pm 4.79 = (270.21, 279.79)$

**(c)**

| C | Interval |
|------|------|
| 90% | 6.03 |
| 95% | 7.16 |
| 99% | 9.58 |

Increasing the confidence increases the width of the interval. This effect demonstrates the tradeoff between precision and confidence.

17.13 (a) $8740 \pm 1.96 \frac{1125}{\sqrt{958}} = 8740 \pm 71.24 = (8668.76, 8811.24)$

Yes, the station's calculation is correct (up to roundoff error).

(b) Although the sample is large, it may not be representative of all the city's citizens. The radio show could only get their listeners to respond and the response was voluntary.

17.15 $n = \left(\frac{z^*\sigma}{m}\right)^2$    $n = \left(\frac{1.96(.0002)}{.0001}\right)^2 = 15.3664$

We would need a sample size of 16 in order to get a margin of error of about .0001.

## LESSON 18

18.1 With 95.00% Confidence, $18.681849 < \mu(\text{D-glucose}) < 70.198151$

18.3 (a)  i. Using ActivStats: $t^* = 2.151$ with 14 degrees of freedom.
         ii. Table B: $t^* = 2.145$ with 14 degrees of freedom.

    (b)  i. Using ActivStats: $t^* = .688$ with 19 degrees of freedom.
         ii. Table B: $t^* = .688$ with 19 degrees of freedom.

    (c)  i. Using ActivStats: $-t^* = -1.312, t^* = 1.312$ with 25 degrees of freedom.
         ii. Table B: $-t^* = -1.316, t^* = 1.316$ with 25 degrees of freedom.

18.5 (a)

| Mean | 5.36667 |
|---|---|
| StdErr | 0.271621 |

    (b) With 90.00% confidence, $1.59809 < \mu(\text{Refractory Period}) < 1.90191$

18.7 (a) Neither the histogram nor the boxplot show outliers, but both graphs suggest that the data is skewed right. Because the sample size is not large the skewed distribution may affect the validity of the $t$ procedures. This will be discussed in further detail in chapter 20.

    (b) With 90.00% Confidence, $223.97485 < \mu(\text{crank}) < 224.02902$

18.9 Neither the histogram nor the normal probability plot show problems with non-normality and there are no clear outliers. The $t$ procedures appear to be appropriate for the data.

With 99.00% Confidence, $5.3345583 < \mu(\text{Earth Density}) < 5.5613038$

18.11 (a) The normal probability plot shows a slightly upward curving trend which is indicative of a slight right skew. The sample size is sufficiently large to compensate for the deviation from normality. The $t$-procedures will still be appropriate.

**(b)** With 90.00% Confidence, $22.209084 < \mu(\text{Active}) < 26.624249$

## REVIEW V

**R1** $\bar{y} \sim N(12, .01/\sqrt{25})$

**R3** $\bar{y} = 22$ nanograms, $m = 2$ nanograms.

The margin of error represents a level of accuracy for given level of confidence. In our example, we are 90% confident that the true mean response lies within 2 nanograms of our estimate.

**R5** **(a)** The abbreviation SEM stands for standard error of the mean.

**(b)** $s = \text{SEM}\sqrt{n} = .01(1.732) = .01732$

**(c)** 90% C.I. $= .84 \pm 1.645(.01) = (.824, .856)$

**R7** **(a)** With 95.00% Confidence, $123.00947 < \mu(\text{Dust Weight}) < 123.19053$

**(b)** With 95.00% Confidence, $121.60952 < \mu(\text{Dust Weight}) < 124.59048$

**(c)** Not knowing $\sigma$ increases the interval.

**R9** $H_0$: $\mu = 18$

$H_a$: $\mu < 18$

**R11** **(a)** The Central Limit Theorem states that even if the distribution of the population is not normally distributed, the sample means will have an approximately normal distribution as long as the sample size is sufficiently large. Since 104 is a "large" sample, we can assume that the $t$ procedures are appropriate for the data.

**(b)** 99% C.I. $= .069 \pm 1.962(.0171) = (.0354, .1026)$

The essential condition that the data must satisfy is that the 104 corporate CEOs were an SRS of all corporate CEOs.

**R13** Although the results were significant, we cannot conclude that the consumers strongly prefer design A. The study was not an experiment which ruled out the possibility of other factors. We can only conclude that the consumers bought significantly more of design A.

**R15** **(a)** 5%

**(b)** It is not surprising because we can expect 5% of the tests to show significant results when there is, in fact, no difference between the two groups. (Five percent of 77 is 3.85)

**R17** **(a)** The margin of error would decrease.

**(b)** The $P$-value would decrease.

## LESSON 19

19.1 (a) If the null hypothesis were true the sampling distribution of the means would be: $\bar{y} \sim N(115, 30/\sqrt{25})$

(b) The result $\bar{y} = 118.6$ is not good evidence that the true population mean is greater than 115, because it is not very far from the specified value in the null hypothesis. The distance is less than 1 standard deviation from the hypothesized value. This indicates that if the true mean were 115 observing a sample mean of 118.6 would not be a rare event. The result $\bar{y} = 125.7$, however, is considerably greater than the hypothesized value which indicates observing this sample mean would be rare if the true mean were 115.

(c) The 27.43% of the area under the curve is shaded to the left.

(d) $\bar{y} = 118.6$; $p$-value $= .2743$    $\bar{y} = 125.7$; $p$-value $= .0375$

19.3 $H_0$: $\mu = 5$ mm

$H_a$: $\mu \neq 5$ mm

19.5 $H_0$: $\mu = 50$

$H_a$: $\mu < 50$

19.7 $H_0$: $\mu = 224$ mm

$H_a$: $\mu \neq 224$ mm

$\bar{y} = 224.002, n = 16, \sigma = .06$

$z = .13, P(Z < -.13 \text{ or } Z > .13) = .4483 + .4483 = .8966$ or 89.66%

Since $P$ is large we have insufficient evidence against the null hypothesis. There is not enough evidence to support the claim that the process mean is different from the target value 224 mm.

19.9 (a) 99% C.I. $= 145 \pm 2.33(2.0656) = (140.19, 149.81)$

(b) $H_0$: $\mu = 140$

$H_a$: $\mu > 140$

$z = 2.42, P(Z > 2.42) = .0078$

Conclusion: Since $P$ is low, there is insufficient evidence to support the null hypothesis. It is likely that the true mean cellulose content is greater than 140 mg/g.

(c) The procedures carried out in (a) and (b) are valid if:

1. The samples are a simple random sample (SRS) of the population of interest.

2. Since $n$ is small (15), we must assume that the population is normally distributed.

19.11 If there were no difference between the true mean level of all church attenders and nonattenders, a difference as high or higher than that found between the two sample groups would occur less than 5% of the time. Therefore the data reveal that the difference in the level of ethnocentrism between church attenders and nonattenders is too high to be attributed to chance alone.

19.13 Although many people use the student's definition as a practical way of interpreting the $p$-value, it is not correct. Assuming the null hypothesis is true, a $p$-value represents the probability of observing an event which is as extreme or more extreme than that of the test statistic. Therefore a $P$-value of $P = .03$ indicates that if the null hypothesis were true, a sample statistic as extreme or more extreme than the test statistic would only occur 3% of the time.

19.15 A $P$-value of less than .1% is highly significant, because it represents the probability of observing differences as extreme or more extreme as the differences observed by our test subjects if there were truly no perceivable differences among the various tobacco companies' advertisements. It is good evidence that there is a perceivable difference among all advertisements, because the sample is representative of all advertisements. As long as the differences are truly perceivable increasing the sample size would only strengthen the conclusion.

19.17 **(a)** $z = 1.64$, $P$-value $= P(\bar{x} > 491.4) = P(Z > 1.64) = .0505$

  Fail to reject $H_0$ at the .05 level of significance. The mean score of the children using the training program was not significantly higher than the mean score of those who did not use the program.

  **(b)** $z = 1.65$, $P$-value $= P(\bar{x} > 491.5) = P(Z > 1.65) = .0495$

  Reject $H_0$ at the .05 level of significance. The mean score of the children using the training program was significantly higher than 475.

19.19 A test of significance answers only: (b) Is the observed effect due to chance?

# LESSON 20

20.1

```
          Summary of Sweetness
          Mean       1.02
          StdDev   1.1961
```

Sweetness: Test $H_0$: $\mu$(Sweetness) $= 0$ vs. $H_a$: $\mu$(Sweetness) $> 0$
$t$-Statistic $= 2.697$ w/9 df
Reject $H_0$ at Alpha $= 0.05$
$p = 0.0123$

20.3 $H_0$:   $\mu = 1.3$

$H_a$:   $\mu > 1.3$
Refractory Period:
Test $H_0$: $\mu$(Refractory Period) $= 1.3$ vs. Ha: $\mu$(Refractory Period) $> 1.3$
$t$-Statistic $= 6.971$ w/3 df
$p = 0.0030$
Reject $H_0$ at Alpha $= 0.05$. The data show evidence that the true mean refractory period for poisoned rats is above 1.3 milliseconds.

20.5 (a) $H_0$:   $\mu_d = 0$

$H_a$:   $\mu_d > 0$
$t = 43.47$
$t_{(.01,199)} = 2.364$ using 100 degrees of freedom.
Reject $H_0$ at the .01 level of significance. The data show evidence that there is an increase in the amount charged after the removal of the annual fee.

(b) 99% C.I. $= (313.95, 350.05)$

(c) The criticism comes primarily from the fact that the bank did not isolate the effects of removing the fee from the other possible factors which could lead to increased spending. In order to avoid this criticism, the bank could perform a comparative experiment which controls for the effects of outside variables. Collect an SRS of 200 people who charge at least $2400 a year. Randomly divide the sample into two groups; Group 1 will retain the annual fee while Group 2 will not be charged the annual fee. By randomly dividing the two groups, we are randomizing across other variables with the only effective difference being the annual fee. Track both groups' expenditures within the same year. By comparing spending patterns within the same year we are holding constant other factors which may vary across different years. At the end of the year compare the two groups. Determining whether there is a statistically significant difference between the two groups will be discussed further in Chapter 21.

20.7 $H_0$:   $\mu = 5.517$

$H_a$:   $\mu \neq 5.517$

Earth Density:

Test $H_0$: $\mu$(Earth Density) $= 5.517$ vs. Ha: $\mu$(Earth Density) $\neq$ 5.517

$t$-Statistic $= -1.683$ w/28 df

$p = 0.1034$

At the .01 level of significance, we must fail to reject the null hypothesis. The data show evidence that there is not a statistically significant difference between Cavendish's results and the modern figure for the density of the earth.

20.9 **(a)** The numbers 5.71 and 2.82 are the mean and standard error of the mean respectively.

**(b)** $H_0$: $\mu = 0$

$H_a$: $\mu < 0$

Test $H_0$: $\mu$(Difference) $= 0$ vs. $H_a$: $\mu$(Difference) $< 0$

$t$-Statistic $= -2.024$ w/13 df

$p = 0.0320$

Reject $H_0$ at the .05 level of significance.

Fail to Reject $H_0$ at the .01 level of significance.

The data are significant at the .05 level of significance, but not at the .01 level of significance.

**(c)** With 90.00% Confidence, $-10.714388 < \mu$(Difference) $< -0.7141836$. We are 90% certain that the interval we have created captures the true mean difference.

20.11 **(a)** The parameter $m$ in this setting is the mean difference in pounds per plant between Variety A and Variety B.

**(b)** $H_0$: $\mu_d = 0$

$H_a$: $\mu_d > 0$

**(c)** $t = 1.29$, $P$-value $= .115$

The large $P$-value suggests that the data do not show a statistically significant difference between the two corn yields.

20.13 The distribution of particulate levels for the city is approximately normal and unimodal. There appears to be one high outlier.

An estimate of the true mean particulate levels for the city is $54.0667 \pm 7.247$. The margin of error, 7.247, constitutes a 95% confidence interval.

20.15 The scatterplot shows a linear, positive, and strong relationship between city and rural particulate levels. There is one potential outlier with a city particulate level of 69 given its rural value of 51. The graph suggests that least-squares regression is appropriate.

The regression results show that 95.1% of the variation in city particulate levels has been explained by the linear relationship.

For a day on which a rural reading was 88, the predicted city particulate level is: City Particulate Levels = $-2.58 + 1.09(88)$ = 93.34.

## LESSON 21

### 21.1

| Summary Statistics: | | | |
|---|---|---|---|
| **Group** | **Total Cases** | **Mean** | **StdDev** |
| Poultry | 17 | 459 | 84.7393 |
| Beef | 20 | 401.15 | 102.435 |

$$SE_{\bar{y}_1 - \bar{y}_2} = 30.77, \quad t = 1.88$$

**21.3 (a)** $H_0$:  $\mu_{\text{Control}} - \mu_{\text{Treatment}} = 0$

$H_a$:  $\mu_{\text{Control}} - \mu_{\text{Treatment}} > 0$
$t = 2.45$, $P$-value $= .0087$
Reject $H_0$ at the .05 and .01 levels of significance. The data show evidence that beta-blockers reduce pulse rates.
**(b)** 99% C.I. $= (.922, 9.277)$

**21.5** The conclusion is that ratio of current assets to current liabilities is significantly higher for healthy businesses than for businesses that went bankrupt a year later. We can conclude this because of the extremely high value of $t = 7.36$.

**21.7 (a)** Single sample.
**(b)** Matched pairs.

**21.9 (a)** The graphical displays of the data show:

Women's SSHA Scores: Right skewed, unimodal, with one potential outlier (200).

Men's SSHA Scores: Approximately normal, unimodal with no outliers.

Although the distribution for women is right skewed and contains one potential outlier, the use of a $t$ procedure is still acceptable, because the sum of the sample sizes is "large."
**(b)** $H_0$:  $\mu 1 - \mu 2 = 0$
$H_a$:  $\mu 1 - \mu 2 > 0$

Women $-$ Men:

Test $H_0$: $\mu$(Women) $- \mu$(Men) $= 0$ vs. $H_a$: $\mu$(Women) $- \mu$(Men) $> 0$

$t$-Statistic $= 2.056$ w/35 df

$p = 0.0236$

Reject $H_0$ at Alpha $= 0.05$. The data show evidence that the true mean for women is higher than that for men.

(c) With 90.00% Confidence, $3.5376862 < \mu$(Women) $- \mu$(Men) $< 36.073425$

21.11 $H_0$:  $\mu 1 - \mu 2 = 0$

$H_a$:  $\mu 1 - \mu 2 > 0$

Active $-$ Passive:

Test $H_0$: $\mu$(Active) $- \mu$(Passive) $= 0$ vs. $H_a$: $\mu$(Active) $- \mu$(Passive) $> 0$

$t$-Statistic $= 4.282$ w/39 df

$p \leq 0.0001$

There is good evidence that the active learning group is superior to the passive learning group ($P$-value $\leq .0001$)

21.13 (a) 95% C.I. $= (-1.215, 7.216)$

(b) We are 95% confident that the interval we created frames the true mean change in sales. Since this interval contains 0 and negative numbers, we cannot be sure the change was positive.

21.15 When conducting multiple tests, each test contains the probability $\alpha$ that a type I error can occur. Therefore, about 5% of tests would show significant results, even if there are no differences between adults and children.

## LESSON 22

22.1 (a) Since the variability of $\hat{p}$ is a function of $\hat{p}$, the variability from state to state will change as $\hat{p}$ changes regardless of the sample size.

(b) Since 1% is not a static number for each state and the sample proportion $\hat{p}$ is different for each state the variability will differ from state to state.

22.3 $\hat{p} = 54/639 = .0845$, 90% C.I. $= (.0664, .1026)$

22.5 (a) $\hat{p} = 132/200 = .66$, 95% C.I. $= (.5943, .7257)$

(b) $H_0$:  $p = .73$

$H_a$:  $p \neq .73$

$z = -2.23$, P-value $= P(Z < -2.23$ or $Z > 2.23) = 2(.0129) = .258$ or 25.8%

The high P-value of .258 indicates that the true proportion of first year students at this university who think being very well-off is important does not differ significantly from the national value.

(c) It is assumed that the procedures used in (a) and (b) are appropriate if $n(1 - p_o) > 10$. Since $200(1 - .73) = 54$, the $z$ procedures for one sample are appropriate.

22.7 (a) $\hat{p} = 22/148 = .15$; 95% C.I. $= (.0925, .2075)$

(b) This is unlikely because the businesses which responded are not an SRS of all small businesses.

22.9 (a) $H_0$:  $p_{\text{acetaminophen}} - p_{\text{ibuprofen}} = 0$

$H_a$:  $p_{\text{acetaminophen}} - p_{\text{ibuprofen}} \neq 0$

$\hat{p}_{\text{acetaminophen}} = 44/650 = .068$

$\hat{p}_{\text{ibuprofen}} \quad = 49/347 = .14$

**pooled sample proportion;** $\hat{p} = (44 + 49)/(650 + 347) = 93/997 = .093$

Checking conditions necessary for $z$ procedures:

$n_1\hat{p} = 650(.093) = 60.45, \quad n_1(1 - \hat{p}) = 650(.907) = 589.55, \quad n_2\hat{p} = 347(.093) = 32.271, \quad n_2(1 - \hat{p}) = 347(.907) = 314.729$

Since all values are over 5, the conditions necessary to perform a two-sample test have been met.

(b) $z = -3.73$, P-value $= P(Z < -3.73$ or $Z > 3.73) = 0$. The data show very strong evidence that the proportions of people experiencing discomfort are not equal for the two types of pain relievers.

22.11 $\hat{p}_{\text{complainers}} = 54/639 = .0845$

$\hat{p}_{\text{control}} \quad = 22/743 = .03$

(a) 90% C.I. $= (.0297, .0793)$

(b) Checking conditions necessary for $z$ procedures:

$n_1\hat{p}_1 = 639(.0845) = 54, \quad n_1(1-\hat{p}_1) = 639(.9155) = 585, \quad n_2\hat{p}_2 = 743(.03) = 22.29, \quad n_2(1 - \hat{p}_2) = 743(.97) = 720.71$

Since all values are over 5, the conditions necessary to perform a two sample confidence interval have been met.

22.13 $H_0$:  $p_{\text{men}} - p_{\text{women}} = 0$

$H_a$:  $p_{\text{men}} - p_{\text{women}} \neq 0$

$\hat{p}_{men}$   $= 15/106 = .14$

$\hat{p}_{women} = 7/42 = .17$

**pooled sample proportion;** $\hat{p} = (15 + 7)/(106 + 42) = 22/148 = .1486$

Checking conditions necessary for $z$ procedures:

$n_1 \hat{p} = 106(.1486) = 15.75$,   $n_1(1 - \hat{p}) = 106(.8514) = 90.25$,   $n_2 \hat{p} = 42(.1486) = 6.24$,   $n_2(1 - \hat{p}) = 42(.8514) = 35.76$

Since all values are over 5, the conditions necessary to perform a two-sample test have been met.

$z = -.46$, P-value $= P(Z < -.46 \text{ or } Z > .46) = 2(.3228) = .6456 \text{ or } 64.56 \%$

The data do not provide strong evidence that the proportions of small business failures in this class are different between the two sexes.

22.15 **(a)** $H_0$:   $p_{men} - p_{women} = 0$

$H_a$:   $p_{men} - p_{women} \neq 0$

$\hat{p}_{men}$   $= 775/840 = .92$

$\hat{p}_{women} = 680/1077 = .63$

99% C.I. $= (.245, .335)$

Since 0 is not contained within the 99% confidence interval of the two means, there is a statistically significant difference between the two proportions at the .01 level of significance.

**(b)** $H_0$:   $\mu_{men} - \mu_{women} = 0$

$H_a$:   $\mu_{men} - \mu_{women} \neq 0$

| Men: | $\bar{y}_1 = 272.4$ | Women: | $\bar{y}_2 = 274.73$ |
|---|---|---|---|
| | $s_1 = 59.2$ | | $s_2 = 57.5$ |
| | $n_1 = 840$ | | $n_2 = 1077$ |

99% C.I. $= (-9.27, 4.61)$

Since 0 is contained within the 99% confidence interval, there is no evidence of a statistically significant difference in NAEP's scores between the two sexes at the .01 level of significance.

22.17 **(a)** Recall: P-value $= .6456$ or $64.56\%$

**(b)** $H_0$:   $p_{men} - p_{women} = 0$

$H_a$:   $p_{men} - p_{women} \neq 0$

$\hat{p}_{\text{men}} = 450/3180 = .14$

$\hat{p}_{\text{women}} = 210/1260 = .16$

**pooled sample proportion;** $\hat{p} = (450 + 210)/(3180 + 1260) = 660/4440 = .1486$

$z = -2.53$, P-value $= P(Z < -2.53 \text{ or } Z > 2.53) = 2(.0057) = .0114 \text{ or } 1.14\%$

Since $.0114 < .05$, reject $H_0$. There is a statistically significant difference between the proportion of failed businesses of men and women at the .05 level of significance.

(c) (a) 95% C.I. $= (.164, .104)$

(b) 95% C.I. $= (-.0536, .0064)$

22.19 $n = \dfrac{\hat{p}(1-\hat{p})}{\left(\frac{m}{z^*}\right)^2} = \hat{p}(1-\hat{p})\left(\frac{z^*}{m}\right)^2 = .44(1 - .44)\left(\frac{1.96}{.03}\right)^2 = .2464(4268.44) = 1051.744$

In order to obtain a margin of (.03), a sample size of 1052 will be needed.

# LESSON 23

23.1 (a) There are three rows and four columns; $r = 3, c = 2$.

(b) More than 12 hours per week appears to be least conducive to successful grades (37.5%). The range which appears to have the highest proportion of successful students is 2–12 hours per week (74.7%). Less than 2 hours per week produces a mid proportion of successful students (55%).

(c) Place the amount of hours per week students put towards extracurricular activities on the horizontal axis. Place the percentage of successful students on the vertical axis. The bar graph should visually show clear differences among the three categories. The highest bar should be 2–12 hours per week followed by less than 2 and more than 12.

(d) $H_0$:  The proportions of success are the same.

$H_a$:  The proportions of success are not the same.
$\chi^2 = 6.942$, P-value $= .0311$

We reject the null hypothesis that the proportions are equal at the .05 level of significance, but we fail to reject the null hypotheses at the .01 level of significance.

(e) The term which contributes most to the $\chi^2$ statistic is the term associated with more than 12 hours of activity. The relationship this points to is that there may be a point when too much activity is associated with declines in academic success.

(f) Although there is evidence to suggest a relationship between time spent on extracurricular activities and academic success, cause and effect are not determined by purely statistical methods.

23.3 (a) NC: .030, MC: .131, NMC: .064. The data show that HMO members with medical complaints had the largest proportion of people who left, followed by non-medical complaints and no complaints.

(b) NC; Left: 22, MC; Left: 721, NC: 743, NC; Stayed: 173, MC: 199, NMC; Stayed: 412, NMC: 440, Left: 76, Stayed: 1306, Total: 1382

(c) $H_0$:  There is an association between complaint status and HMO retention levels.

$H_a$:  There is no association between complaint status and HMO retention levels.

$\chi^2 = 31.76$, $P$-value $= 0$

Reject $H_0$ at the .01 level of significance. There is a significant association between complaint status and HMO retention levels.

(d) Since all expected counts are greater than 5, the chi-squared test is assumed to be appropriate.

23.5 (a) $H_0$:  There is an association between sex and choice of major.

$H_a$:  There is no association between sex and choice of major.

$\chi^2 = 10.83$, $P$-value $= .0127$

Reject the null hypothesis that there is no relationship between sex and choice of major at the .05 level of significance, but fail to reject at the .01 level of significance.

(b) The most marked disparity between women and men is in Administration. Women in Administration comprise 69% of the respondents while men only comprise 31%. The other majors differ slightly, but are on average split 50% for men and women.

(c) The two cells which have the largest component of the chi-squared statistic are the differences between observed and expected counts for men and women in Administration. Their components comprise 6.73 of the 10.83 units in the chi-squared statistic.

(d) Although two of the cell counts are small, their individual expected counts are both over one. The one value which is slightly under five (4.59) comprises less than 20% of the expected counts. The chi-squared test is appropriate.

(e) The percent which did not respond is: $1 - 386/722 = .47$ or 47%.

23.7 (a) $H_0$:    There is an association between city and medical releases.

$H_a$:    There is no association between city and medical releases.
$\chi^2 = 10.83$, $P$-value $= .0127$
Reject the null hypothesis at the .01 level of significance. There is evidence of an association between city and parental medical releases.

(b) The data should be gathered throughout all economic sectors—not just the poor areas, in such a way that a good SRS of all cities can be gathered.

23.9 (a) $H_0$:    There is an association between season and response levels.

$H_a$:    There is no association between season and response levels.

(b) The percent of successful interviews appears higher in the colder months and lower in the warmer months. The statistical significance of this variation is that the surveys in the warmer months may not be as valid as the surveys taken when a higher percentage is successful. A possible reason for this type of variation could be due to warmer weather bringing more people outdoors.

23.11 (a) The null hypothesis states that the proportion of women in the four majors are equal.

(b) The proportion of students who are women is 58.29% . The expected number of women in accounting is: $.5829(124) = 72.28$. The expected number of women in administration is: $.5829(131) = 76.36$. These are the same values computed by software.

(c) The expected counts, using the formula (row total) (column total)/table total, are: 72.28, 16.36, 6.41, 69.95, 51.72, 54.64, 4.59, 50.05

23.13 (a) Choose an SRS of 8000 people who have previously had strokes. Randomly divide the sample into four equal groups; Placebo, Aspirin, Dipyridamole and Both. Administer the drugs or placebo at the recommended dosage. After a set time (two years) compare the members who are still participating in the four groups to determine if there is a significantly different proportion of strokes.

(b) $H_0$:    There is an association between medication and repeat strokes.

$H_a$:    There is no association between medication and repeat strokes.

$$\chi^2 = 24.243, P\text{-value} = 0$$

There is evidence of a relationship between type of medication and incidence of repeat strokes.

The percentage who have repeat strokes is roughly equal for both Aspirin and Dipyridamole, but is higher for the Placebo group. The group with the lowest percentage of strokes is the Both group.

(c) $\chi^2 = 10.442$, P-value $= .005$. The results are still significant as low as the .005 level of significance.

(d) $\chi^2 = 7.466$, P-value $= .006$. The results are still significant as low as the .006 level of significance.

## LESSON 24

24.1 (a) The scatterplot shows a linear, positive, and strong relationship between gas used and degree days. The regression explains 99.1% of the observed variation in gas used.

(b) There appears to be no important deviations from normality in the normal probability plot of residuals. There appears to be no systematic pattern in the scatterplot of residuals vs. the explanatory variable.

(c) 95% C.I. $= b \pm t_{(\alpha/2, n-2)}SE_b = .19 \pm 2.145(.004934) = (.18, .20)$

(d) Fit = 99.323, 95.0% CI = (94.905, 103.741)

24.3 (a) The scatterplot shows a linear, positive, and strong relationship. There is one unusual point which corresponds to the Rolls Royce Silver Spur.

(b) Removing the Rolls Royce Silver Spur has little influence on the regression.

(c) The normal probability plot shows no important deviations from normality. The scatterplot of residuals vs. the explanatory variable shows no clear systematic pattern. The assumptions are met reasonably well.

(d) The slope of the true regression line represents the true average increase in highway mileage for every additional city mile per gallon for all 1997 model cars.

95% C.I. $= b \pm t_{(\alpha/2, n-2)}SE_b = 1.2 \pm 2.00(.0524) = (1.0952, 1.3048)$

(e) Fit = 25.534, 95.0% PI = (22.886, 28.182)

24.5 The relationship is linear, positive, and strong with one outlier. The normal probability plot of the residuals shows a right skew. There appears to be a weak negative association between the residuals and the explanatory variable. The assumptions may be violated.

`Fit = 52.095, 95.0% CI = (50.279, 53.912)`

24.7 (a) The outlier in weight is not an outlier in the scatterplot of width vs. length.

(b) The regression equation is: $\hat{y} = -.883 + .298(Length)$

(c) `Fit = 6.5549, 95.0% CI = (6.4255, 6.6842)`

(d) The normal probability plot of residuals shows no important deviations from normality. The scatterplot of residuals vs. length, however, show a fan pattern which is the classic sign of non-constant variance. The predicted values for longer fish will be less accurate.

24.9 (a) $L = -.790, U = -.3575$

(b) The student's work did not make sense, because the relationship is not linear. In addition, the student's interpretation of a confidence interval is incorrect.

24.11 (a) The regression equation is: $\hat{y} = -3.66 + 1.1969(Femur)$

(b) $S_{Regression} = \sqrt{\dfrac{\sum_{i=1}^{n}(y_i - \hat{y})^2}{n-2}} = \sqrt{\dfrac{11.79}{3}} = \sqrt{3.93} = 1.982$

(c) $SE_b = \dfrac{S_{Regression}}{\sqrt{\sum_{i-1}^{n}(x_i - \bar{x})^2}} = \dfrac{1.982}{\sqrt{696.8}} = .075084, t =$

$\dfrac{b}{SE_b} = \dfrac{1.1969}{.07584} = 15.78$

## REVIEW VI

R1 Both the control and the nitrite group have a roughly normal and unimodal distribution. Neither group have any outliers.

$H_0$: $\mu1 - \mu2 = 0$

$H_a$: $\mu1 - \mu2 > 0$

Control − Nitrite:

Test $H_0$: $\mu(Control) - \mu(Nitrite) = 0$ vs. $H_a$: $\mu(Control) - \mu(Nitrite) > 0$

$t$-Statistic $= 0.8909$ w/56 df

$p = 0.1884$

Fail to reject $H_0$. The data show insufficient evidence that the amino acid uptake is less for the nitrite group than for the control group.

R3 $H_0$: $p = .10$

$H_a$: $p < .10$

$z = -3.34$, P-value $= P(Z < -3.34) = .0004$

The data show significant evidence that less than 10% of the patients who took the pain reliever suffered adverse symptoms.

R5 (a) The term s.e. stands for standard error. Total calories; Drivers: $y = 2821, s = 435.578$, Total calories; Conductors: $y = 2844, s = 437.3$, Alcohol; Drivers: $y = .24, s = .594$, Alcohol; Conductors: $y = .39, s = 1.002$.

(b) $H_0$: $\mu_{drivers} - \mu_{conductors} = 0$

$H_a$: $\mu_{drivers} - \mu_{conductors} \neq 0$
$z = -.35$, P-value $= .7264$

Fail to reject $H_0$ at the .05 level of significance. The data show insufficient evidence that drivers and conductors consume different number of calories per day.

(c) $H_0$: $\mu_{drivers} - \mu_{conductors} = 0$

$H_a$: $\mu_{drivers} - \mu_{conductors} \neq 0$
$z = -1.20$, P-value $= .2302$

The data show no significant difference between the mean level of alcohol consumed by drivers and conductors up to the .2302 level of significance.

(d) 90% C.I. $= (.208, .572)$

(e) 80% C.I. $= (-.311, .011)$

R7 $H_0$: $p = .105$

$H_a$: $p < .105$
$z = -1.64$, P-value $= .0505$

There is insufficient evidence that blacks were under-represented at the .01 level of significance.

R9 (a) $H_0$: $p_{high-tech} - p_{non-high-tech} = 0$

$H_a$: $p_{high-tech} - p_{non-high-tech} > 0$

$\hat{p}_{high-tech} = 73/91 = .80$

$\hat{p}_{non-high-tech} = 75/109 = .69$

**pooled sample proportion**; $\hat{p} = (73 + 75)/(91 + 109) = .74$

Checking conditions necessary for $z$ procedures:

$n_1 \hat{p} = 73(.74) = 54.02$, $n_1(1 - \hat{p}) = 73(.26) = 18.98$, $n_2 \hat{p} = 109(.74) = 80.66$, $n_2(1 - \hat{p}) = 109(.26) = 28.34$

Since all values are over 5, it is assumed that the conditions necessary to perform a two-sample test have been met.

$z = 1.77$, P-value $= .0384$ or 3.84%

Reject $H_0$ at the .05 level of significance. There is evidence that the proportion of high-tech companies that offer stock options is higher than non-high-tech companies.

(b) 95% C.I. = $(-.01, .23)$

R11 (a) $H_0$:  $\mu_A - \mu_B = 0$

$H_a$:  $\mu_A - \mu_B \neq 0$

$z = -1.48$, $P$-value = .1388

There is insufficient evidence that there is a real difference between the two proposals.

(b) Since the sum of the sample sizes is "large" and the two-sample $t$ procedure is used, the distributions of the data do not affect the validity of the test.

(c) No, the bank's study is not an experiment, because the treatment was not deliberately imposed on the subjects. As a result we cannot conclude that the differences in the two plans were the only variables which affected the results.

R13 (a) 99% C.I. = $(1016.55, 1069.45)$

(b) The large sample sizes.

R15 $H_0$:  $p = .488$

$H_a$:  $p > .488$

$\hat{p} = 273/555 = .49$

$z = .09$, $P$-value = .4641

The data show insufficient evidence that the proportion of females born to chemists is higher than that of the overall proportion of Washington State.

R17 (a) 95% C.I. = $(.214, .286)$

(b) $H_0$:  $p_{\text{sickle-cell}} - p_{\text{non-sickle-cell}} = 0$

$H_a$:  $p_{\text{sickle-cell}} - p_{\text{non-sickle-cell}} < 0$

$\hat{p}_{\text{sickle-cell}} = 36/136 = .26$

$\hat{p}_{\text{non-sickle-cell}} = 152/407 = .37$

**pooled sample proportion;** $\hat{p} = (36 + 152)/(136 + 407) = .35$

Checking conditions necessary for $z$ procedures:

$n_1 \hat{p} = 136(.35) = 47.6$,   $n_1(1 - \hat{p}) = 136(.65) = 88.4$,   $n_2 \hat{p} = 407(.35) = 142.45$,   $n_2(1 - \hat{p}) = 407(.65) = 264.55$

Since all values are over 5, the conditions necessary to perform a two-sample test have been met.

$z = -2.38$, $P$-value = .0087

Reject $H_0$ at the .01 level of significance. There is evidence that the proportion of children with the sickle-cell trait who have malaria is less than the non-sickle-cell children.

R19 $H_0$: The proportions are equal.

$H_a$: The proportions are not equal.
$\chi^2 = 3.277$, P-value $= .513$
There is insufficient evidence that the proportions are different across the different types of industry.

R21 $H_0$: The proportions are equal.

$H_a$: The proportions are not equal.
$\chi^2 = 3.955$, P-value $= .412$
There is insufficient evidence that the proportions are different across the different cities.

R23 Although Thailand's scores are higher than Sweden's, the variability is likely to be much higher for Thailand than for Sweden. The increased variability will reduce the significance of a test.

R25 (a) The boxplot of IQs show that both sexes have roughly symmetric distributions. There are two low outliers for male IQs and one low outlier for female IQs. The median IQ is slightly higher for males than females. The spreads as measured by IQRs are about equal for both sexes.

The boxplot of GPAs shows that both sexes have roughly symmetric distributions. There is one low outlier for male GPAs. Females have no clear outliers. The medians are roughly equal for both sexes. The spreads as measured by IQRs are also roughly equal.

(b) With the exception of male GPA, which has slight right skew, the four normal probability plots show a roughly normal distribution. The large sample sizes allow even skewed distributions, but the outliers may present a problem. The fact that it is not known if this an SRS is the most serious problem with the data.

# Data Table Index

# Index